Lecture Notes in Physics

Springer
Berlin
Heidelberg
New York
Barcelona
Budapest
Hong Kong
London
Milan
Paris
Santa Clara
Singapore
Tokyo

The Editorial Policy for Proceedings

The series Lecture Notes in Physics reports new developments in physical research and teaching – quickly, informally, and at a high level. The proceedings to be considered for publication in this series should be limited to only a few areas of research, and these should be closely related to each other. The contributions should be of a high standard and should avoid lengthy redraftings of papers already published or about to be published elsewhere. As a whole, the proceedings should aim for a balanced presentation of the theme of the conference including a description of the techniques used and enough motivation for a broad readership. It should not be assumed that the published proceedings must reflect the conference in its entirety. (A listing or abstracts of papers presented at the meeting but not included in the proceedings could be added as an appendix.)

When applying for publication in the series Lecture Notes in Physics the volume's editor(s) should submit sufficient material to enable the series editors and their referees to make a fairly accurate evaluation (e.g. a complete list of speakers and titles of papers to be presented and abstracts). If, based on this information, the proceedings are (tentatively) accepted, the volume's editor(s), whose name(s) will appear on the title pages, should select the papers suitable for publication and have them refereed (as for a journal) when appropriate. As a rule discussions will not be accepted. The series editors and Springer-Verlag will normally not interfere with the detailed editing except in fairly obvious cases or on technical matters.

Final acceptance is expressed by the series editor in charge, in consultation with Springer-Verlag only after receiving the complete manuscript. It might help to send a copy of the authors' manuscripts in advance to the editor in charge to discuss possible revisions with him. As a general rule, the series editor will confirm his tentative acceptance if the final manuscript corresponds to the original concept discussed, if the quality of the contribution meets the requirements of the series, and if the final size of the manuscript does not greatly exceed the number of pages originally agreed upon. The manuscript should be forwarded to Springer-Verlag shortly after the meeting. In cases of extreme delay (more than six months after the conference) the series editors will check once more the timeliness of the papers. Therefore, the volume's editor(s) should establish strict deadlines, or collect the articles during the conference and have them revised on the spot. If a delay is unavoidable, one should encourage the authors to update their contributions if appropriate. The editors of proceedings are strongly advised to inform contributors about these points at an early stage.

The final manuscript should contain a table of contents and an informative introduction accessible also to readers not particularly familiar with the topic of the conference. The contributions should be in English. The volume's editor(s) should check the contributions for the correct use of language. At Springer-Verlag only the prefaces will be checked by a copy-editor for language and style. Grave linguistic or technical shortcomings may lead to the rejection of contributions by the series editors. A conference report should not exceed a total of 500 pages. Keeping the size within this bound should be achieved by a stricter selection of articles and not by imposing an upper limit to the length of the individual papers. Editors receive jointly 30 complimentary copies of their book. They are entitled to purchase further copies of their book at a reduced rate. As a rule no reprints of individual contributions can be supplied. No royalty is paid on Lecture Notes in Physics volumes. Commitment to publish is made by letter of interest rather than by signing a formal contract. Springer-Verlag secures the copyright for each volume.

The Production Process

The books are hardbound, and the publisher will select quality paper appropriate to the needs of the author(s). Publication time is about ten weeks. More than twenty years of experience guarantee authors the best possible service. To reach the goal of rapid publication at a low price the technique of photographic reproduction from a camera-ready manuscript was chosen. This process shifts the main responsibility for the technical quality considerably from the publisher to the authors. We therefore urge all authors and editors of proceedings to observe very carefully the essentials for the preparation of camera-ready manuscripts, which we will supply on request. This applies especially to the quality of figures and halftones submitted for publication. In addition, it might be useful to look at some of the volumes already published. As a special service, we offer free of charge LATEX and TEX macro packages to format the text according to Springer-Verlag's quality requirements. We strongly recommend that you make use of this offer, since the result will be a book of considerably improved technical quality. To avoid mistakes and time-consuming correspondence during the production period the conference editors should request special instructions from the publisher well before the beginning of the conference. Manuscripts not meeting the technical standard of the series will have to be returned for improvement.

For further information please contact Springer-Verlag, Physics Editorial Department II, Tiergartenstrasse 17, D-69121 Heidelberg, Germany

Hildegard Meyer-Ortmanns
Andreas Klümper (Eds.)

Field Theoretical Tools
for Polymer
and Particle Physics

Springer

Editors

Hildegard Meyer-Ortmanns
Fachbereich Physik
Bergische Universität Wuppertal
Gaußstrasse 20
D-42097 Wuppertal, Germany

Andreas Klümper
Institut für Theoretische Physik
Universität zu Köln
Zülpicher Strasse 77
D-50937 Köln, Germany

Cataloging-in-Publication Data applied for.

Die Deutsche Bibliothek - CIP-Einheitsaufnahme

Field theoretical tools for polymers and particle physics /
Hildegard Meyer-Ortmanns ; Andreas Klümper (ed.). - Berlin ;
Heidelberg ; New York ; Barcelona ; Budapest ; Hong Kong ;
London ; Milan ; Paris ; Santa Clara ; Singapore ; Tokyo : Springer,
1998
   (Lecture notes in physics ; 508)
   ISBN 3-540-64308-7

ISSN 0075-8450
ISBN 3-540-64308-7 Springer-Verlag Berlin Heidelberg New York

Typesetting: Camera-ready by the authors/editors
Cover design: *design & production* GmbH, Heidelberg
SPIN: 10644157   55/3144-543210 - Printed on acid-free paper

Preface

This volume of lecture notes is based on lectures presented at the Workshop on *Field Theoretical Tools for Polymer and Particle Physics* within the framework of the *Graduiertenkolleg* (college for graduate students) at the University of Wuppertal. The workshop took place in June 1997. The *Graduiertenkolleg* of the Theory Department in Wuppertal was founded in 1992. It is devoted to *Field Theoretical and Numerical Methods in Statistical Physics and Elementary Particle Physics*. The workshop was addressed to advanced graduate students who are in a stage of specializing on a specific topic in their diploma or Ph.D. thesis. The topics of the lectures have been selected in a way that the presented methods as well as the models have applications in both areas, those of particle physics and of polymer physics. The lectures may serve as some guide through more recent research activities and illustrate the applicability of joint methods in different contexts. Unavoidably, their style varies from *rigorous* to *phenomenological* and from *detailed* to *review-like*. We summarize the main topics.

– Analytical Tools

Random Walk models originally were introduced in theoretical polymer physics in the context of phase transitions. Later representations of Euclidean lattice field theories in terms of random walks played an important role in the triviality issue of the Φ^4-theory in dimensions ≥ 4, particularly in $d = 4$.

Familiar examples for *polymer expansions* in statistical physics are high- and low temperature expansions of the Ising model. Because of the intimate relation between statistical mechanics and Euclidean quantum field theory, the expansion techniques have been applied to investigate the phase structure of lattice gauge theories, to discuss the thermodynamic limit as well as the continuum limit.

The *bosonization technique* is particularly suited for 2-dimensional systems such as the Thirring model. In 2 dimensions it leads to an equivalent bosonic theory (the Sine-Gordon model in case of the Thirring model) which is local without an additionally introduced truncation of nonlocal terms. In higher dimensions, in field theoretical models like effective models for QCD, the same technique leads to nonlocal effective actions, but their truncation may be justified in a certain parameter range such as low-momenta and large number of colours. The bosonic action then is no longer equivalent to the original fermionic model, but still representative for aspects on a certain (energy) scale.

The topic of *stochastic classical many-body systems* is defined in terms of a master equation describing the time evolution of configuration probabilities. It is this evolution equation that can be cast in an operator form reminiscent of the formulation of quantum mechanics. For general diffusion-reaction models the associated "quantum Hamiltonian" does not possess hermiticity properties. This however, does not limit the usefulness of this approach as to date many quantum many body systems in $d = 1 + 1$ are known to be integrable for hermitian as well as non-hermitian nearest neighbour interactions. This serves for new non-perturbative results for phase diagrams and correlation functions of stochastic systems.

– Numerical Tools

The *Hybrid Monte Carlo Method* is one of the most established algorithms to perform Monte Carlo simulations for QCD with dynamical fermions. Recently it has been successfully applied to Fourier accelerated simulations of polymer chains.

Langevin dynamics was used to study the phase conversion mechanism as a function of the strength of first order phase transitions. Nucleation or large domain coarsening, well known conversion mechanisms in condensed matter systems, here are discussed in the background of phase transitions in the early universe.

Certain aspects of neural networks can be described by methods of statistical mechanics. One of such aspects is the phase structure of the Hopfield model. More general devices, so called multilayered neural networks, nowadays are used in the particle data analysis of high energy experiments at large colliders.

– Common Models: The Gross-Neveu Model

In 3 dimensions the *Gross-Neveu model* serves as an effective model for the chiral phase transition in QCD. The chiral transition in 2-flavour QCD has led to a controversy about its universality class. A similar controversy arises in the Gross-Neveu model on the universality class of the corresponding parity-breaking transition. The controversy is based on plausibility arguments. In the Gross-Neveu model it can be solved by a combined application of large-N expansions, the method of dimensional reduction, and linked cluster expansions on a lattice, while the type of universality class of the chiral transition in QCD is still under debate.

In 2 dimensions the model provides a qualitative (to some extent even a quantitative) description of the finite-density phase transition in *polyacetylene*. Polyacetylene is the prototype of conjugate polymers. Recently conjugate polymers have attracted much attention. Similarly to semiconductors they become electrically conducting after a procedure called *doping*.

The various facets of the Gross-Neveu model are complemented by a confrontation with lab-experiments on polyacetylene. The experimentally observed particle-like excitations may be identified with the theoretically predicted solitons in the 2-dimensional Gross-Neveu model.

The full programme of the workshop can be found under the *List of Participants*.

We would like to thank the speakers of the workshop for their interesting talks and the authors for writing up their contribution in the form of lectures rather than proceedings. We are indebted to our colleagues at the Theory Department in Wuppertal for their support. Financial support of the *Deutsche Forschungsgemeinschaft* (DFG) is gratefully acknowledged.

Wuppertal/Köln, Germany,   H. Meyer-Ortmanns
December 1997   A. Klümper

Contents

Polymers, Spin Models and Field Theory
Bo Söderberg 68

Reaction-Diffusion Mechanisms and Quantum Spin Systems
Gunter M. Schütz 78

List of Participants

I. Analytical Methods for Polymer and Particle Physics

E. Eisenriegler:
Random Walks in Polymer Physics I
Random Walks in Polymer Physics II
A. Pordt:
Random Walks in Euclidean Quantum Field Theory
Polymer Expansions in Particle Physics
B. Söderberg:
Polymers, Polyelectrolytes and Field Theory
G. Schütz:
Reaction-Diffusion Mechanisms and Quantum Spin Systems
B. Derrida:
The Asymmetric Exclusion Model for Diffusion-Reaction Systems
T. Giamarchi:
Bosonization in Condensed Matter Physics
D. Ebert:
Bosonization in Particle Physics
Hadronization in Particle Physics
F. Niedermayer:
Finite-Size Scaling Analysis in the Presence of Goldstone Bosons

II. Numerical Methods for Polymer and Particle Physics

F. Karsch:
Finite-Size Scaling Analysis for QCD
T. Lippert:
The Hybrid-Monte Carlo Algorithm for QCD
A. Irbäck:
The Hybrid-Monte Carlo Method for Polymer Chains
Simulations of Toy Proteins
B. A. Berg:
Confidence Limit from Neural Network Data Analysis

III. Models with Applications in Polymer and Particle Physics

Random Walks in Polymer Physics

Erich Eisenriegler

Institut für Festkörperforschung, Forschungszentrum Jülich,
D–52425 Jülich, Germany

Abstract. Long flexible polymer chains can be modelled as self-avoiding random walks and their properties analyzed by exploiting the polymer-magnet analogy and using field theoretical tools. We discuss the behavior of chains in an infinite unbounded space and the interaction of chains with impenetrable boundaries, such as those of mesoscopic spheres and cylinders. This is relevant for colloidal particles immersed in a solution of free nonadsorbing polymers, and the understanding of polymer–depletion effects is advanced considerably if we use properties of the field theory such as the local character of renormalization, various operator expansions for short distances, and conformal invariance.

1 Introduction

Dilute solutions of long flexible polymer chains in good solvent show a behavior that is independent of most of the details of the chemical microstructure of the chains and of the (low molecular weight) solvent. As an example consider the mean square end–to–end distance or the radius of gyration $\mathcal{R}^2$ of a single isolated chain. It follows a power law [1, 2, 3]

$$\mathcal{R}^2 \sim N^{2\nu} \tag{1.1a}$$

as the number N of repeat units (monomers) along the chain becomes large. Although the chemical structure of repeat units is quite different for different types of polymer chains [4], the exponent ν is the same. In three dimensions $d = 3$ and in the case of a good solvent (effective repulsion between monomers)

$$\nu = 0.588 \ . \tag{1.1b}$$

Another example is the osmotic pressure Π of polymer chains. It obeys a scaling law [1, 2, 3]

$$\Pi = k_B T \, c_p \, X(c_p \mathcal{R}^d) \ , \tag{1.2}$$

where c_p is the number density of polymer chains and $c_p \mathcal{R}^d$ characterizes the degree of overlap between chains. Although it is assumed that N is large and that the monomer density $c_p N$ is much smaller than in a dense polymer melt, the overlap may be either large (semidilute solution) or small (dilute solution). Eq. (1.2) contains two nontrivial statements. First, for a given substance, e.g. PMM in acetone, the dependence of $\Pi/(k_B T c_p)$ upon the two variables c_p and $\mathcal{R}$ is only via the product $c_p \mathcal{R}^d$. Second, the same (universal) scaling function X appears for different substances [5].

This poses two questions: How does the universal scaling behavior come about, and how can one calculate the universal quantities (such as ν or X) ? A first hint comes from an oversimplified treatment that treats the polymer chain configurations as *random* walks where each of the N steps is statistically independent. For such an 'ideal' polymer chain, scaling and universality follow from the central limit theorem that tells us that the scaling function of the end–to–end distance distribution is a Gaussian and that $\nu_{id} = 1/2$, independent of the detailed form of the probability distribution for one step. For a chain at the θ point [1, 2, 3] in $d = 3$ this is indeed essentially realized. However, for a chain in good solvent the repulsive (excluded volume) interaction between monomers is important. For such a *'self avoiding* random walk' (SAW) ν is larger, and the end–to–end distribution is non–Gaussian. A sufficiently general framework to explain the universality in this case is provided by the renormalization group. This is known from other critical systems, such as lattice spin models or field theories. Here we use the result that the SAW can be mapped onto a special type of near–critical field theory in order to calculate physical properties of SAW's.

First polymers are considered in an infinite unbounded space and then the interaction of polymers with boundaries. With applications to mesoscopic colloidal particles in mind, we consider not only planar but also curved boundaries, such as surfaces of spheres and cylinders.

2 Modelling a Polymer Chain in a Good Solvent and Relation to Field Theory

Since many details are irrelevant, there is much freedom in choosing a model. One of the most convenient models for analytical calculations is the 'bead and spring'–model with the single chain partition function

$$Z_N(r_B, r_A) = \int dr_{N-1} \ldots \int dr_1\, P(r_B, r_{N-1}) \ldots P(r_1, r_A)$$
$$\cdot \hat{\prod_{(i,i')}} \left[1 - bl^d\, \delta(r_i - r_{i'})\right] \ . \tag{2.1}$$

Here the product of N normalized Gaussians

$$P(r, r') = (4\pi l^2)^{-d/2}\, e^{-(r-r')^2/(4l^2)} \tag{2.2}$$

defines the chain structure with fixed end points at r_A, r_B and $N-1$ internal beads at $r_1, \ldots, r_{N-1}$, and it introduces a characteristic size l per monomer. The product $\prod$ is over the $\binom{N-1}{2}$ pairs (i, i') of internal beads. Since $b > 0$, it decreases the probability weight for those configurations where the beads overlap. The hat on $\prod$ means that only those products of δ-functions are retained where every bead position r_i occurs not more than once. This makes the model well defined [3].

To become familiar with the model, first consider its ideal–chain limit $b = 0$. Then $Z_N = Z_N^{(0)}$ is a convolution of P's, which is handled easily by introducing the Fourier transform $\tilde{Z}_N^{(0)}$, given by

$$\int d(r_B - r_A)e^{ip(r_B - r_A)}Z_N^{(0)}(r_B, r_A) = (\tilde{P}(p))^N , \qquad (2.3)$$

where

$$\tilde{P}(p) = e^{-p^2 l^2} \qquad (2.4)$$

is the Fourier transform of P. Then $Z_N^{(0)}$ follows from P simply on replacing l^2 by $l^2 N$ and is given by

$$Z_N^{(0)}(r_B, r_A) = (4\pi N l^2)^{-d/2} e^{-(r_B - r_A)^2/(4N l^2)} . \qquad (2.5)$$

This particularly simple coarse graining behavior arises, of course, from choosing the macroscopic Gaussian shape of the ideal chain end–to–end distance distribution already at the microscopic level.

The mean square end–to–end distance defined by

$$\mathcal{R}_e^2 = \int dr_B (r_B - r_A)^2 Z_N(r_B, r_A) / \int dr_B Z_N(r_B, r_A)$$
$$\equiv d \mathcal{R}_x^2 \qquad (2.6)$$

follows for $b = 0$ from Eqs (2.3)–(2.5) with the result

$$(\mathcal{R}_x^2)_0 = 2l^2 N , \qquad (2.7)$$

consistent with $\nu_{id} = 1/2$. Here we have used $\int dr_B Z_N^{(0)} = 1$ and $\int dr_B (r_B - r_A)^2 Z_N^{(0)} = (-\Delta_p \tilde{Z}_N^{(0)})_{p=0}$, with Δ the Laplacian, and the form of $\tilde{Z}_N^{(0)}$ from (2.3) and (2.4).

Next consider the first–order in b contribution $Z_N^{(1)}$ of Z_N. Since $\hat{\prod}$ in (2.1) is given to first order by

$$1 - bl^d \sum_{\substack{i,i' \\ (N>i>i'>0)}} \delta(r_i - r_{i'}) ,$$

one finds, on using the above simple coarse graining properties, that

$$Z_N^{(1)}(r_B, r_A) = -bl^d \sum \int dr Z_{N-i}^{(0)}(r_B, r) Z_{i-i'}^{(0)}(r, r) Z_{i'-0}^{(0)}(r, r_A)$$
$$= -bl^d \sum Z_{i-i'}^{(0)}(0, 0) Z_{N-i+i'-0}(r_B, r_A) , \qquad (2.8)$$

with the double sum $\sum$ as above. Manipulations similar to those near (2.7) then lead to the first–order expressions

$$\int dr_B \, Z_N^{(1)}(r_B, r_A) = -bl^d \sum Z_{i-i'}^{(0)}(0,0) \tag{2.9}$$

and

$$\int dr_B (r_B - r_A)^2 Z_N^{(1)}(r_B, r_A) = -bl^d \sum Z_{i-i'}^{(0)}(0,0) \cdot 2dl^2 (N-i+i') \ , \tag{2.10}$$

which we need for the denominator and numerator in (2.6). Using the form of $Z_{i-i'}^{(0)}$ that follows from (2.5), one then finds

$$\mathcal{R}_x^2 / (\mathcal{R}_x^2)_0 = 1 + b\,\sigma(N) \tag{2.11}$$

to first order. Here

$$\sigma = (4\pi)^{-d/2} \sum_{j=1}^{N-2} j^{1-\frac{d}{2}} \left(1 - \frac{j+1}{N}\right) \tag{2.12}$$

has an asymptotic behavior for large chain length N which depends in a crucial way on the space dimension d. Apart from N–independent prefactors,

$$\sigma \to (1, \quad \ln N, \quad N^{(4-d)/2}) \tag{2.13a}$$

as $N \to \infty$ for

$$(d > 4, \quad d = 4, \quad d < 4) \ . \tag{2.13b}$$

This suggests that for $d > 4$ the excluded volume interaction ($b > 0$) does not change the exponent $\nu = \nu_{id}$. It also shows that for $d \leq 4$ a perturbation expansion in b cannot be used to determine the $N \to \infty$ behavior. A different tool (renormalization group) is needed.

The peculiarity of dimension 4 can also be seen in a more intuitive way: Due to $\nu_{id} = 1/2$ a random walk has Hausdorff dimension 2 and behaves like a surface or plane with internal dimension 2. Thus in $d = 4$ two remote mesoscopic parts of the walk practically do not overlap, since two 2–dimensional planes in $d = 4$ intersect only at a point.

To see the relation between chain conformation statistics and field theory, consider the (discrete) Laplace transform

$$G_t(r_B, r_A) = \sum_{N=1}^{\infty} l^2 e^{-Nl^2 t} Z_N(r_B, r_A) \tag{2.14}$$

of the chain partition function Z_N in (2.1). This converts the convolution with respect to bead labels of the partition functions $Z^{(0)}$ of chain pieces which appears in the perturbation expansion of Z_N into a product of their Laplace transforms. For example the contribution from the first order expression (2.8) becomes

$$G_t^{(1)}(r_B, r_A) = -bl^{d-4} \int dr \, G_t^{(0)}(r_B, r) G_t^{(0)}(r, r) G_t^{(0)}(r, r_A) \ . \tag{2.15}$$

The rhs of (2.15) has the structure of the first–order contribution to the correlation function $\langle \phi(r_B)\phi(r_A) \rangle$ in a Landau–Ginzburg–Wilson $(\phi^2)^2$ field theory [1, 2, 3], where the strength of the $(\phi^2)^2$ interaction is $\sim bl^{d-4}$. Note that $G_t^{(0)}$, which plays the role of the unperturbed (Gaussian) propagator $\langle \phi \, \phi \rangle_0$, contains an effective upper cutoff ($\sim l^{-1}$) for wavevectors, which makes the field theory well defined. This follows from its Fourier transform

$$\int d(r_B - r_A)\, e^{ip(r_B - r_A)}\, G_t^{(0)}(r_B, r_A) = \sum_{N=1}^{\infty} l^2\, e^{-Nl^2 t}\, e^{-p^2 N l^2}$$

$$= \frac{l^2}{e^{l^2(t+p^2)} - 1}, \quad (2.16)$$

(see Eqs. (2.14), (2.3) and (2.4)), which tends to zero for $t + p^2 \gg l^{-2}$ and reproduces the usual Gaussian propagator expression $1/(t + p^2)$ for $t + p^2 \ll l^{-2}$.

To make the relationship precise [6], consider a field $\phi = (\phi_1, \phi_2, \ldots, \phi_n)$ with n components. With the explicit form

$$\mathcal{H}' = bl^{d-4} \int dr\, \frac{1}{2} \sum_{\alpha,\beta=1}^{n} \frac{1}{2}\phi_\alpha^2(r)\, \frac{1}{2}\phi_\beta^2(r) \quad (2.17)$$

of the $(\phi^2)^2$ interaction and with the Gaussian propagator

$$\langle \phi_\alpha(r)\phi_\beta(r') \rangle_0 = \delta_{\alpha,\beta}\, G_t^{(0)}(r, r') \quad (2.18)$$

consider the $\mathcal{M}^{\text{th}}$ order expression $\langle \phi_1(r_B)\phi_1(r_A)\frac{1}{\mathcal{M}!}(-\mathcal{H}')^{\mathcal{M}} \rangle_{0,\text{conn}}$ to the correlation function $\langle \phi_1(r_B)\phi_1(r_A) \rangle$, which may be easily Wick–factorized. Introducing the diagrammatic notation shown in Fig. 1 for $\mathcal{H}'$ in (2.17), one obtains the Wick contributions for $\mathcal{M} = 1$ in Fig. 2. The first contribution is

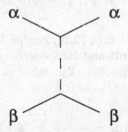

Fig. 1. Diagrammatic notation for $\mathcal{H}'$ in Eq. (2.17).

a 'chain–diagram', which leads to a result independent of n. The second is a diagram with a loop and is proportional to n (since the loop contains a free component index α to be summed from 1 to n). For $\mathcal{M} = 2$ the contributions

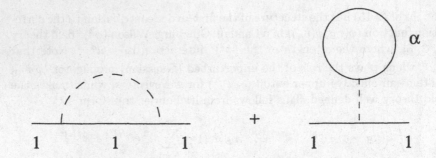

Fig. 2. The two first-order contributions to the correlation function.

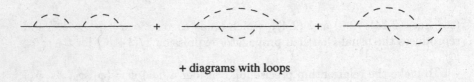

+ diagrams with loops

Fig. 3. Second-order contributions to the correlation function.

are shown in Fig. 3. It is easy to see that for each of the chain diagrams in Figs 2 and 3 the Wick combinatorial factor is precisely cancelled by the numerical (symmetry) factors appearing in (2.17) and that these contributions are equal to the (Laplace transform of the) corresponding contributions in a perturbation expansion of the chain partition function in (2.1). Thus we require that diagrams with loops vanish, which is achieved by formally setting $n = 0$ in the perturbation expansion. This reasoning can be extended to arbitrary order and implies the relation

$$\sum_N l^2 e^{-Nl^2 t} Z_N(r_B, r_A) = \langle \phi_1(r_B) \phi_1(r_A) \rangle \,|_{n=0} \ . \tag{2.19}$$

A slightly different derivation of (2.19) can be found in [7]. See also [8] and the contributions of A. Pordt and B. Soederberg in this book.

Besides the partition function Z_N, other properties of a polymer chain are of interest. One example is the fraction

$$F_m(r) \, dr = \frac{1}{N-1} \sum_{j=1}^{N-1} \delta(r - r_j) \, dr \tag{2.20a}$$

of beads (monomers) in a volume element dr around a given point r in space. Due to its simple normalization

$$\int dr \, F_m(r) = 1 \ , \tag{2.20b}$$

the fraction density F_m is less dependent on microscopic details and conventions [9] than the monomer density $(N-1)F_m$. A further advantage [9] is that in the scaling limit of large N the scaling dimension of F_m equals its (naive) inverse length dimension d. For a long chain with ends fixed at r_A and r_B the configurational average $\{F_m\}_{A,B}$ of F_m is given by [10]

$$\{F_m(r)\}_{B,A} = \frac{(Nl^2)^{-1}\mathcal{L}\langle\frac{1}{2}\phi^2(r)\cdot\phi_1(r_B)\phi_1(r_A)\rangle|_{n=0}}{\mathcal{L}\langle\phi_1(r_B)\phi_1(r_A)\rangle|_{n=0}} \qquad (2.21)$$

in terms of the field theory. Here the dot following $\phi^2(r)$ denotes a cumulant average, and $\mathcal{L} = \mathcal{L}_{t \to Nl^2}$ is the inverse of the Laplace transform operation on the lhs of Eq (2.19). Thus the denominator on the rhs of (2.21) equals the partition function $Z_N(r_B, r_A)$. To establish the validity of (2.21), one needs to show that the product $\{F_m\}_{B,A} Z_N(r_B, r_A)$, that is, the numerator in the chain configuration average $\{F_m\}$, equals the numerator on the rhs of (2.21). The simplest case is the ideal chain limit $b = 0$, where $N - 1$ times the product is given by

$$\int dr_{N-1}\ldots\int dr_1\, P(r_B, r_{N-1})\ldots P(r_1, r_A) \sum_{j=1}^{N-1} \delta(r - r_j)$$

$$= \sum_{j=1}^{N-1} Z_{N-j}^{(0)}(r_B, r)\, Z_{j-0}^{(0)}(r, r_A) \ , \qquad (2.22)$$

which is indeed the inverse Laplace transform of

$$l^{-2}\langle\frac{1}{2}\phi^2(r)\cdot\phi_1(r_B)\phi_1(r_A)\rangle_0 = l^{-2}\, G_t^{(0)}(r_B, r)\, G_t^{(0)}(r, r_A) \ , \qquad (2.23)$$

as one may verify on applying the direct Laplace transform operation of (2.19) to the rhs of (2.22). Fig. 4 shows a useful diagrammatic notation for this ideal chain expression. The wiggly line can be interpreted either as $\sum \delta(r - r_j)$ or as $\frac{1}{2}\phi^2(r)$. To first order in b the numerator on the rhs of (2.21) times

Fig. 4. Diagrammatic notation for the expressions in Eqs. (2.22) or (2.23).

N leads to three contributions shown in Fig. 5. Diagrams with a closed loop have been discarded since they vanish for $n = 0$. The same three contributions are obtained from the numerator in the chain configurational average of $(N - 1)F_m$. This follows from the lhs of (2.22) when the second term of the expression above Eq. (2.8) is inserted. The three contributions are due to

Fig. 5. First–order expressions for the numerator in Eq. (2.21).

$j > i > i'$, $i > j > i'$ and $i > i' > j$, respectively. The two terms $j = i$ and $j = i'$ of the j–sum (2.20a) can be neglected for large N. For a general proof of (2.21) see Refs. [6, 7].

3 The Universal Scaling Behavior of a Long Polymer Chain

Starting from the universal critical behavior of the Landau–Ginzburg–Wilson $(\phi^2)^2$ field theory [11], we can use relations such as (2.19) or (2.21) to infer [1, 2, 6] the corresponding behavior for long polymer chains. Note that the Laplace variable t in Eq. (2.19) is proportional to the deviation of the temperature from the (mean field or Gaussian) critical temperature of the field theory. Eq. (2.19) gives an expansion of the correlation function on its rhs in a power series in $e^{-l^2 t}$ which converges and is analytic for large enough t (paramagnetic region). Decreasing t towards the true critical value t_c where the correlation function is singular is equivalent to approaching the radius of convergence of the series. Thus the singular behavior for $t \searrow t_c$ is related to the behavior of Z_N for $N \to \infty$.

Consider the polymer partition function

$$Z_N(r_A) = \int dr_B \, Z_N(r_B, r_A) \tag{3.1}$$

with one end fixed. Eq. (2.19) gives

$$\sum_N l^2 e^{-N l^2 t} \, Z_N(r_A) \sim l^2 \left[(t - t_c) l^2 \right]^{-\gamma} \tag{3.2}$$

with the susceptibility exponent γ, since the integral $\int dr_B$ of the correlation function on the rhs of (2.19) is the susceptibility. Eq. (3.2) implies that

$$Z_N(r_A) \sim e^{N l^2 t_c} \, N^{\gamma - 1} \tag{3.3}$$

as $N \to \infty$. While t_c and thus the exponential N–dependence depend on details [12], one finds the universal values

$$\gamma - 1 = 0, \quad 0.161, \quad 11/32 \tag{3.4a}$$

for the power law exponent from the known [11] values of γ in space dimensions

$$d = 4, \quad 3, \quad 2 . \tag{3.4b}$$

Next consider $\mathcal{R}_x^2$ in (2.6). For the numerator we need

$$\int dr_B \, r_B^2 \langle \phi_1(r_B)\phi_1(0)\rangle \sim \xi^{d+2-2x} , \tag{3.5}$$

where ξ is the correlation length

$$\xi \sim (t - t_c)^{-\nu} \tag{3.6}$$

of the field theory. Eq. (3.5) follows from the scaling form

$$\langle \phi_1(r)\phi_1(0)\rangle \sim \xi^{-2x} \, \mathcal{X}(r/\xi) \tag{3.7}$$

of the correlation function where x is the scaling dimension of the order parameter. $\mathcal{X}$ is a universal scaling function. Calculating for the denominator the susceptibility also from Eq. (3.7) yields

$$d - 2x = \gamma/\nu , \tag{3.8}$$

and from (2.19)

$$\mathcal{R}_x^2 \sim N^{(d+2-2x)\nu-1}/N^{(d-2x)\nu-1} = N^{2\nu} . \tag{3.9}$$

Thus the polymer exponent ν in Eq. (1.1a) coincides with the correlation length [11] exponent ν in (3.6) for the case $n = 0$, and [13]

$$\nu = \frac{1}{2}, \quad 0.588, \quad \frac{3}{4}, \quad 1 \tag{3.10a}$$

for

$$d = 4, \quad 3, \quad 2, \quad 1 . \tag{3.10b}$$

If r_A and r_B in Eq. (2.19) coincide, the partition function on its lhs becomes $Z_N(0,0)$, which is the partition function of a phantom ring polymer, i.e. a ring polymer where we sum over all possible knot structures, and the quantity on its rhs becomes $\langle \phi_1^2(0)\rangle$, which is proportional to the energy density of the field theory. For the $N \to \infty$ behavior

$$Z_N(0,0) \sim e^{Nl^2 t_c} N^{-\nu d} \tag{3.11}$$

of the ring polymer partition function, the power law exponent is negative, whereas the exponent was positive in case of (3.3). This reflects the strong constraint due to the coinciding ends of the chain. It also implies that the main contribution to the Laplace transform on the lhs of (2.19) for t near t_c

is now from N's of order 1, where corrections to (3.11) have to be taken into account. This is avoided if we consider the second derivative

$$\sum_N e^{-Nl^2(t-t_c)}\, N^2\, e^{-Nl^2 t_c}\, Z_N(0,0) \sim \frac{d^2}{dt^2}\langle\phi_1^2(0)\rangle|_{n=0} , \qquad (3.12)$$

since the factor N^2 suppresses the corrections. The exponent in (3.11) follows from the known behavior $\sim (t-t_c)^{-(3-\nu d)}$ of the rhs of (3.12).

Finally consider the normalized end–to–end distance distribution. For $|r_B - r_A|$ and $\mathcal{R}_x$ large on a microscopic scale

$$Z_N(r_A, r_B)/Z_N(r_A) = \mathcal{R}_x^{-d}\, \mathcal{Y}(|r_B - r_A|)/\mathcal{R}_x) , \qquad (3.13)$$

which is consistent with the scaling law (3.7) and with (3.3), (3.8). It is interesting to consider the distribution for distances $|r_B - r_A|$ much smaller than $\mathcal{R}_x$. In this case one finds

$$\mathcal{Y}(y) \sim y^\theta , \quad y \ll 1 , \qquad (3.14)$$

with the short distant exponent [2]

$$\theta = \frac{\gamma - 1}{\nu} \equiv d - 2x - \frac{1}{\nu} \equiv x_{\phi^2} - 2x , \qquad (3.15)$$

where we have introduced the critical exponent

$$x_{\phi^2} = d - \frac{1}{\nu} \qquad (3.16)$$

of the energy density. The form of θ follows from (3.3), (3.9), (3.13) and the plausible assumption that the $N \to \infty$ dependence of $Z_N(r_B, r_A)$ with finite and fixed $|r_B - r_A|$ is the same as that of $Z_N(0,0)$ in (3.11). Using Eqs. (3.4) and (3.10), one finds

$$\theta = 0, \quad 0.27, \quad 11/24 \qquad (3.17a)$$

for

$$d = 4, \quad 3, \quad 2 . \qquad (3.17b)$$

θ vanishes at the upper critical dimension $d = 4$ where the excluded volume interaction is only marginally effective and where the distribution is a Gaussian. For $d < 4$, θ is positive and leads to a distribution that *increases* with distance $|r_B - r_A|$ if $|r_B - r_A| \ll \mathcal{R}_x$. It decreases, of course, for $|r_B - r_A| \gg \mathcal{R}_x$. This is a consequence of the excluded volume interaction, which suppresses configurations with close ends.

Eq. (3.14) is a special case of a general short-distance relation of the field theory, which can be written in the operator form [11]

$$\phi(r_B)\phi(r_A) \sim |r_B - r_A|^{x_{\phi^2} - 2x}\phi^2\left(\frac{r_B + r_A}{2}\right) \qquad (3.18a)$$

for distances

$$l \ll |r_B - r_A| \ll \text{other lengths} . \tag{3.18b}$$

Eq. (3.18a) holds in any [14] correlation function if $|r_B - r_A|$, while being large on the microscopic scale l, is much smaller than the other lengths (such as the correlation length ξ or the distances from $|r_B + r_A|/2$ to the positions of other operators) that appear in the correlation function. In all these cases the same exponent (3.15) appears, and its form follows from the requirement that both sides of (3.18a) have the same scaling dimension. The factor of proportionality in (3.18a) is also independent of the particular correlation function.

The behavior of the distribution (3.13) for $|r_B - r_A|$ much larger than $\mathcal{R}_x$ is also interesting as has been discussed by Fisher (1966), compare the last reference in [6].

4   Polymers Interacting with Boundaries

Consider the situation where the polymer chain can only move in a part of the space due to an impenetrable boundary [15, 16, 17]. One example is a half space with a planar boundary. Another one is the exterior of a spherical mesoscopic (colloidal) particle, in which case the boundary is curved. In this article we assume inert impenetrable boundaries and disregard the possibility of an energy gain per chain–monomer at the boundary (attractive impenetrable boundary), which can lead to adsorption [15, 16, 17]. In this article, the boundary acts on the monomers like a repelling hard wall. Again it is possible to map [16] the problem onto a field theory, and relations of the form (2.19) and (2.21) again apply. However, it is now a field theory with a boundary [18, 19], which satisfies the Dirichlet condition

$$\phi = 0 \qquad \text{at the boundary} . \tag{4.1}$$

The main effect of the boundary is to generate a boundary–layer which is depleted of chain–monomers since the number of chain conformations, and thus the entropy, is strongly reduced near the boundary. Here we only consider the case of non–overlapping chains (dilute solution), and the width of the depletion layer is of the order of the unperturbed chain–extension $\mathcal{R}_x$. The depletion effect shows up most directly in the density profiles ρ_m and ρ_e of chain–monomers and chain–ends.

5   Planar Boundary and Density–Force Relations

For a dilute and monodisperse solution of free polymers in the half space with a planar boundary (wall), the bulk–normalized density profiles $\rho_e/\rho_e^{(bulk)}$ for ends or $\rho_m/\rho_m^{(bulk)}$ for monomers have the scaling form

$$\rho(z)/\rho^{(bulk)} = Y(z/\mathcal{R}_x) \ , \tag{5.1}$$

where z, the distance from the wall, is large on the microscopic scale and where $Y = Y_e$ or Y_m is a universal scaling function. For $z \gg \mathcal{R}_x$, $Y \to 1$. For

$$\text{microscopic distances} \ll z \ll \mathcal{R}_x \ , \tag{5.2}$$

ρ_e, ρ_m and thus Y_e, Y_m have a power law behavior in z. The power law exponents are positive, in accordance with the depletion phenomenon, and are known as 'surface exponents'. In the case of ρ_e the exponent is new and is no simple combination of the bulk exponents ν and γ. In the case of ρ_m the exponent is $1/\nu$.

The reason is that ρ_m near the wall can be related to the *force* that the polymers exert onto the wall [20]. Using in the bulk of the dilute solution $\rho_m^{(bulk)}$ and the monomer number N per chain as independent variables, the force per area is given by

$$\frac{f}{\mathcal{A}} = k_B T \, \rho_m^{(bulk)}/N \ , \tag{5.3}$$

since it equals the chain osmotic pressure in the bulk, which by the ideal gas law is $k_B T$ times the bulk density of chains $\rho_m^{(bulk)}/N$. From the plausible assumption [21] that ρ_m for $z \ll \mathcal{R}_x$ is proportional to $f/\mathcal{A}$, which by means of (5.3) and (3.9) is proportional to $\rho_m^{(bulk)}/\mathcal{R}_x^{1/\nu}$, one finds from (5.1) that

$$Y_m(z/\mathcal{R}_x) \to B \cdot (z/\mathcal{R}_x)^{1/\nu} \tag{5.4a}$$

for $z \ll \mathcal{R}_x$, with B a universal amplitude. On using (5.1), (5.3) this can also be written as

$$\frac{\mathcal{R}_x^{1/\nu}}{N}\rho_m(z) \to \frac{f}{\mathcal{A}k_B T} \cdot B\, z^{1/\nu} \tag{5.4b}$$

for microscopic distances $\ll z \ll \mathcal{R}_x$. Note that the lhs of (5.4b) is defined and can be determined for a given solution in an unambiguous way (similar to the fraction F_m for a single chain considered in Eqs. (2.20)). Its inverse length dimension $d - 1/\nu$ equals its scaling dimension. Eq. (5.4b) relates this modified monomer density to the force via the universal amplitude B.

These largely phenomenological density–force considerations can be supplemented by a field–theoretic analysis which allows one to calculate B and to show that $B z^{1/\nu}$ occurs not only in the present context but also in a variety of other situations.

Since Eq. (2.21) also applies for a chain in the presence of a boundary [16, 19], the z–dependence of the half space profile ρ_m is proportional to

$$\rho_m(z) \sim \mathcal{L} \int dr_B \int dr_A \langle \phi^2(r_\parallel, z) \cdot \phi_1(r_A)\phi_1(r_B)\rangle_w \ , \tag{5.5}$$

where $\langle\,\rangle_w$ denotes the half space average with the Dirichlet condition (4.1) (w stands for wall). The behavior of $\phi^2(r)$ on approaching the wall follows from the short distance relation [19]

$$\phi^2(r_\parallel, z) \sim z^{x_{\phi_\perp^2} - x_{\phi^2}} \cdot \frac{1}{2}\phi_\perp^2(r_\parallel) \ . \tag{5.6}$$

This is analogous to the bulk relation (3.18a) for the case in which one operator approaches another one. The surface operator

$$\frac{1}{2}\phi_\perp^2(r_\parallel) = \frac{1}{2}\left[(\partial_z \phi(r_\parallel, z))\right]_{z=0} \tag{5.7}$$

on the rhs of (5.6) is the operator of lowest inverse length dimension that is even in ϕ and nonvanishing at the Dirichlet boundary, and $x_{\phi_\perp^2}$ is its scaling dimension [22]. Actually $\frac{1}{2}\phi_\perp^2(r_\parallel)$ equals $T_{\perp\perp}(r_\parallel, 0)$, the component of the stress tensor at the boundary perpendicular to the boundary. Quite generally the stress tensor generates coordinate transformations. In particular, integrating $\frac{1}{2}\phi_\perp^2 = T_{\perp\perp}$ over the planar boundary generates a shift away from the surface [23, 24, 25]. For example

$$\int d^{d-1}r_\parallel \langle T_{\perp\perp}(r_\parallel, 0) \cdot \phi_1(r_A)\phi_1(r_B)\rangle_w = (\partial_{z_A} + \partial_{z_B})\langle \phi_1(r_A)\phi_1(r_B)\rangle_w \tag{5.8}$$

if $z_A, z_B > 0$. Thus the scaling dimension $x_{\phi_\perp^2}$ of $\frac{1}{2}\phi_\perp^2 = T_{\perp\perp}$ equals its (naive) inverse length dimension d. Together with (3.16) and (5.5), (5.6) this is the field theoretic explanation of the $z^{1/\nu}$-behavior of ρ_m.

A more explicit form of (5.6) follows from introducing, instead of ϕ^2, the quantity

$$\Psi(r_\parallel, z) = \mathcal{R}_x^{1/\nu}(Nl^2)^{-1}\frac{1}{2}\phi^2(r_\parallel, z) \ , \tag{5.9}$$

which appeared in Eq. (2.21) and whose scaling dimension equals its inverse length dimension $d - 1/\nu = x_{\phi^2}$. This leads to the short distance relation

$$\Psi(r_\parallel, z) \to B\,z^{1/\nu} \cdot T_{\perp\perp}(r_\parallel, 0) \tag{5.10}$$

valid for microscopic distances $\ll z \ll$ other lengths. One can show [26] that B is the universal amplitude introduced in Eqs. (5.4) for free chains in the half space. For a single chain in the half space with the ends fixed at r_B, r_A, Eq. (5.10), the half space counterparts of (2.21), (2.19), and the shift identity (5.8) imply

$$\int d^{d-1}r_\parallel\,\mathcal{R}_x^{1/\nu}\{F_m(r_\parallel, z)\}_{B,A}^{(w)} \to B\,z^{1/\nu} \cdot (\partial_{z_B} + \partial_{z_A})\ln Z_N^{(w)}(r_A, r_B) \ . \tag{5.11}$$

Here the derivative of the logarithm of the single chain partition function $Z^{(w)}$ in the half space on the rhs of (5.11) is the force per $k_B T$ which the chain exerts onto the wall. Thus $Bz^{1/\nu}$ determines the density–force relation also for this situation. The same expression $Bz^{1/\nu}$ appears in other cases of interest [26]. For a single chain trapped between two parallel walls it relates the monomer density near one wall to the disjoining pressure, and for a mesoscopic particle immersed in a polymer solution in the half space it relates the change in monomer density near the wall to the attractive depletion force between the particle and the wall.

For $d = 4 - \varepsilon$ close to the upper critical dimension 4, the value of B for chains with self repulsion is given [26] by $2(1 - 0.075\,\varepsilon)$ and deviates only slightly from the d–independent value $B_{id} = 2$ for ideal (random walk) chains. For $d = 2$ it is shown in the Appendix that B can be expressed in the form

$$B = (nC_T)\,[2^{-4/3}a/n]^{1/2} \cdot \left\{ \left[(\mathcal{R}_x^2/\hat{\mathcal{R}}^2) \frac{\sqrt{\pi}}{2} U_2/U_0 \right]^{2/3} \kappa \right\} \qquad (5.12)$$

in terms of other known universal amplitudes. Here $\hat{\mathcal{R}}^2 = \hat{\mathcal{R}}_x^2 + \hat{\mathcal{R}}_y^2$ is the mean square radius of gyration of a self avoiding polygon (ring polymer) with the same number of links as the open chain that defines $\mathcal{R}_x^2$ and the ratio of the two mean square extensions is given by $\mathcal{R}_x^2/\hat{\mathcal{R}}^2 \approx 6.85$ [27]. The remaining quantities are universal amplitudes of the two–dimensional $O(n)$ model for $n \to 0$. Here $U_2 = 20/9$, $U_0 = 4\pi^2/3$ and $\kappa \approx 0.226630$ are noncritical universal bulk amplitudes introduced in Ref. [28], and $nC_T \to 8\pi^2/5$ and $a/n \to 4/(3\sqrt{3})$ are half space amplitudes at bulk criticality that are explained in the Appendix. Inserting their numerical values, one finds $B \approx 2.01$ for $d = 2$, which is again very close to $B_{id} = 2$. This suggests a similar behavior for $d = 3$.

6   Spherical and Cylindrical Particles and Expansion for Small Radius

It is interesting to compare the densities ρ_e and ρ_m near a planar boundary with the corresponding densities in the vicinity of a spherical or cylindrical particle with radius R that is immersed in a dilute and monodisperse polymer solution. For $\mathcal{R}_x/R \ll 1$ surface curvature effects are negligible, and one again finds the planar wall behavior. For $\mathcal{R}_x/R$ of order 1 or larger the polymer chains sense the global shape of the obstacle (particle). Fig. 6 shows the bulk normalized end density $\rho_e/\rho_e^{(bulk)}$ for ideal chains as a function of the (radial) distance $r - R$ from the particle surface [29]. For a given distance, $\rho_e/\rho_e^{(bulk)}$ is the ratio of the partition function of a chain with only one end fixed at that distance (as in Eq. (3.1)) and of the corresponding chain in the absence of the particle. It is obvious that for a fixed distance from the particle surface, ρ_e for a cylinder is smaller than for a sphere of equal radius and that ρ_e (for either cylinder or sphere) decreases with increasing R. In particular, ρ_e for a planar wall is smaller than ρ_e for a cylinder or sphere. Fig. 7 shows the

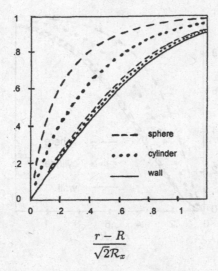

$$\frac{r - R}{\sqrt{2}\mathcal{R}_x}$$

Fig. 6. Bulk-normalized density of chain ends for $\sqrt{2}\mathcal{R}_x/R$ equal to 5 (upper curves) and equal to 0.2 (lower curves).

corresponding monomer densities $\rho_m/\rho_m^{(bulk)}$ [29], for which similar effects can be observed.

Here we consider polymers in the presence of a 'small' spherical particle, where R, albeit being large on the microsopic scale, is much smaller than $\mathcal{R}_x$ and other characteristic lengths. The effect of the particle on the chain can be described in terms of a δ-function potential, located at the center r_s of the particle, that weakly repels the monomers. This means that the Boltzmann weight W_s for the chain that arises from the presence of the sphere tends for small R to [29]

$$W_s[r_j] \to 1 - A\,R^{d-\frac{1}{\nu}}\mathcal{R}_x^{\frac{1}{\nu}}F_m(r_s) \ , \qquad (6.1)$$

where the $[r_j]$–dependent chain monomer fraction density $F_m(r)$ is defined in Eq. (2.20a). The product $\mathcal{R}_x^{1/\nu}F_m$ is the modified monomer density (that we have encountered in the different context of Eq. (5.11)) with a scaling dimension equal to its inverse length dimension $d - 1/\nu$. The exponent of R in (6.1) follows from comparing scaling dimensions, and the amplitude A is dimensionless and universal [30]. The potential is 'weak' for small R since $d - \frac{1}{\nu} > 0$.

As an application of Eq. (6.1) and in order to see the relation to the field theory, consider the fraction

$$\mathcal{F}(r_A, r_B) = Z_N^{(\text{with sphere})}(r_A, r_B)/Z_N^{(bulk)}(r_A, r_B)$$
$$\equiv \mathcal{L}\langle\phi_1(r_A)\phi_1(r_B)\rangle_{\text{with sphere}}/\mathcal{L}\langle\phi_1(r_A)\phi_1(r_B)\rangle_{bulk} \qquad (6.2)$$

of chain configurations with fixed ends at r_A, r_B that survive on immersing a spherical particle. For $R \ll \mathcal{R}_x, |r_A - r_s|, |r_B - r_s|$, Eqs. (6.1) and (2.21)

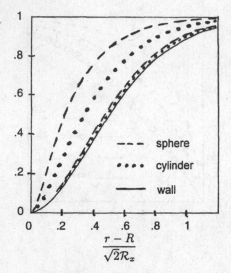

Fig. 7. Bulk-normalized density of chain monomers.

lead to

$$\mathcal{F}(r_A, r_B) \to 1 - AR^{d-\frac{1}{\nu}} \mathcal{R}_x^{\frac{1}{\nu}} \{F_m(r_s)\}_{A,B}^{(bulk)}$$
$$\equiv 1 - AR^{d-\frac{1}{\nu}} \mathcal{L} \langle \Psi(r_s) \cdot \phi_1(r_A)\phi_1(r_B) \rangle_{bulk} / \mathcal{L} \langle \phi_1(r_A)\phi_1(r_B) \rangle_{bulk} \ , \quad (6.3)$$

with $\Psi(r)$ related to $\phi^2(r)$ as in Eq. (5.9). Comparison of the last expressions in Eqs. (6.2) and (6.3) shows that the Boltzmann factor $\exp(-\mathcal{H}_s)$ for the field ϕ that arises from the presence of the sphere tends for small R to

$$\exp(-\mathcal{H}_s[\phi]) \sim 1 - AR^{d-\frac{1}{\nu}}\Psi(r_s) \ . \quad (6.4)$$

Thus, as expected in the Landau Ginzburg Wilson field theory (or in a spin model for magnetism [18, 19]), a small mesoscopic sphere with Dirichlet condition (4.1) (free spin or 'ordinary' boundary condition) at its surface acts for distant points like a temperature that is enhanced (or a spin–spin coupling that is diminished) in a microscopic region around r_s. A more quantitative discussion has been given in Refs. [31, 32, 33], where on conformally mapping the half space onto the exterior of a sphere it was shown that $A\Psi$ on the rhs of (6.4) can be written as

$$A\Psi(r_s) = (a/B_{\phi^2})^{1/2}\phi^2(r_s) \quad (6.5)$$

with the universal and nonuniversal amplitudes a and B_{ϕ^2} defined in (A2) and (A3). On comparing with a similar relationship

$$\frac{1}{B}\Psi(r_\parallel, z) = \left[(aB_{\phi^2})^{1/2} 2^{-x_{\phi^2}} C_T\right]^{-1} \phi^2(r_\parallel, z) \quad (6.6)$$

that follows from the two forms (5.10) and (A1), (A2) of the short distance expansion near a planar wall, one infers, for arbitrary d, the identity

$$A B = 2^{-(d-\frac{1}{\nu})} a\, C_T \qquad (6.7)$$

between the universal amplitudes A, B of polymers and the universal amplitudes a, C_T of the field theory at the bulk critical point, which are defined in (A1-3).

The proportionality between $R^d \{F_m\}_{bulk}$ and $(1 - \mathcal{F})(R/\mathcal{R}_x)^{1/\nu}$ contained in Eq. (6.3) is expected intuitively. Both quantities are proportional to the mean fraction of monomers of an unperturbed chain (in the absence of the particle) in the region S around r_s which after the immersion is occupied by the particle. This is obvious for the first quantity. It also holds for the second, since $1 - \mathcal{F}$ is the probability that an unperturbed chain visits S and since the monomer fraction inside S of a typical visiting chain is proportional to $(R/\mathcal{R}_x)^{1/\nu}$.

Another example where the small R expansion (6.1) can be usefully applied is the free energy cost F_s of immersing a small sphere in a dilute solution of free chains with or without self avoidance. The result is [29]

$$F_s/(k_B T n_b) = A\, R^{d-\frac{1}{\nu}}\, \mathcal{R}_x^{\frac{1}{\nu}} \ . \qquad (6.8)$$

Here n_b is the number density of chains (before immersing the particle), and $k_B T\, n_b$ is their pressure (n_b is identical to the quantity denoted by $\rho_m^{(bulk)}/N$ below Eq. (5.3)). Also for end and monomer density profiles around a single small sphere in a free chain solution, Eq. (6.1) can be applied and leads to simple explicit expressions [29].

While power laws such as (6.8) are known from previous work [34, 35, 36], the present approach reveals a universal amplitude A that appears not only for polymers interacting with a single isolated small sphere but also where polymers mediate a depletion interaction between a small sphere and a planar wall [29] or between two or more small spheres. In the latter cases the free energy of interaction is a product of powers of $A\, R^{d-\frac{1}{\nu}}$ and pair or multiple correlation functions of 'monomer densities' $\mathcal{R}_x^{\frac{1}{\nu}} F_m$ in the absence of obstacles, which allows a systematic discussion of the non pairwise additive depletion interaction [37].

For ideal chains $A = A_{id} = 2\pi^{d/2}/\Gamma\left(\frac{d}{2}-1\right)$ is exactly known [29] and takes the values $A_{id} = 2\pi^2$ and 2π in $d = 4$ and 3, respectively. For chains with self avoidance it can be estimated from its form

$$A = 2\pi^2 \left\{1 - \varepsilon\, A_1 + \mathcal{O}(\varepsilon^2)\right\} \qquad (6.9a)$$

in $d = 4 - \varepsilon \lesssim 4$, where

$$A_1 = \frac{1}{2}\ln \pi + \frac{1}{8}\ln 2 + \frac{3}{8}C_E - \frac{1}{4} = 0.624 \qquad (6.9b)$$

and $C_E = 0.577$ is Euler's constant, and from its value

$$A = (a/n)^{1/2}/\{\,\} = 2\pi^2 \cdot 0.193 \tag{6.10}$$

in $d = 2$, where the curly bracket is defined in Eq. (5.12). Eqs. (6.9) and (6.10) follow from Eq. (6.7). In the case of (6.10) the necessary numerical information has been given below Eq. (5.12). In the case of (6.9) one needs the ε–expansion $a/n = \frac{1}{2} + \mathcal{O}(\varepsilon^2)$, that of B in Ref. [26] and that of C_T (see Ref. [25] where C_T is called C_I).

There is also a small R expansion for a self avoiding chain near a cylindrical particle with radius R [29]. In this case the δ–function potential is smeared over the cylinder axis, the R–dependent factor is $R^{2-\frac{1}{\nu}}$ in three dimensions (instead of $R^{3-\frac{1}{\nu}}$ for the sphere) and the corresponding universal amplitude A_{cyl} can also be estimated [37]. This is qualitatively different from the corresponding behavior of an ideal chain, in which case $2 - \frac{1}{\nu}$ vanishes and there is a logarithmic dependence for small R [38].

This is consistent with the general trend that a chain with excluded volume interaction shows *weaker* depletion effects than an ideal chain (see also Table I in Ref. [29]). To understand this trend consider the fraction $\mathcal{F}$ in Eq. (6.2). Due to the depletion effect the numerator $Z_N^{(\text{with sphere})}$ contains fewer configurations where the chain intersects itself and is reduced less by turning on the excluded volume interaction than the denominator $Z_N^{(bulk)}$. Thus the fraction $\mathcal{F}$ increases.

Finally consider a small sphere that *touches* a planar wall w at the point $r = 0$. In this case the Boltzmann weight from the sphere $\exp(-\mathcal{H}_{ts})$ tends to

$$\exp(-\mathcal{H}_{ts}[\phi]) \sim 1 - A_{ts}\, R^d\, T_{\perp\perp}(0) \ . \tag{6.11}$$

This should be compared with Eq. (6.4), which holds for a sphere in the presence of a *distant* wall (distance from wall to sphere much larger than R). Note that the small touching sphere (ts) is represented by the surface operator $T_{\perp\perp}(0)$ in the wall w, as introduced in and below Eq. (5.7). The universal amplitude A_{ts} can be determined by calculating the critical energy density profile $\langle -\phi^2 \rangle_{w,ts}$ of the field theory in the (w, ts) geometry on mapping conformally onto the space (w, w) between two parallel walls [32,33], with the result

$$A_{ts} = S_d \left[(1 - d)\,\Delta C_T\right]/x_{\phi^2} \ . \tag{6.12}$$

Here S_d is the surface of the d–dimensional unit sphere, C_T is explained in Eq. (A1) and Δ is the critical Casimir amplitude, for large separation ϑ between the two walls w, in the term $\Delta\vartheta^{1-d}$ of the free energy per $k_B T$ and area $\mathcal{A}$ of the d–dimensional $O(n)$ model. The quantity in square brackets is known as a Fisher–de Gennes–amplitude [24,25,39,40]. Here ΔC_T is to be evaluated for $n \to 0$. In particular for ideal chains in $d \geq 1$

$$(A_{ts})_{id} = 2\pi^{d/2}(d - 1)\,\zeta(d) \,/\, \Gamma(d/2) \ , \tag{6.13}$$

with ζ the Riemann zeta function, leading to the values $(A_{ts})_{id} = [\pi^6/15,$ $8\pi\zeta(3),\ \pi^3/3,\ 2]$ in $d = [4,3,2,1]$. In $d = 2$ the value of A_{ts} is, to a large extent, model–independent [24,31,39] and for chains with excluded volume interaction is also given by $\pi^3/3$. For $d = 1$ the immersion of the touching 'sphere' simply implies a shift by $2R$ of the boundary point, and the value $A_{ts} = 2$ should apply quite generally. Eq. (6.11) may be used to calculate, e.g., the free energy cost F_{ts} of immersing a sphere with $R \ll \mathcal{R}_x$ so that it touches the wall confining a polymer solution in the half space. The result is

$$F_{ts}/(k_BTn_b) = A_{ts}R^d \qquad (6.14)$$

and is, as expected, much smaller than the corresponding quantity (6.8) for immersion in a bulk solution. One may compare (6.14) with the correspond-ing quantity, given [41] by $ABR^d(D/R)^{1/\nu}$, for immersing the sphere at a distance D from the confining wall with $R \ll D \ll \mathcal{R}_x$. Both quantities are independent of $\mathcal{R}_x$ and are the limits of an $\mathcal{R}_x$–independent universal scaling behavior $R^d\,J(D/R)$ valid for $R, D \ll \mathcal{R}_x$.

The free energy cost $F(D)$ of immersing a single sphere at a distance D from the wall and the corresponding bulk quantity $F_s = F(\infty)$ have a mean-ing for a mixture of spherical particles and polymer chains with sufficiently small bulk densities c_b and n_b, respectively. For arbitrary ratios of the meso-scopic lengths R, D, $\mathcal{R}_x$ they determine for the deviation $c(D) - c_b$ of the particle density profile from its bulk value the contribution of order c_bn_b in the form $c_b(F_s - F(D))/k_BT$. This implies an enrichment of particles near the wall. The bulk quantity F_s represents an increase in chemical potential of the particles due to the presence of polymers and determines the decrease of $c_b = c_b^{(0)}(1 - F_s/k_BT)$, i.e. of particle solubility, for increasing n_b and fixed particle density $c_b^{(0)}$ in a coexisting ideal gas phase.

If the chains are ideal the relation (6.14) can be readily generalized to the case where the particle touching the wall has the form of a cylinder, with the axis parallel to the wall, provided its radius R is small and its length Λ is large so that $R \ll \mathcal{R}_x \ll \Lambda$. Then the leading contribution to the free energy cost per k_BTn_b of immersing the cylinder is again independent of $\mathcal{R}_x$ and is in the case of three dimensions $(d = 3)$ given by $\Lambda R^2\,\pi^3/3$. The reason is that in the ideal polymer partition function the dependence on the component of the end–to–end distance parallel to the wall and cylinder–axis factors off and is normalized to 1 (compare Eq. (2.5)). One then is left with the two–dimensional problem where a polymer moves in the half plane in the presence of a disc touching the boundary line and where Eqs. (6.11) and (6.13) with $d = 2$ apply.

Instead of a sphere touching a wall one may consider a sphere touching another sphere with the same small radius R. The Boltzmann factor arising from the two spheres has in leading order the form of the rhs of Eq. (6.4) with the amplitude A replaced by a different amplitude mA. The anisotropy of the two-sphere configuration shows up in higher order corrections only.

Thus the free energy of immersing the two touching spheres is for $R \ll \mathcal{R}_x$ given by the rhs of Eq. (6.8) with A replaced by mA. An argument similar to that given below Eq. (6.11) relates m to the critical energy density profile between two parallel walls with distance ϑ : The quantity m equals the ratio of the value of the profile at a point halfway between the two walls and the value of the half–space profile at a distance $\vartheta/2$ from the wall. This leads for ideal chains in $d > 2$ to $m = 2(1 - 2^{3-d})\zeta(d - 2)$ implying $m = 2\ln 2$ in $d = 3$. For self-avoiding chains in $d = 2$ one finds $m = (\pi/2)^{2/3}$. Note that $1 < m < 2$ as expected.

The above free energy expressions for small particle radius R such as (6.8), (6.14) and the expression in [41] are complementary to expressions of the Asakura–Oosawa [42] and Deriagin [26] type that best apply for R large compared to $\mathcal{R}_x$. Work is in progress to obtain the free energy of the depletion interaction in the intermediate regime where R and $\mathcal{R}_x$ are of the same order [43].

7   Summary

Hopefully this contribution will convince the reader that field theoretical methods are useful in polymer physics. After a brief introduction to the 'polymer–magnet analogy' the emphasis was on short distance expansions which can be used both for polymers in unbounded space [Eqs. (3.18)] and for polymers interacting with boundaries. In Sec. 5 the scaling behavior of the monomer density near a planar boundary was discussed in terms of the expansion (5.10), which leads to a general relation between the density and the force onto the boundary in which a universal amplitude B appears. In Sec. 6 the behavior of a small spherical particle was considered. Here too a short distance relation [Eq. (6.1) or (6.4)] applies for quite general conditions, and it contains a universal amplitude A. The two numbers A and B in three dimensions $d = 3$, which appear in a variety of different situations of interest, encompassing the polymer induced depletion–interaction between walls and particles, can be estimated by calculating them in an ε–expansion (close to $d = 4$) and in $d = 2$. A small spherical particle touching a planar wall [Eqs. (6.11-14)] or touching another spherical particle of equal size was also considered.

Acknowledgments: It is a pleasure to thank A. Bringer, S. Dietrich, A. Hanke and, in particular, T. Burkhardt for useful discussions.

A   Density–Force Amplitude in $d = 2$

The amplitude C_T in the short distance expansion [14]

$$\phi^2(r_\|, z)/\langle -\phi^2(r_\|, z)\rangle_{w,crit} \to C_T \, z^d \, T_{\perp\perp}(r_\|, z) \qquad (A.1)$$

of the energy density near a wall at the 'ordinary transition' is universal and in two dimensions is given by $C_T = 4\pi x_{\phi^2}/c$ with c the conformal charge [24, 39]. Since we make a subtraction so that $\langle\phi^2\rangle_{bulk,crit} = 0$, here $\langle-\phi^2\rangle_{w,crit}$ is proportional to the excess energy density, over the bulk value, in the half space at the bulk critical point and decays with the power law behavior

$$\langle-\phi^2(r_\|, z)\rangle_{w,crit} = (a B_{\phi^2})^{1/2} \cdot (2z)^{-x_{\phi^2}} \tag{A.2}$$

from positive values to zero as z increases. While a is a universal amplitude, B_{ϕ^2} is the nonuniversal amplitude in the critical bulk correlation function

$$\langle\phi^2(r) \cdot \phi^2(0)\rangle_{bulk,crit} = B_{\phi^2}\, r^{-2x_{\phi^2}} \tag{A.3}$$

and should not be confused with the universal density–force amplitude B. For the $O(n)$ model in two dimensions the explicit form of a can be found, e.g., in Eq. (4.4) of Ref. [40] where $a^{1/2}$ is called $\mathcal{A}_O$. The $n \to 0$ behavior of C_T and a is given below Eq. (5.12).

The form (5.12) of B follows from the defining equations (5.9), (5.10) on using (A1), (A3) provided we identify the curly bracket in (5.12) with

$$\{\ \} = \left[\frac{1}{n}\langle\Psi(r)\Psi(0)\rangle_{bulk,crit}\, r^{2x_{\phi^2}}\right]^{1/2} \tag{A.4}$$

as $n \to 0$.

To establish (A4), we switch to the notation of Ref. [28]. Note that Ψ in Eq. (5.9) can be written as

$$\Psi(r) = \mathcal{R}_x^{1/\nu}\,\lambda^{-1}\,\tau(r)\ , \tag{A.5}$$

where $-\lambda$ is the exponent in the Laplace transform that relates the chain and field theoretical expressions and τ is the local thermal perturbation of the critical Hamiltonian of the field theory. In our notation

$$\lambda = Nl^2(t - t_c) \tag{A.6}$$

$$\tau(r) = (t - t_c)\frac{1}{2}\phi^2(r)\ , \tag{A.7}$$

and in the notation of Ref. [28]

$$\lambda = N(x_c - x)/x_c \tag{A.8}$$

$$\tau(r) = (x_c - x)\mathcal{E}(r)\ . \tag{A.9}$$

The expression in the square bracket of (A4) may then be written as the product

$$\frac{1}{n}\langle\Psi(r)\Psi(0)\rangle_{bulk,crit}\, r^{2x_{\phi^2}} = (\mathcal{R}_x^2/\hat{\mathcal{R}}^2)^{1/\nu} \cdot \left(\lambda^{-2}\hat{\mathcal{R}}^{2/\nu}/\xi^{2/\nu}\right) \cdot$$
$$\cdot \left(\xi^{2/\nu}\frac{1}{n}\langle\tau(r)\tau(0)\rangle_{bulk,crit}\, r^{2x_{\phi^2}}\right) \tag{A.10}$$

of three universal factors [44]. Of course, the critical average $\langle\rangle_{bulk,crit}$ only refers to the product of the two operators ϕ^2 or $\mathcal{E}$ in the angular bracket and does not affect factors $t - t_c$ or $x_c - x$. The quantity ξ in (A10) is a correlation length corresponding to the thermal perturbation. One may choose any type of ξ (true, second moment, etc), but we choose the one of Ref. [28]. Then the second and third universal factors on the rhs of (A10) are given by $\left(\frac{\sqrt{\pi}}{2} U_2/U_0\right)^{1/\nu}$ and κ^2, respectively, as follows from Eqs. (5), (24) and (82), (86) of Ref. [28]. This leads to Eq. (A4).

References

[1] de Gennes, P.G. (1979): *Scaling Concepts in Polymer Physics* (Cornell University Press)

[2] des Cloizeaux, J., Jannink, G. (1990): *Polymers in Solution* (Clarendon)

[3] Schäfer, L.: *The Physics of Polymer Solutions as Explained by the Renormalization Group* (Springer)

[4] For polyethylene the repeat unit is CH_2. For polystyrene (PS) a CH_2 alternates with CH where a benzene ring is attached, and for polydimethylsiloxane (PDMS) an oxygen alternates with $Si(CH_3)_2$. Another frequently used polymer is polymethylmetacrylate (PMM), where the repeat unit is even more complicated.

[5] For an explicit comparison between PMM in acetone and PDMS in cyclohexane see e.g. Ref. [3].

[6] de Gennes, P.G. (1972), Phys. Lett. **38A**, 339; des Cloizeaux, J. (1974), Phys. Rev. A5, 1665, and J. de Physique (Paris) **36**, 281 (1975). Similarities between the generating function (2.14) of self–avoiding walks and the magnetic pair correlation function have been noticed earlier by Fisher, M.E., Sykes, M.F. (1959), Phys. Rev. **114**, 45, and by Fisher, M.E. (1966), J. Chem. Phys. **44**, 616

[7] Emery, V.J. (1975), Phys. Rev. **B11**, 399; Duplantier, B. (1980), C.R. Hebd. Séan. Acad. Sci., Paris **290B**, 199

[8] Fernandez, R., Fröhlich, J., Sokal, A. (1991): *Random Walks, Critical Phenomena, and Triviality in Quantum Field Theory* (Springer)

[9] For a real polymer chain the value of N depends on which chemical piece is considered as a monomer. Note also that the dimensionless quantity N has, in accordance with Eq. (1.1a), the nonvanishing scaling dimension $-(1/\nu)$.

[10] Since we are interested in the long chain limit $N \gg 1$, we do not distinguish here between $N - 1$ and N.

[11] See e.g. Zinn–Justin, J. (1989): *Quantum Field theory and Critical Phenomena*, Oxford. For exponents of the $O(n)$ model in the unbounded plane $(d = 2)$ see Cardy, J.L., Hamber, H.W. (1980), Phys. Rev. Lett. **45**, 499, and Nienhuis, B. (1982), Phys. Rev. Lett. **49**, 1062 and J. Stat. Phys. **34**, 731 (1984)

[12] Since in our model (2.1) the partition function $Z_N^{(0)}(r_A)$ without self–repulsion ($b = 0$) is equal to 1 and $t_c^{(0)} = 0$, the quantity $\exp(l^2 t_c)$ in (3.3) corresponds to the ratio μ/μ_0 of 'connective constants' with and without self repulsion. Although μ/μ_0 is nonuniversal, one expects $\mu < \mu_0$. This is consistent with a decrease $t_c < 0$ in the critical temperature due to the $(\phi^2)^2$ interaction.

[13] This shows that the Flory approximation [1, 2, 3] $\nu_F = 3/(d+2)$ for the exponent ν is amazingly good.

[14] The operator equation gives the leading behavior for all connected correlation functions except for a few low–order correlations where the rhs is augmented by operator–free terms that are analytic in $t - t_c$.

[15] de Bell, K., Lookman, T. (1993), Rev. Mod. Phys. **65**, 87

[16] Eisenriegler, E. (1993): *Polymers near Surfaces* (World Scientific, Singapore)

[17] Fleer, G., Cohen Stuart, M., Scheutjens, J., Crosgrove T., Vincent, B. (1993): *Polymers at Interfaces* (Chapman Hall)

[18] Binder, K. (1983): *Phase Transitions and Critical Phenomena*, edited by C. Domb and J.L. Lebowitz (Academic Press, London), Vol. 8, p. 1

[19] (a) Diehl, H.W. (1986): *Phase Transitions and Critical Phenomena*, edited by C. Domb and J.L. Lebowitz (Academic Press, London), Vol. 10, p. 75.
(b) Diehl, H.W. (1997): Int. J. Mod. Phys. B **11**, 3503

[20] For definiteness assume that the wall is a semipermeable membrane, impermeable for the chains, but permeable for the solvent.

[21] Joanny, J.F., Leibler, L., de Gennes, P.G. (1979), J. Polym. Sci. Polym. Phys. Ed. **17**, 1073

[22] Since $x_{\phi_1^2} = d$ is the lowest scaling dimension for even surface operators both for $d = 4 - \varepsilon$ and $d = 2$, one can expect this also for $d = 3$.

[23] Diehl, H.W., Dietrich, S., Eisenriegler, E. (1983), Phys. Rev. B**27**, 2937

[24] Cardy, J.L. (1990), Phys. Rev. Lett. **65**, 1443

[25] Eisenriegler, E., Krech, M., Dietrich, S. (1996), Phys. Rev. B**53**, 14377

[26] Eisenriegler, E. (1997), Phys. Rev. E**55**, 3116. In a Monte–Carlo simulation of a self–avoiding chain between two parallel repulsive walls Milchev, A. and Binder, K. (preprint 1997) found effective density–force amplitudes with the predicted order of magnitude. However, simulations with still larger chain lengths and distances between the walls are needed for a test of universality and an accurate estimate of B.

[27] Cardy, J.L., Guttmann, A.J. (1993), J. Phys. (London) A**26**, 2485

[28] Cardy, J., Mussardo, G. (1993), Nucl. Phys. B**410**, 451

[29] Eisenriegler, E., Hanke, A., Dietrich, S. (1996), Phys. Rev. E**54**, 1134

[30] In Ref. [29] the spatial dimension d is denoted by D and the amplitude A in Eq. (6.1) by $-\tilde{A}_K/D^{1-(1/2\nu)}$, with K denoting a sphere.

[31] Burkhardt, T.W., Eisenriegler, E. (1985), J. Phys. (London) A**18**, L83

[32] Burkhardt, T.W., Eisenriegler, E. (1995), Phys. Rev. Lett. **74**, 3189

[33] Eisenriegler, E., Ritschel, U. (1995), Phys. Rev. B51, 13717

[34] de Gennes, P.G. (1979), C.R. Acad. Sci. Paris 288B, 359

[35] Odijk, T. (1996) Macromol. 29, 1842; J. Chem. Phys. 106, 3402

[36] Sear, R.P. (1997) J. Phys. II (Paris) 7, 877

[37] Hanke, A., Eisenriegler, E., Dietrich, S., in preparation.

[38] Joanny, J.F. (1988), J. Phys. (Paris) 49, 1981

[39] Burkhardt, T.W., Xue, T. (1991), Phys. Rev. Lett. 66, 895

[40] Burkhardt, T.W., Eisenriegler, E. (1994), Nucl. Phys. B424, 487

[41] This expression follows from the more general [29] expression $Y_m(D/\mathcal{R}_x) \cdot F_s/(k_B T n_b)$ valid for $R \ll D, \mathcal{R}_x$ on expanding Y_m for small argument using Eq. (5.4a).

[42] Compare, e.g., Lekkerkerker, H.N.W. (1997), Physica A224, 227

[43] For ideal chains in the presence of spheres compare also the computer simulation studies by Meijer, E.J., Frenkel, D. (1994), J. Chem. Phys. 100, 6873

[44] That $\lambda^{-2}\mathcal{R}_x^{2/\nu}/\xi^{2/\nu}$ is a universal ratio follows e.g. from calculating $\mathcal{R}_e^2$ from (2.6) on inverting the Laplace transform (2.19) and on using the scaling form (3.7) of the order–parameter correlation function.

Random Walks in Field Theory

Andreas Pordt *

Institut für Theoretische Physik I, Universität Münster,
Wilhelm-Klemm-Str. 9, D-48149 Münster, Germany,
e-mail: pordt@uni-muenster.de

Abstract. We present basic definitions and various examples for random walk models. The random walk representation by Brydges, Fröhlich and Spencer for the Green functions of the N-component field theory is discussed. As an application it is shown that the continuum limit of the Φ^4 model is trivial in $D > 4$ dimensions.

1 Introduction and Definition of a Random Walk Model

In the fifties random walk models were introduced in theoretical polymer physics studying the behavior of phase transitions. Later in the eighties they were used as a new tool for investigating critical phenomena of classical spin systems and continuum limits of Euclidean field theories. The main object was the so-called triviality problem of the Φ^4-theory in larger or equal to four dimensions. Many results were found by using the random walk representation of Euclidean lattice field theory. The triviality in $D > 4$ dimensions was shown by Aizenman (1981,1982) and Fröhlich (1982). But the principal question of triviality of the Φ^4 model in four dimensions was and is until now not answered.

This talk follows in main parts the book by Fernando, Fröhlich, and Sokal (1992). For an extensive list of literature we refer the reader to this book.

The talk is organized as follows. We start with the basic definitions of a random walk model. Some classes of random walk models are presented in the second section. In the next three sections we give a short introduction to the random walk representation of Euclidean lattice field theory (N-component model). For that we recall the basic definitions of a lattice field theory in section 3. In the fourth section we show that the Gaussian model (free field theory) is equivalent to a simple random walk model. In the fifth section we will prove the Brydges-Fröhlich-Spencer (BFS) representation of classical lattice spin systems (Brydges, Fröhlich, and Spencer (1982)). A variant of the BFS representation is Symanzik's complete polymer representation (Symanzik (1966,1969)) presented in section 6. In section 7 we derive three inequalities for BFS random walks which will be used in a proof of triviality. The triviality of the Φ^4_D-theory in $D > 4$ dimensions is proven in the last section 8.

---

* Supported by Deutsche Forschungsgemeinschaft

We start here with basic definitions of random walk models. First of all we specify the space Λ where the walks "live". The space Λ is a denumerable set[1]. For example Λ is the D-dimensional hypercubic lattice $\mathbb{Z}^D$ or the finite torus $(\mathbb{Z}/L\mathbb{Z})^D = \mathbb{Z}_L^D$. *Walks* $\omega = (\omega(0), \omega(1), \ldots, \omega(k))$ are ordered sequences of elements in Λ. ω consists of k steps $(\omega(i-1), \omega(i))$, $i = 1, \ldots, k$. The number of steps in ω, $|\omega| = k$, is called the *length* of ω. The initial (or final) site of ω is denoted by $b(\omega) := \omega(0)$ (or $e(\omega) := \omega(k)$). For two lattice sites $x, y \in \Lambda$ which are the initial resp. final site of a walk ω, $b(\omega) = x$, $e(\omega) = y$ we write $\omega : x \to y$. The *support* of a walk ω consists of all elements $x \in \Lambda$ "visited" by ω

$$\text{supp}(\omega) := \{x \in \Lambda \,|\, \omega(j) = x \text{ for some } j\} \ . \tag{1.1}$$

Denote by Ω the set of all walks. A random walk model is specified by non-negative functionals $\rho^{[n]} : \Omega^n \to \mathbb{R}_+ \cup \{0\}$ which are called *statistical weights*, for all $n \in \mathbb{N}$. For sequences of n walks $\omega_1, \ldots, \omega_n$ the statistical weight is $\rho(\omega_1, \ldots, \omega_n)$. We define $\rho^{[0]} = 1$ and omit in the following the superscript $[n]$ since it is redundant by the number of arguments of ρ.

Decompose the set $\{x_1, \ldots, x_{2n}\} \subseteq \Lambda$ into n pairs $(x_1 x_2), \ldots (x_{2n-1} x_{2n})$ and define its "kernel" as the sum over all walks $\omega_i : x_{2i-1} \to x_{2i}$

$$K(x_1 x_2 | \cdots | x_{2n-1} x_{2n}) := \sum_{\substack{\omega_1 : x_1 \to x_2 \\ \vdots \\ \omega_n : x_{2n-1} \to x_{2n}}} \rho(\omega_1, \ldots, \omega_n) \ . \tag{1.2}$$

K is called the *generating function* of the random walk model.

Let P_{2n} be the set of all permutations of the set $\{1, \ldots, 2n\}$. Denote by Q_{2n} the set of all permutations $\pi \in P_{2n}$ such that $\pi(2k - 1) < \pi(2k)$, $k = 1, \ldots n$, and $\pi(2l - 1) < \pi(2l + 1)$, $l = 1, \ldots, n - 1$. An element $\pi \in Q_{2n}$ can be considered as a "pairing" $(\pi(1)\pi(2)), \ldots, (\pi(2n - 1)\pi(2n))$ of the set $\{x_1, \ldots, x_{2n}\}$. There are $(2n-1)(2n-3)\cdots 3 = \frac{(2n)!}{2^n n!} = |Q_{2n}|$ of such pairings. The *Green function* of a random walk model is defined by

$$S(x_1, \ldots, x_{2n}) := \sum_{\pi \in Q_{2n}} K(x_{\pi(1)} x_{\pi(2)} | \cdots | x_{\pi(2n-1)} x_{\pi(2n)}) \ . \tag{1.3}$$

A random walk model is called *symmetric* if ρ is symmetric, i. e.

$$\rho(\omega_1, \ldots, \omega_n) = \rho(\omega_{\pi(1)}, \ldots, \omega_{\pi(n)}) \tag{1.4}$$

for all permutations π. Then we see that the Green function S_{2n} is also symmetric and given by the sum over all permutations devided by $2^n n!$

$$S(x_1, \ldots, x_{2n}) = \frac{1}{2^n n!} \sum_{\pi \in S_{2n}} K(x_{\pi(1)} x_{\pi(2)} | \cdots | x_{\pi(2n-1)} x_{\pi(2n)}) \ . \tag{1.5}$$

---

[1] Λ could also be the continuum $\mathbb{R}^D$, which is not denumerable. But we consider here only random walk models on a lattice consisting of a denumerable number of lattice sites.

2 Classification of Random Walk Models

The random walk models studied in this talk are specified by non-negative
activities $J_{xy} \geq 0$ for all steps (xy), $x, y \in \Lambda$ and the *interaction energy*
$U_n(\omega_1, \ldots, \omega_n)$ for n walks $\omega_1, \ldots, \omega_n$. For a given activity J and interaction
energy U define the random walk model by the statistical weight

$$\rho(\omega_1, \ldots, \omega_n) := \prod_{i=1}^{n} J^{\omega_i} \exp\{-U_n(\omega_1, \ldots, \omega_n)\}, \qquad (2.6)$$

where

$$J^{\omega} := \prod_{k=0}^{|\omega|-1} J_{\omega(k)\omega(k+1)} \qquad (2.7)$$

for a walk $\omega = (\omega(0), \ldots, \omega(k))$.

We call a random walk model specified by (2.6) a (general) *polymer-chain
model*. If the interaction energy U is independent of the activity J we say
that ρ depends *trivially* on J. For random walk representation of field theories
we will see that the interaction energy U will depend on the activity J (*deep
J-dependence*).

We call a random walk model specified by (2.6) *repulsive* if

$$U_{m+n}(\omega_1, \ldots, \omega_{m+n}) \geq U_m(\omega_1, \ldots, \omega_m) + U_n(\omega_{m+1}, \ldots, \omega_{m+n}) . \qquad (2.8)$$

Replacing the inequality in (2.8) by an equality if the supports of $(\omega_1, \ldots, \omega_m)$
and $(\omega_{m+1}, \ldots, \omega_{m+n})$ are disjoint we call the random walk model a *generalized contact-interacting (CIW) model*. Thus a CIW-model has the property
that

$$U_{m+n}(\omega_1, \ldots, \omega_{m+n}) = U_m(\omega_1, \ldots, \omega_m) + U_n(\omega_{m+1}, \ldots, \omega_{m+n}) \qquad (2.9)$$

if $\bigcup_{i=1}^{m} \text{supp}(\omega_i) \cap \bigcup_{j=1}^{n} \text{supp}(\omega_{m+j}) = \emptyset$.

For a walk ω and a lattice site $x \in \Lambda$ define the *visitation function* $n_x(\omega)$
by the number of times ω "visits" the site x, i. e.

$$n_x(\omega) := |\{i \mid \omega(i) = x\}| . \qquad (2.10)$$

For a family of walks $\omega_1, \ldots, \omega_n$ define the *total visitation function* by

$$n_x(\omega_1, \ldots, \omega_n) := \sum_{j=1}^{n} n_x(\omega_j) . \qquad (2.11)$$

Simple CIW models have the property that the interaction energy U obeys

$$\exp\{-U_n(\omega_1, \ldots, \omega_n)\} = \prod_{x \in \Lambda} c_x(n_x(\omega_1, \ldots, \omega_n)) \qquad (2.12)$$

for some functions $c_x : \mathbb{N} \to \mathbb{R}_+ \cup \{0\}$, $c_x(0) = 1$, $x \in \Lambda$. A simple CIW model is called *repulsive* if

$$c_x(m+n) \leq c_x(m)\, c_x(n) \ . \tag{2.13}$$

In the following we present four examples of simple CIW models which are specified by the functions c_x, $x \in \Lambda$.

Simple Random Walk.

$$c_x(n) = \tau^{-n}, \qquad \tau > 0 \ . \tag{2.14}$$

Self-avoiding Walk (SAW).

$$c_x(n) = \begin{cases} 1 & : \ \text{if } n = 0, 1, \\ 0 & : \ \text{if } n \geq 2 \ . \end{cases} \tag{2.15}$$

This is equivalent to

$$U_m(\omega_1, \ldots, \omega_m) = \begin{cases} 0 & : \ \text{if supp } \omega_i \cap \text{supp } \omega_j = \emptyset \ \forall i, j, \ i \neq j, \\ \infty & : \ \text{otherwise} \ . \end{cases} \tag{2.16}$$

Domb-Joyce Model.

$$c_x(n) = e^{-vn(n-1)/2}, \qquad v \geq 0 \ . \tag{2.17}$$

If $v = 0$ then the Domb-Joyce model is the simple random walk model (with $\tau = 1$) and if $v = \infty$ it is the SAW model.

Edwards Model.

$$c_x(n) = \int_0^\infty d\nu_n(t)\, \mathcal{Z}(t), \tag{2.18}$$

where

$$d\nu_n(t) = \begin{cases} \delta(t)\, dt & : \ \text{if } n = 0, \\ \frac{t^{n-1}}{(n-1)!}\, dt & : \ \text{if } n \geq 1 \end{cases} \tag{2.19}$$

and

$$\mathcal{Z}(t) = \exp\{-\lambda t^2 - \tau t\}, \qquad \lambda \geq 0 \ . \tag{2.20}$$

For $\lambda = 0$ the Edwards model is the simple random walk model. If $\lambda \to \infty$ and $\tau = -\left(2\lambda \ln(\lambda/\pi)\right)^{1/2}$ then it is a SAW model.

All the above examples of simple CIW models fulfill (2.13) and are therefore repulsive.

For Ising models there is an alternative way of rewriting correlation functions which is called the *Aizenman random walk (ARW) representation* (Aizenman (1982) and (1985)). We will not go into details here concerning this "random current representation" and refer the reader to the original articles or the book by Fernando, Fröhlich, and Sokal (1992).

Brydges, Fröhlich, and Spencer (1982) found a random walk model (BFS model) which writes correlation functions of classical spin models as sums over random walks. This will be the subject of section 5.

Before showing that Euclidean lattice field theories can be represented by random walk models we introduce the basic definitions of Euclidean field theory on a lattice in the following section.

3   Euclidean Lattice Field Theory

For N-component fields $\Phi : \Lambda \to \mathbb{R}^N$ on a lattice Λ define a single-spin measure by

$$dP_x(\Phi_x) = g_x(\Phi_x^2)\, d\Phi_x \qquad (3.21)$$

for all $x \in \Lambda$, where $d\Phi_x$ is the Lebesgue-measure on $\mathbb{R}^N$ and

$$g_x(\Phi_x^2) = \mathrm{e}^{-f_x(\Phi_x^2)}, \qquad (3.22)$$

where f_x is a convex function and growing at least linearly for large Φ_x^2. Define a ferromagnetic non-local action by

$$S(\Phi) = - \sum_{<xy>} J_{xy}\Phi_x\Phi_y, \qquad J_{xy} \ge 0 \ . \qquad (3.23)$$

The sum goes over all bonds $< xy >$ (unoriented pairs of lattice sites $x, y \in \Lambda$, $x \ne y$).

The basic definition of a field theory is the *generating functional of correlation functions*

$$Z[h] = \int \prod_{x \in \Lambda} dP_x(\Phi_x) \exp\{S(\Phi) + \sum_{x \in \Lambda} h_x\Phi_x\} \ . \qquad (3.24)$$

$h : \Lambda \to \mathbb{R}^N$ is called the *external source* and

$$\sum_x h_x\Phi_x = \sum_x \sum_{a=1}^{N} h_x^a\Phi_x^a \qquad (3.25)$$

is called the *source term*. The *n-point correlation functions* are given by differentiation of the generating function $Z[h]$ with respect to the external source h

$$< \Phi_{x_1} \cdots \Phi_{x_n} > = \frac{\partial^n}{\partial h_{x_1} \cdots \partial h_{x_n}} \frac{Z[h]}{Z[0]}\Big|_{h=0} \qquad (3.26)$$

for all $x_1, \ldots, x_n \in \Lambda$. The *n-point connected (truncated) correlation functions* are given by differentiation of the logarithm of the generating functional $\ln Z[h]$

$$< \Phi_{x_1} \cdots \Phi_{x_n} >^c = \frac{\partial^n}{\partial h_{x_1} \cdots \partial h_{x_n}} \ln Z[h]|_{h=0} \, . \qquad (3.27)$$

The relation between these two types of correlation functions is given by the following equation

$$< \Phi_{x_1} \cdots \Phi_{x_n} > = \sum_{\{I\}: \{1,\dots,n\}=\sum I} \prod_I < \prod_{i \in I} \Phi_{x_i} >^c, \qquad (3.28)$$

where the sum goes over all partitions of the set $\{1, \dots, n\}$ into disjoint nonempty subset I.

The definition (3.21) of the single-spin distribution dP_x and the generating functional (3.24) implies invariance under sign-changing $\Phi \to -\Phi$ (global $\mathbb{Z}_2$-symmetry). This implies that the n-point (connected) correlation functions vanish if n is odd. We consider in the following only $\mathbb{Z}_2$-symmetric models. For random walk models with magnetic fields (which are not $\mathbb{Z}_2$-symmetric) we refer the reader to Fernando, Fröhlich, and Sokal (1992), Chapter 9.

Special examples of relation (3.28) for $n = 2, 4$ are given by

$$< \Phi_{x_1} \Phi_{x_2} > = < \Phi_{x_1} \Phi_{x_2} >^c, \qquad (3.29)$$

$$< \Phi_{x_1} \Phi_{x_2} \Phi_{x_3} \Phi_{x_4} > = < \Phi_{x_1} \Phi_{x_2} \Phi_{x_3} \Phi_{x_4} >^c + < \Phi_{x_1} \Phi_{x_2} >^c < \Phi_{x_3} \Phi_{x_4} >^c$$

$$+ < \Phi_{x_1} \Phi_{x_3} >^c < \Phi_{x_2} \Phi_{x_4} >^c$$

$$+ < \Phi_{x_1} \Phi_{x_4} >^c < \Phi_{x_2} \Phi_{x_3} >^c \, . \qquad (3.30)$$

In the generic case the generating functional $Z[h]$ and the correlation functions cannot be calculated in closed form. There is one exception. For a single-spin measure defined by

$$g(\Phi_x^2) = e^{-\frac{\mu}{2} \Phi_x^2}, \qquad \mu > 0 \qquad (3.31)$$

the integral over the field Φ on the right hand side of (3.24) is Gaussian. Gaussian integrals can be calculated explicitly. A field theory defined by a single-spin measure (3.31) is called a *Gaussian (or free) field theory*. Its correlation functions and the relation to random walks are presented in the next section.

4   Gaussian Model and Simple Random Walks

The generating functional of the Gaussian model specified by the single-spin distribution (3.31) is

$$Z[h] = \int \prod_{x \in \Lambda} d\Phi_x \exp\{-\frac{1}{2} \sum_{x,y \in \Lambda} \Phi_x w_{xy} \Phi_y + \sum_x h_x \Phi_x\}, \qquad (4.32)$$

where

$$w_{xy} = \begin{cases} \mu & : \text{ if } x = y \\ -J_{xy} & : \text{ if } x \neq y . \end{cases} \qquad (4.33)$$

This Gaussian integral is explicitly

$$Z[h] = \mathcal{N} \exp\{\frac{1}{2} \sum_{x,y} h_x v_{xy} h_y\}, \qquad (4.34)$$

where v_{xy} is the kernel of the operator

$$v = (\mu \mathbb{1} - J)^{-1}, \qquad (4.35)$$

and where the operator J is defined by

$$(J\Phi)_x = \sum_{y \in \Lambda} J_{xy} \Phi_y . \qquad (4.36)$$

v is called the *(free) propagator* of the Gaussian model. By the definition (3.27) of the truncated correlation function and the explicit result (4.34) for the generating functional we obtain for the connected n-point correlation function

$$< \Phi_{x_1} \cdots \Phi_{x_n} >^c = \begin{cases} v_{xy} & : \text{ if } n = 2, \\ 0 & : \text{ if } n \neq 2 . \end{cases} \qquad (4.37)$$

By (3.28) and (3.29) we obtain for the $2n$-point correlation function

$$< \Phi_{x_1} \cdots \Phi_{x_{2n}} >^c = \sum_{\pi \in Q_{2n}} < \Phi_{x_{\pi(1)}} \Phi_{x_{\pi(2)}} > \cdots < \Phi_{x_{\pi(2n-1)}} \Phi_{x_{\pi(2n)}} >, \quad (4.38)$$

where Q_{2n} denotes the set of all pairings of $\{1, \ldots, 2n\}$ (see section 1). Thus the sum on the right hand side of (4.38) goes over all pairings of $x_1, \ldots, x_{2n}$. Suppose that

$$\|J\| := \sup_x \sum_y |J_{xy}| < \mu . \qquad (4.39)$$

Then the Neumann series expansion of the free propagator

$$v = (\mu - J)^{-1} = \mu^{-1} (\mathbb{1} - \mu^{-1} J)^{-1} = \mu^{-1} \sum_{n \geq 0} (\mu^{-1} J)^n \qquad (4.40)$$

is convergent. The corresponding expansion of the kernel of the propagator is

$$v_{xy} = \sum_{n \geq 0} \mu^{-(n+1)} \sum_{x_1, \ldots, x_{n-1} \in \Lambda} J_{xx_1} J_{x_1 x_2} \cdots J_{x_{n-1} y}$$

$$= \sum_{\omega : x \to y} J^\omega \prod_{z \in \Lambda} \mu^{-n_z(\omega)} . \qquad (4.41)$$

By (4.37, 4.38, 4.41), the definitions (1.2, 1.3, 2.6) and the definition of the simple random walk model in section 2 we see that

$$< \Phi_{x_1} \cdots \Phi_{x_{2n}} >= S_{2n}(x_1, \ldots, x_{2n}), \tag{4.42}$$

where $S_{2n}(x_1, \ldots, x_{2n})$ is the Green function for the simple random walk model specified by

$$c_x(n) = \mu^{-n}, \qquad \tau > 0 . \tag{4.43}$$

We will show in the next section that the fundamental identity (4.42) holds also for non-Gaussian models.

5 Random Walk Representation and Euclidean Field Theory

We consider here a random walk representation of a 1-component Euclidean field theory defined in section 3. The restriction to 1-components is only for notational convenience. A generalization to N-components is obvious.

It was shown by Brydges, Fröhlich, and Spencer (1982) that there exists a statistical weight ρ of a random walk model such that the fundamental identity

$$< \Phi_{x_1} \cdots \Phi_{x_{2n}} >= S_{2n}(x_1, \ldots, x_{2n}) \tag{5.44}$$

holds for all $x_1, \ldots, x_{2n} \in \Lambda$. For k walks $\omega_1, \ldots, \omega_k$ the statistical weight ρ defining the Green function S_{2n} of the random walk model is

$$\rho(\omega_1, \ldots, \omega_k) = J^{\omega_1 + \cdots + \omega_k} \int \mathcal{Z}(\underline{t}_1 + \cdots + \underline{t}_k) \prod_{i=1}^{k} d\nu_{\omega_i}(\underline{t}_i) \tag{5.45}$$

where

$$\mathcal{Z}(\underline{t}) := \frac{1}{Z[0]} \int [\prod_{x \in \Lambda} d\Phi_x \, g_x(\Phi_x^2 + 2t_x)] \, e^{\frac{1}{2} \sum_{x,y} \Phi_x J_{xy} \Phi_y} \tag{5.46}$$

for all $\underline{t} : \Lambda \to \mathbb{R}_+ \cup \{0\}$, and

$$d\nu_\omega(\underline{t}) := \prod_{x \in \Lambda} d\nu_{n_x(\omega)}(t_x) . \tag{5.47}$$

$d\nu_n(t)$ is defined by (2.19).

We will prove the fundamental identity (5.44). Let us use the abbreviations

$$(\Phi, U\Phi) := \sum_{x,y} \Phi_x U_{xy} \Phi_y, \tag{5.48}$$

$$[d\Phi] := \prod_{x \in \Lambda} d\Phi_x . \tag{5.49}$$

By the fundamental theorem of calculus we have

$$\int [d\Phi] \frac{\partial}{\partial \Phi_y} \left[F(\Phi) e^{-\frac{1}{2}(\Phi, U\Phi)} \right] = 0 \ . \tag{5.50}$$

Thus

$$\int [d\Phi] \left[\frac{\partial F}{\partial \Phi_y}(\Phi) - \sum_z{}' U_{yz}\Phi_z F(\Phi) \right] e^{-\frac{1}{2}(\Phi, U\Phi)} = 0 \ . \tag{5.51}$$

Suppose that U is invertible with a positive-definite real part and F is a polynomially bounded function. Multiplication of both sides of (5.51) by U^{-1} yields the Gaussian integration-by-parts formula

$$\int [d\Phi]\, \Phi_x F(\Phi) e^{-\frac{1}{2}(\Phi, U\Phi)} = \sum_y (U^{-1})_{xy} \int [d\Phi]\, \frac{\partial F}{\partial \Phi_y} e^{-\frac{1}{2}(\Phi, U\Phi)} \ . \tag{5.52}$$

Insertion of the Fourier transform

$$g_z(\Phi_z^2) = \int_{-\infty}^{\infty} da_z\, \widetilde{g}_z(a_z)\, e^{-ia_z \Phi_z^2} \tag{5.53}$$

into the term

$$< \Phi_x F(\Phi) > = \frac{1}{Z[0]} \int [\prod_{z\in\Lambda} d\Phi_z\, g_z(\Phi_z^2)]\, \Phi_x F(\Phi)\, e^{\frac{1}{2}\sum_{x,y} \Phi_x J_{xy}\Phi_y} \tag{5.54}$$

and using the Gaussian integration-by-parts formula (5.52) yields

$$< \Phi_x F(\Phi) > =$$

$$\frac{1}{Z[0]} \sum_y \int [d\Phi] \int_{\Im a_z = -C < 0} [da_z \widetilde{g}_z(a_z)]\, (2ia - J)^{-1}_{xy}$$

$$\frac{\partial F}{\partial \Phi_y} e^{\frac{1}{2}(\Phi,(J-2ia)\Phi)} \tag{5.55}$$

The interchange of the order of integration and the shift of the path of integration to $\Im a_z = -C = 0$ is allowed since $\widetilde{g}_z$ is an entire analytic function decaying faster than any inverse power as $|\Re a_z| \to \infty$. This follows from the supposed properties of the single-spin distribution dP_x. For $C = -\Im a_z$ sufficiently large the Neumann series of $(2ia - J)^{-1}$ converges. Thus

$$(2ia - J)^{-1} = \sum_{\omega:\, x\to y} J^{\omega} \prod_{z\in\Lambda} (2ia)^{-n_z(\omega)} \tag{5.56}$$

and

$$< \Phi_x F(\Phi) > =$$

$$\frac{1}{Z[0]} \sum_y \sum_{\omega:\, x\to y} J^{\omega} \int [d\Phi] \int_{\Im a_z = -C < 0} [da_z \widetilde{g}_z(a_z)]$$

$$\left[\prod_z (2ia_z)^{-n_z(\omega)} \right] \frac{\partial F}{\partial \Phi_y} e^{\frac{1}{2}(\Phi,(J-2ia)\Phi)} \ . \tag{5.57}$$

Using the identity

$$b^{-n} = \int d\nu_n(t)\, e^{-bt}, \qquad \Re b > 0 \tag{5.58}$$

we obtain

$$< \Phi_x F(\Phi) > =$$

$$\frac{1}{Z[0]} \sum_y \sum_{\omega:\, x \to y} J^\omega \int d\nu_\omega(\underline{t})$$

$$\int [d\Phi] \prod_z g_z(\Phi_z^2 + 2t_z) \frac{\partial F}{\partial \Phi_y} \, e^{\frac{1}{2}(\Phi, J\Phi)} \ . \tag{5.59}$$

We have derived the equation

$$\int [d\Phi] \prod_z g_z(\Phi_z^2)\, \Phi_x F(\Phi) e^{\frac{1}{2}(\Phi, J\Phi)} =$$

$$\sum_y \sum_{\omega:\, x \to y} J^\omega \int d\nu_\omega(\underline{t})$$

$$\int [d\Phi] \prod_z g_z(\Phi_z^2 + 2t_z) \frac{\partial F}{\partial \Phi_y} \, e^{\frac{1}{2}(\Phi, J\Phi)} \ . \tag{5.60}$$

For $\Phi_{x_1} F(\Phi) = \Phi_{x_1} \cdots \Phi_{x_{2n}}$ we derive by repeated application of relation (5.60) the fundamental identity (5.44). $\square$

The random walk model defined by the statistical weight (5.45) were introduced by Brydges, Fröhlich, and Spencer (1982). It is called the *Brydges-Fröhlich-Spencer model (BFS)*.

We have shown that the Green function for the random walk model is equal to the correlation function of the field theoretic model. In the next section we will introduce a random walk representation due to Symanzik (1966) and (1969) of the correlation functions where the statistical weights are represented by sums over "loops".

6   Symanzik's Complete Polymer Representation

There is an alternative way of writing the partition and correlation functions. It is a fully expanded high-temperature expansion. The partition function is thereby represented as a gas of closed random walks by Symanzik (1966) and (1969)(see also Brydges (1982)). For that, we consider an N-component field theory and present a random walk representation such that the fundamental identity

$$< \Phi_{x_1}^a \cdots \Phi_{x_{2n}}^a > = S_{2n}(x_1, \ldots, x_{2n}) \tag{6.61}$$

holds for all indices $a \in \{1, \ldots, N\}$. S_{2n} is the Green function for a random walk model defined by the statistical weight

$$\rho(\omega_1, \ldots, \omega_k) =$$

$$J^{\omega_1 + \cdots + \omega_k} \frac{1}{Z[0]} \sum_{n=0}^{\infty} \frac{1}{n!} \left(\frac{N}{2}\right)^n \sum_{x_1, \ldots x_n \in \Lambda} \sum_{\substack{\underline{\omega}_1 : x_1 \to x_1 \\ \vdots \\ \underline{\omega}_n : x_n \to x_n}}$$

$$\frac{J^{\underline{\omega}_1 + \cdots + \underline{\omega}_n}}{|\underline{\omega}_1| \cdots |\underline{\omega}_n|} \, e^{-U^N_{n+k}(\omega_1, \ldots \omega_k, \underline{\omega}_1, \ldots, \underline{\omega}_n)}, \qquad (6.62)$$

where

$$e^{-U^N_k(\omega_1, \ldots \omega_k)} = \prod_{x \in \Lambda} c_x(n_x(\omega_1, \ldots \omega_k)), \qquad (6.63)$$

and

$$c_x(n) := \frac{\int d\nu_{n+\frac{N}{2}}(t) \, g_x(2t)}{\int d\nu_{\frac{N}{2}}(t) \, g_x(2t)} . \qquad (6.64)$$

The partition function is

$$Z[0] = \sum_{n=0}^{\infty} \frac{1}{n!} \left(\frac{N}{2}\right)^n \sum_{x_1, \ldots x_n \in \Lambda} \sum_{\substack{\underline{\omega}_1 : x_1 \to x_1 \\ \vdots \\ \underline{\omega}_n : x_n \to x_n}}$$

$$\frac{J^{\underline{\omega}_1 + \cdots + \underline{\omega}_n}}{|\underline{\omega}_1| \cdots |\underline{\omega}_n|} \, e^{-U^N_k(\underline{\omega}_1, \ldots, \underline{\omega}_n)} . \qquad (6.65)$$

For a proof see Brydges, Fröhlich, and Spencer (1982).

For $N = 0$ there is a relation between the N-component model and the polymer-chain (simple CIW) model (de Gennes (1970)). In the limit $N \to 0$ there are no loops in (6.62, 6.65)

$$\lim_{N \to 0} Z[0] = 1, \qquad (6.66)$$

$$\lim_{N \to 0} \rho(\omega_1, \ldots, \omega_k) = J^{\omega_1 + \cdots + \omega_k} \exp\{-U_k(\omega_1, \ldots, \omega_k)\} \qquad (6.67)$$

where $U_k = \lim_{N \to 0} U^N_k$ is a simple contact interaction defined by (2.12) with

$$c_x(n) := \frac{\int d\nu_n(t) \, g_x(2t)}{g_x(0)} . \qquad (6.68)$$

Thus the N-component model in the limit $N \to 0$ corresponds to a simple CIW model and ρ depends trivially on the activity J.

In the case where the N-component fields Φ "live" on the $(N-1)$-dimensional sphere (*non-linear σ-model*), i. e. the single-spin distribution obeys

$$g_x(\Phi^2_x) \propto \delta(\Phi^2_x - N) \qquad (6.69)$$

we see by (6.68) that in the limit $N \to 0$

$$c_x(n) = \begin{cases} 1 & : \quad \text{if } n = 0, 1, \\ 0 & : \quad \text{if } n > 1 \ . \end{cases} \qquad (6.70)$$

Thus the N-component model on the sphere with radius N corresponds to a SAW model in the limit $N \to 0$.

7   Inequalities

In this section we will prove some inequalities for random walk models in the BFS class (cp. Fernando, Fröhlich, and Sokal (1992)). These inequalities are useful for proving properties of Euclidean field theories. We will use the inequalities in section 8 to show that the continuum limit of the Φ_D^4-theory in $D > 4$ dimensions is trivial.

Let us suppose that the statistical weight ρ is of the form (5.45) and the single-spin distribution dP_x is given by (3.21, 3.22). Denote by ω_I for an index set $I = \{i_1, \ldots i_n\} \subset \mathbb{N}$ and the family of random walks $\omega_{i_1}, \ldots, \omega_{i_n}$. The following inequalities hold :

I. For all families of walks $\omega_1', \ldots, \omega_l'$ we have

$$\sum_{\pi \in Q_{2k}} \sum_{\substack{\omega_j : x_{\pi(2j-1)} \to x_{\pi(2j)} \\ j=1,\ldots,k}} \rho(\omega_1, \ldots, \omega_k, \omega_1', \ldots, \omega_l') \qquad (7.71)$$

$$\leq \left(\sum_{\pi \in Q_{2k}} \sum_{\substack{\omega_j : x_{\pi(2j-1)} \to x_{\pi(2j)} \\ j=1,\ldots,k}} \rho(\omega_1, \ldots, \omega_k) \right) \rho(\omega_1', \ldots, \omega_l') \ . \quad (7.72)$$

II. For all families of walks $\omega_1, \ldots, \omega_k$ and $\omega_1', \ldots, \omega_l'$ sharing no commmon site,

$$\left(\bigcup_{i=1}^{k} \text{supp } \omega_i \right) \cap \left(\bigcup_{j=1}^{l} \text{supp } \omega_j' \right) = \emptyset \qquad (7.73)$$

we have

$$\rho(\omega_1, \ldots, \omega_k, \omega_1', \ldots, \omega_l') \geq \rho(\omega_1, \ldots, \omega_k) \, \rho(\omega_1', \ldots, \omega_l') \ . \qquad (7.74)$$

III. For all lattice sites $z \in \Lambda$

$$\sum_{\substack{\omega : x \to y \\ x \in \text{supp } \omega}} \rho(\omega) \leq \sum_{z' \in \Lambda} J_{zz'} \sum_{\substack{\omega_1 : x \to z \\ \omega_2 : z' \to y}} \rho(\omega_1, \omega_2) + \sum_{\omega_1 : x \to y} \rho(\omega_1) \, \delta_{yz} \qquad (7.75)$$

Proof of inequality I : We will prove (7.71) only for the case $k = l = 1$. Generalization is straightforward. By definition (5.45) we have

$$\rho(\omega_1, \omega_2) = J^{\omega_1 + \omega_2} \int \mathcal{Z}(\underline{t}_1 + \underline{t}_2) \, d\nu_{\omega_1}(\underline{t}_1) d\nu_{\omega_2}(\underline{t}_2)$$

$$= J^{\omega_1} \int d\nu_{\omega_1}(\underline{t}_1) \, \mathcal{Z}(\underline{t}_1) \int d\nu_{\omega_2}(\underline{t}_2) \frac{\mathcal{Z}(\underline{t}_1 + \underline{t}_2)}{\mathcal{Z}(\underline{t}_1)}$$

$$= J^{\omega_1} \int d\nu_{\omega_1}(\underline{t}_1) \, \mathcal{Z}(\underline{t}_1) \, \rho(\omega_2)_{\underline{t}_1} \qquad (7.76)$$

where $\rho_{\underline{t}}$ is defined in the same way as ρ where each $g_x(\Phi_x^2)$ is replaced by $g_x(\Phi_x^2 + 2t_x)$. Summing over all $\omega_2 : x \to y$ we obtain

$$\sum_{\omega_2 : x \to y} \rho(\omega_1, \omega_2) = J^{\omega_1} \int d\nu_{\omega_1}(\underline{t}_1) \, \mathcal{Z}(\underline{t}_1) < \Phi_x \Phi_y >_{\underline{t}}, \qquad (7.77)$$

where $< \cdot >_{\underline{t}}$ denotes the expectation value of a model defined by the single-spin distribution $dP_x = g_x(\Phi_x^2 + 2t_x) \, d\Phi_x$. Furthermore

$$\frac{\partial}{\partial t_y} < \Phi_x \Phi_y >_{\underline{t}} = -2 < \Phi_x \Phi_y; f_y'(\Phi_y^2 + 2t_y) >_{\underline{t}} \qquad (7.78)$$

where we have used the notation

$$< A; B > = < A \cdot B > - < A > < B > . \qquad (7.79)$$

Since f_y is a convex function we see that $f_y'(\Phi_y^2 + 2t_y)$ is non-decreasing for $\Phi_y \geq 0$. By Griffiths second inequality (see Glimm and Jaffe (1987)) follows

$$< \Phi_x \Phi_y; f_y'(\Phi_y^2 + 2t_y) >_{\underline{t}} \geq 0 . \qquad (7.80)$$

Thus

$$\frac{\partial}{\partial t_y} < \Phi_x \Phi_y >_{\underline{t}} \leq 0 \qquad (7.81)$$

and therefore

$$< \Phi_x \Phi_y >_{\underline{t}} \leq < \Phi_x \Phi_y > . \qquad (7.82)$$

Thus (7.77,7.82) and the fundamental identity (5.44) imply

$$\sum_{\omega_2 : x \to y} \rho(\omega_1, \omega_2) \leq \rho(\omega_1) \sum_{\omega_2 : x \to y} \rho(\omega_2) \qquad \square \qquad (7.83)$$

Proof of inequality II : We will prove (7.74) only for the case $k = l = 1$. Generalization is straightforward. The inequality

$$\rho(\omega_1, \omega_2) \geq \rho(\omega_1) \, \rho(\omega_2), \qquad \text{supp } \omega_1 \cap \text{supp } \omega_2 = \emptyset \qquad (7.84)$$

follows immediately from

$$\mathcal{Z}(\underline{t} + \underline{t}') \geq \mathcal{Z}(\underline{t}) \, \mathcal{Z}(\underline{t}'), \qquad \text{if } t_x t_x' = 0 \text{ for all } x \in \Lambda . \qquad (7.85)$$

Therefore it remains to prove the inequality (7.85). We have for all $\alpha \geq 0$

$$\frac{d}{d\alpha} \mathcal{Z}(\underline{t} + \alpha \underline{t}') = -2 \sum_x t'_x < f'_x(\Phi_x^2 + 2t_x + 2\alpha t'_x) >_{\underline{t}+\alpha \underline{t}'}$$

$$= -2 \sum_x t'_x < f'_x(\Phi_x^2 + 2\alpha t'_x) >_{\underline{t}+\alpha \underline{t}'}$$

$$\geq -2 \sum_x t'_x < f'_x(\Phi_x^2 + 2\alpha t'_x) >_{\alpha \underline{t}'}$$

$$= \frac{d}{d\alpha} \mathcal{Z}(\alpha \underline{t}'), \tag{7.86}$$

where the second equality follows from $t_x t'_x = 0$ for all $x \in \Lambda$, and the inequality follows in the same way as inequality (7.81) by using the convexity of f_x and Griffiths second inequality. Finally, by the fundamental theorem of calculus and (7.86) we obtain

$$\ln \mathcal{Z}(\underline{t} + \underline{t}') = \ln \mathcal{Z}(\underline{t}) + \int_0^1 d\alpha \, \frac{d}{d\alpha} \ln \mathcal{Z}(\underline{t} + \alpha \underline{t}')$$

$$\geq \ln \mathcal{Z}(\underline{t}) + \int_0^1 d\alpha \, \frac{d}{d\alpha} \ln \mathcal{Z}(\alpha \underline{t}')$$

$$= \ln \mathcal{Z}(\underline{t}) + \ln \mathcal{Z}(\underline{t}') \ . \tag{7.87}$$

The last equality follows by $\mathcal{Z}(\underline{0}) = 1$. $\square$

Proof of inequality III : By the definition of the statistical weight ρ and the visitation function n_x we have

$$\sum_{\substack{\omega: \ x \to y \\ z \in \text{supp } \omega}} \rho(\omega) =$$

$$\sum_{\omega': \ x \to z} \chi(n_z(\omega') = 1) \left(\rho(\omega') \delta_{zy} + \sum_{z' \in \Lambda} J_{zz'} \sum_{\omega'': \ z' \to z} \rho(\omega', \omega'') \right) \tag{7.88}$$

where $\chi(n_z(\omega') = 1)$ is the indicator function which is equal to one if the argument is true and zero otherwise. The trick to obtain (7.88) is to split the walk ω the *first* time it visits the lattice site z. Since $\rho(\omega) \geq 0$ and $J_{zz'} \geq 0$ we obtain the inequality (7.75) by (7.88). $\square$

8   Triviality

We consider in this section the Φ^4-theory on the lattice $\Lambda = \mathbb{Z}^D$ defined by (3.21,3.23,3.24) and the weight

$$g(\Phi^2) = \exp\{-\frac{\lambda}{4!} \Phi^4 - \frac{\mu}{2} \Phi^2\} \tag{8.89}$$

for $\lambda > 0$. Furthermore, suppose that

$$J_{xy} = \begin{cases} J & : \quad \text{if } \|x - y\| = 1, \\ 0 & : \quad \text{if } \|x - y\| \neq 1, \end{cases} \tag{8.90}$$

where $\|x\| := \sum_{\mu=1}^{D} |x^\mu|$, $\frac{\mu}{2D} \geq J \geq 0$. For

$$m^2 = \mu - 2JD \geq 0 \tag{8.91}$$

we have

$$Z[h] = \int [d\Phi]\, e^{-\frac{1}{2}(\Phi,(J(-\Delta)+m^2)\Phi) - \sum_x \left(\frac{\lambda}{4!}\Phi_x^4 - h_x \Phi_x\right)} . \tag{8.92}$$

For Θ, $1 \leq \Theta < \infty$ and $x_1, \ldots, x_n \in \Theta^{-1}\mathbb{Z}^D$ define scaled correlation functions by

$$G_\Theta(x_1, \ldots, x_n) := \alpha(\Theta)^{2n} < \Phi_{\Theta x_1} \cdots \Phi_{\Theta x_{2n}} >_{\mu(\Theta), \lambda(\Theta), h(\Theta)}, \tag{8.93}$$

where $\alpha(\Theta), \mu(\Theta), \lambda(\Theta), h(\Theta)$ are chosen such that the limit

$$G^*(x_1, \ldots, x_n) := \lim_{\Theta \to \infty} G_\Theta(x_1, \ldots, x_n) \tag{8.94}$$

exists for all $x_1, \ldots, x_n \in \mathbb{R}^D$ and

$$0 < \lim_{\Theta \to \infty} G_\Theta(x, y) = G^*(x, y) < \infty \tag{8.95}$$

holds for all $x, y \in \mathbb{R}^D$, $0 < \|x - y\| < \infty$. This limit is called the *continuum limit* of the Φ^4-theory.

Since by Griffiths second inequality (see Glimm and Jaffe (1987))

$$\frac{\partial}{\partial \lambda} < \Phi_x \Phi_y > = -\frac{1}{4!} \sum_z < \Phi_x \Phi_y; \Phi_z^4 > \leq 0 \tag{8.96}$$

we see that the *infrared bound* holds

$$< \Phi_x \Phi_y >_\lambda \leq < \Phi_x \Phi_y >_{\lambda=0} = (J(-\Delta) + m^2)_{xy}^{-1} \leq const\, \|x - y\|^{2-D} . \tag{8.97}$$

In order that (8.95) holds we have to suppose

$$\alpha(\Theta) \geq const\, \Theta^{\frac{D}{2}-1}, \qquad \text{if } \Theta \to \infty . \tag{8.98}$$

By the infrared bound (8.97) follows

$$G^*(x, y) \leq const\, \|x - y\|^{2-D} \tag{8.99}$$

for all $x, y \in \mathbb{R}^D$.

The continuum limit is called *trivial* if all correlation functions can be expressed in terms of 2-point correlation functions

$$G^*(x_1, \ldots, x_{2n}) = \sum_{\pi \in Q_{2n}} \prod_{j=1}^{n} G^*(x_{\pi(2j-1)}, x_{\pi(2j)}) \tag{8.100}$$

and the n-point correlation functions vanish for odd n. This is equivalent to

$$G^{*c}(x_1, \ldots, x_n) = 0, \qquad \text{unless } n \neq 2 . \tag{8.101}$$

We will show that if the continuum limit exists in $D > 4$ dimensions for the Φ^4-theory *then* it is trivial.

It can be shown (see the contribution *Polymer Expansion in Particle Physics*, section 3, of this volume) that the connected correlation functions are related to the (not connected) correlation functions by

$$< \Phi_{x_1} \cdots \Phi_{x_{2n}} >^c =$$

$$\sum_{k=1}^{n}(-1)^{k-1}(k-1)! \sum_{I_1 + \cdots + I_k = \{1, \ldots, 2n\}} \prod_{a=1}^{k} < \prod_{i \in I_a} \Phi_{x_i} >, \tag{8.102}$$

where the last sum on the right hand side of (8.102) goes over all partitions of $\{1, \ldots, 2n\}$ into disjoint nonempty subsets $I_1, \ldots, I_k$. By (1.2,1.3) we have

$$< \Phi_{x_1} \cdots \Phi_{x_{2n}} > = \sum_{\pi \in Q_{2n}} \sum_{\substack{\omega_1 : x_{\pi(1)} \to x_{\pi(2)} \\ \vdots \\ \omega_n : x_{\pi(2n-1)} \to x_{\pi(2n)}}} \rho(\omega_1, \ldots, \omega_n) . \tag{8.103}$$

By (8.102)

$$< \Phi_{x_1} \cdots \Phi_{x_{2n}} >^c = \sum_{\substack{\omega_1 : x_{\pi(1)} \to x_{\pi(2)} \\ \vdots \\ \omega_n : x_{\pi(2n-1)} \to x_{\pi(2n)}}} \rho^T(\omega_1, \ldots, \omega_n), \tag{8.104}$$

where

$$\rho^T(\omega_1, \ldots, \omega_n) = \sum_{k=1}^{n}(-1)^{k-1}(k-1)! \sum_{I_1 + \cdots + I_k = \{1, \ldots, n\}} \prod_{a=1}^{k} \rho(\omega_{I_a}) . \tag{8.105}$$

The inverse relation of (8.105) is

$$\rho(\omega_1, \ldots, \omega_n) = \sum_{\sum I = \{1, \ldots, n\}} \prod_{I} \rho^T(\omega_I) . \tag{8.106}$$

The BFS model obeys

$$\rho(\omega_I, \omega_{I'}) = \rho(\omega_I)\,\rho(\omega_{I'}), \qquad \text{if } \operatorname{supp} \omega_I \cap \operatorname{supp} \omega_{I'} = \emptyset . \tag{8.107}$$

For a family of walks $\omega_1, \ldots, \omega_n$ define a diagram $\gamma(\omega_1, \ldots, \omega_n)$ consisting of n vertices $1, \ldots, n$ and edges (ab) iff $\operatorname{supp} \omega_a \cap \operatorname{supp} \omega_b \neq \emptyset$. The above factorization property (8.107) and the definition of ρ^T yields

$$\rho^T(\omega_1, \ldots, \omega_n) = 0, \qquad \text{unless } \gamma(\omega_1, \ldots, \omega_n) \text{ is connected} . \tag{8.108}$$

Recall that a diagram γ is connected iff for all $a, b \in \{1, \ldots, n\}$, $a \neq b$, there exists vertices $a_1, \ldots, a_{k-1}$ such that the edges $(aa_1), (a_1a_2), \ldots, (a_{k-1}b)$ are contained in γ. By the property (8.108) and (8.102, 8.103) follows

$$
|< \Phi_{x_1} \cdots \Phi_{x_{2n}} >^c | \leq
$$
$$
\sum_{\substack{\pi \in Q_{2n} \\ \gamma(\omega_1, \ldots, \omega_n) \text{ connected}}} \sum_{\omega_j : x_{\pi(2j-1)} \to x_{\pi(2j)}}
$$
$$
\sum_{k=1}^n (k-1)! \cdot \sum_{\sum_{a=1}^k I_a = \{1, \ldots, 2n\}} \prod_{a=1}^k \rho(\omega_{I_a}) \ . \tag{8.109}
$$

Using inequality I of section (7) we obtain

$$
|< \Phi_{x_1} \cdots \Phi_{x_{2n}} >^c | \leq
$$
$$
K_n \sum_{\substack{\pi \in Q_{2n} \\ \gamma(\omega_1, \ldots, \omega_n) \text{ connected}}} \sum_{\omega_j : x_{\pi(2j-1)} \to x_{\pi(2j)}} \prod_{j=1}^n \rho(\omega_j) \ . \tag{8.110}
$$

The sum over k, the factorial $(k-1)!$, and the partitions of $\{1, \ldots, n\}$ on the right hand side of (8.109) is bounded by the constant K_n.

We call a diagram γ with vertices $1, \ldots, n$ a *tree* iff γ is connected and consists of $n - 1$ edges (or equivalently contains no loops). A diagram γ' is called a *subdiagram* of γ, $\gamma' \subseteq \gamma$ iff all vertices and edges of γ' are contained in γ. For a connected diagram γ there exists trees $\tau \subseteq \gamma$ with the same number of vertices. τ is called a *spanning tree* of γ. For each connected diagram γ we select a special spanning tree $\tau_*(\gamma)$. We may write the right hand side of (8.110) as a sum over all trees τ with vertex set $\{1, \ldots, n\}$

$$
|< \Phi_{x_1} \cdots \Phi_{x_{2n}} >^c | \leq
$$
$$
K_n \sum_{\pi \in Q_{2n}} \sum_{\tau} \sum_{\substack{\omega_j : x_{\pi(2j-1)} \to x_{\pi(2j)}, \ \tau = \tau_*(\gamma) \\ \gamma(\omega_1, \ldots, \omega_n) \text{ connected}}} \prod_{j=1}^n \rho(\omega_j) \ . \tag{8.111}
$$

For each edge $(ab) \in \tau$ there is associated a pair of walks ω_a, ω_b such that supp $\omega_a \cap$ supp $\omega_b \neq \emptyset$. For each spanning tree τ we define a linear order relation $\prec_\tau$ on $\{1, \ldots, n\}$ such that for all $b \in \{2, \ldots, n\}$ there exists an element $a \in \{1, \ldots, n-1\}$, $a \prec_\tau b$, obeying $(ab) \in \tau$. For each edge $l = (ab) \in \tau$ there exists a lattice site $z_l \in \Lambda$ such that $z_l \in$ supp $\omega_a \cap$ supp ω_b. Thus

$$
|< \Phi_{x_1} \cdots \Phi_{x_{2n}} >^c | \leq
$$
$$
K_n \sum_{\pi \in Q_{2n}} \sum_{\tau} \sum_{\substack{z_l \in \Lambda \\ l = (ab) \in \tau}} \sum_{\substack{\omega_j : x_{\pi(2j-1)} \to x_{\pi(2j)} \\ \tau = \tau_*(\gamma), \ z_l \in \text{supp } \omega_a \cap \text{supp } \omega_b \\ \gamma(\omega_1, \ldots, \omega_n) \text{ connected}}} \prod_{j=1}^n \rho(\omega_j) \ . \tag{8.112}
$$

Consider now the special case $n = 2$. We have

$$| < \Phi_{x_1} \cdots \Phi_{x_4} >^c | \le K_2 \sum_{\pi \in Q_4} \sum_{z \in \Lambda} \sum_{\substack{\omega_1 : x_{\pi(1)} \to x_{\pi(2)}, \ \omega_2 : x_{\pi(3)} \to x_{\pi(4)} \\ z \in \text{supp}\, \omega_1 \cap \text{supp}\, \omega_2}} \rho(\omega_1)\, \rho(\omega_2) \ .$$

(8.113)

Using inequality III of section 7 we obtain

$$\sum_{z \in \Lambda} \sum_{\substack{\omega_1 : x_{\pi(1)} \to x_{\pi(2)}, \ \omega_2 : x_{\pi(3)} \to x_{\pi(4)} \\ z \in \text{supp}\, \omega_1 \cap \text{supp}\, \omega_2}} \rho(\omega_1)\, \rho(\omega_2) \le$$

$$\sum_{z \in \Lambda} \left(\sum_{z_1 \in \Lambda} J_{zz_1} \sum_{\substack{\omega_1' : x_1 \to z \\ \omega_1'' : x_1 \to x_2}} \rho(\omega_1')\, \rho(\omega_1'') + \sum_{\omega_1' : x_1 \to x_2} \rho(\omega_1') \delta_{x_2 z} \right)$$

$$\left(\sum_{z_2 \in \Lambda} J_{zz_2} \sum_{\substack{\omega_2' : x_3 \to z \\ \omega_2'' : x_2 \to x_4}} \rho(\omega_2')\, \rho(\omega_2'') + \sum_{\omega_2' : x_3 \to x_4} \rho(\omega_2') \delta_{x_4 z} \right) \ . \quad (8.114)$$

Using inequality I of section 7 again and the fundamental identity for the 2-point Green function (5.44)

$$\sum_{z \in \Lambda} \sum_{\substack{\omega_1 : x_{\pi(1)} \to x_{\pi(2)}, \ \omega_2 : x_{\pi(3)} \to x_{\pi(4)} \\ z \in \text{supp}\, \omega_1 \cap \text{supp}\, \omega_2}} \rho(\omega_1)\, \rho(\omega_2) \le$$

$$\sum_{z, z_1, z_2 \in \Lambda} J_{zz_1} J_{zz_2} < \Phi_{x_1} \Phi_z > < \Phi_{z_1} \Phi_{x_2} > < \Phi_{x_3} \Phi_z > < \Phi_{z_1} \Phi_{x_4} >$$

$$+ \, \mathcal{E}, \quad (8.115)$$

where

$$\mathcal{E} = \sum_{z, z_1 \in \Lambda} J_{zz_1} \delta_{x_4 z} < \Phi_{x_1} \Phi_z > < \Phi_{z_1} \Phi_{x_2} > < \Phi_{x_3} \Phi_{x_4} >$$

$$+ \sum_{z, z_2 \in \Lambda} J_{zz_2} \delta_{x_2 z} < \Phi_{x_3} \Phi_z > < \Phi_{z_2} \Phi_{x_4} > < \Phi_{x_1} \Phi_{x_2} >$$

$$+ \sum_{z \in \Lambda} \delta_{x_2 z} \delta_{x_4 z} < \Phi_{x_1} \Phi_{x_2} > < \Phi_{x_3} \Phi_{x_4} > \ . \quad (8.116)$$

Using the bounds (8.113,8.115) and the definition (8.93) we obtain for the scaled correlation functions

$$|G_\Theta(x_1, \ldots, x_4)| \le \alpha(\Theta)^4 K_2$$
$$\left(\sum_{z, z_1, z_2 \in \Lambda} J_{zz_1} J_{zz_2} < \Phi_{\Theta x_1} \Phi_z > < \Phi_{z_1} \Phi_{\Theta x_2} > < \Phi_{\Theta x_3} \Phi_z > < \Phi_{z_1} \Phi_{\Theta x_4} > \right.$$
$$+ \, \mathcal{E}) + \text{Perm.} \ . \quad (8.117)$$

Using the upper bound (8.99) for $G^*(x, y)$, the lower bound (8.98) for $\alpha(\Theta)$, $\Theta \to \infty$ and the definition (8.93) of the scaled 2-point correlation function we obtain

$$\alpha(\Theta)^4 \sum_{z, z_1, z_2 \in \Lambda} J_{zz_1} J_{zz_2}$$
$$< \Phi_{\Theta x_1} \Phi_z > < \Phi_{z_1} \Phi_{\Theta x_2} > < \Phi_{\Theta x_3} \Phi_z > < \Phi_{z_1} \Phi_{\Theta x_4} >$$
$$\leq const\, \Theta^{4-D} \tag{8.118}$$

for $\Theta \to \infty$. Similarly we obtain

$$\alpha(\Theta)^4 \mathcal{E}_\Theta \leq const\, \Theta^{2-D} . \tag{8.119}$$

This implies, using the bound (8.117),

$$|G_\Theta(x_1, \ldots, x_4)| \leq const\, \Theta^{4-D} . \tag{8.120}$$

Thus in $D > 4$ dimensions

$$\lim_{\Theta \to \infty} G_\Theta(x_1, \ldots, x_4) = G^*(x_1, \ldots, x_4) = 0\ ! \tag{8.121}$$

To show that

$$G^*(x_1, \ldots, x_n) = 0 \leq const\, \Theta^{4-D} \tag{8.122}$$

for all $n \in \mathbb{N}$ we have to go back to the bound (8.112). Let us consider a special spanning tree τ and the corresponding linear order $\prec_\tau$ of $\{1, \ldots, n\}$. The estimate of the right hand side of (8.112) is organized as follows. Consider a "maximal" edge of τ, that is an edge $l = (ab) \in \tau$ such that b is a maximal element of $\{1, \ldots, n\}$ with respect to the linear order relation $\prec_\tau$. Then consider the term

$$\sum_{z_{l=(ab)} \in \Lambda} \sum_{\substack{\omega_a : \Theta x_{\pi(2a-1)} \to \Theta x_{\pi(2a)}, \\ \omega_b : \Theta x_{\pi(2b-1)} \to \Theta x_{\pi(2b)}}} \rho(\omega_a)\, \rho(\omega_b) . \tag{8.123}$$

Estimation of this term yields in the same way as above a term proportional to Θ^{4-D}. Now consider a maximal edge $l' = (a'b')$ of the tree $\tau' := \tau - \{l\}$ (τ' is the tree τ "trimmed" by l). Doing the same estimation as above for the term corresponding to the edge l' we obtain again a factor Θ^{4-D}. Repeat this procedure ("trimming of a tree") until the remaining tree is empty. Then we see that for $n \geq 2$

$$|G_\Theta(x_1, \ldots, x_{2n})| \leq const\, (\Theta^{4-D})^{n-1}, \qquad \text{if } \Theta \to \infty . \tag{8.124}$$

This proves that the continuum limit for the Φ^4-theory is trivial in $D > 4$ dimensions.

References

Aizenman, M. (1981) : Proof of the Triviality of ϕ_d^4 Field Theory and some Mean-Field Features of Ising Models, Phys. Rev. Lett. **47**, 1

Aizenman, M. (1982) : Geometric Analysis of ϕ_4^4 Fields and Ising Models. Parts I and II. Commun. Math. Phys. **86**, 1

Aizenman, M. (1985) : Rigorous Studies of Critical Behavior. In L. Garrido, editor, *Applications of Field Theory to Statistical Mechanics*, pages 125-139. Lecture Notes in Physics Vol. **216**, Springer

Aragao de Carvalho, C., Caracciolo, S., Fröhlich, J. (1983) : Polymers and $g\phi^4$ Theory in Four Dimensions. Nucl. Phys. **B215[FS7]**, 209-248

Brydges, D. (1982) : Field Theories and Symanzik's Polymer Representation. In P. Dita, V. Georgescu, and R. Purice, editors, *Gauge Theories: Fundamental Interactions and Rigorous esults*. 1981 Brasov lectures, Birkhäuser, Boston-Basel-Stuttgart

Brydges, D., Fröhlich, J., Spencer, T. (1982) : The Random Walk Representation of Classical Spin Systems and Correlation Inequalities. Commun. Math. Phys. **83**, 123-150

Fernando, R. , Fröhlich, J. , Sokal, A. D. (1992): Random Walks, Critical Phenomena, and Triviality in Quantum Field Theory, Springer

Fröhlich, J. (1982) : On the Triviality of $\lambda\phi_4^4$ Theories and the Approach to the Critical Point in $d \geq 4$ Dimensions. Nuclear Physics **B200[FS4]**, 281-296

de Gennnes, P. G. (1970) : Exponents for the Excluded Volume Problem as Derived by the Wilson Method, Phys. Lett. **A38**, 339

de Gennnes, P. G. (1979) : Scaling Concepts in Polymer Physics Cornell University Press

Glimm, J., Jaffe, A. (1987) : Quantum Physics, Second Edition, Springer Verlag, Heidelberg

Symanzik, K. (1966) : Euclidean Quantum Field Theory. I. Equations for a Scalar Model. J. Math. Phys. **7**, 510-525

Symanzik, K. (1969) : Euclidean Quantum Field Theory. In R. Jost, editor, *Local Quantum Theory*. 1968 Varenna Lectures, Academic Press, New York-London

Polymer Expansion in Particle Physics

Andreas Pordt *

Institut für Theoretische Physik I, Universität Münster,
Wilhelm-Klemm-Str. 9, D-48149 Münster, Germany,
e-mail: pordt@uni-muenster.de

Abstract. Polymer expansion is a useful tool in statistical mechanics and Euclidean field theory. Various examples of polymer systems including high and low temperature expansions of the Ising model, N-component field theory and lattice gauge field theory are presented. We discuss the concepts of Kirkwood-Salsburg equations, Moebius transform, cluster expansion formula of the free energy and thermodynamic limit.

1 Introduction and Definitions

The concept of polymer systems is a generalization of expansion methods used in statistical mechanics (Ruelle (1969)). These methods include Mayer expansions, high and low temperature expansions of partition functions and correlation functions. The main idea is to decompose an infinite (or large) system into contributions of independent finite systems (clusters). Then the properties of these finite subsystems are studied and the physical objects of the infinite system are expressed by series expansions (cluster expansion). The equivalence of statistical mechanics and Euclidean field theory allows to apply these expansion methods also for field theory (see Itzykson and Drouffe (1989) or Glimm and Jaffe (1987)). There is a wide range of applications for polymer expansion techniques. Using polymer systems and cluster expansion techniques the existence of infinite volume limits, mass gaps and phase transitions of lattice (gauge) field theories can be shown. Also the existence of continuum limits of field theories are proven by using polymer expansion methods.

Combining renormalization group methods with polymer expansions one obtains a powerful tool to describe the behavior of a field theory at different scales. These methods are out of the scope of this talk (see Pordt (1994) and references cited therin).

For bosonic models perturbative expansions are generically divergent, whereas polymer expansions are convergent for certain ranges of parameters.

In this talk we present basic facts about polymer systems. We start this section with an abstract definition of polymers and define the Moebius transformation. The second section contains various examples of polymer systems showing the wide applicability of polymer systems. In the third section

---

* Supported by Deutsche Forschungsgemeinschaft

a generalized exponential function and its inverse, the logarithm, is introduced. These functions are important concepts for polymer representations and they are used to present the cluster expansion formula for the free energy in the fifth section. The Kirkwood-Salsburg equations are the starting point for series expansions of expectation values. The Mayer-Montroll equations represent correlation functions in terms of polymer activites. They are both presented in the fourth section. In the sixth section two convergence criteria of polymer expansions are given. Polymer expansions for Green functions and expectation values of N-component and gauge field theory on a lattice are derived in the seventh section. This talk ends with the definition of the thermodynamic limit using polymer representations.

We start with some general definitions (cp. Kotecký and Preiss (1986)). Let $\mathbf{K}$ be a denumerable set and $\iota \subseteq \mathbf{K} \times \mathbf{K}$ a reflexive and symmetric relation, i.e. $(\omega, \omega) \in \iota$ and $(\omega_1, \omega_2) \in \iota$ if $(\omega_2, \omega_1) \in \iota$ for all $\omega, \omega_1, \omega_2 \in \mathbf{K}$. We call the elements of $\mathbf{K}$ *polymers*. Two polymers $\omega_1, \omega_2 \in \mathbf{K}$ are called *incompatible* if $(\omega_1, \omega_2) \in \iota$ (or in infix notation $\omega_1 \iota \omega_2$). If $(\omega_1, \omega_2) \notin \iota$ we say that ω_1 and ω_2 are *compatible*. For all subsets $\mathbf{L} \subseteq \mathbf{K}$ we denote by $\mathcal{B}(\mathbf{L})$ the set of all finite subsets of $\mathbf{L}$ and by $\mathcal{D}(\mathbf{L}) \subseteq \mathcal{B}(\mathbf{L})$ the set of all finite subsets of $\mathbf{L}$ which consists of mutually compatible polymers. We write $\mathcal{B} = \mathcal{B}(\mathbf{K})$ and $\mathcal{D} = \mathcal{D}(\mathbf{K})$. For $\mathbf{P} \in \mathcal{B}$ we denote by $|\mathbf{P}|$ the number of polymers in $\mathbf{P}$ and call $\mathbf{P}$ a *cluster* if it is not decomposable into two nonempty subsets $\mathbf{P} = \mathbf{P}_1 + \mathbf{P}_2$ such that all pairs of polymers ω_1 and ω_2 with $\omega_1 \in \mathbf{P}_1$, $\omega_2 \in \mathbf{P}_2$ are compatible. A polymer functional $\rho : \mathcal{B} \to \mathbb{R}$ defines for all $\Omega \in \mathcal{D}$ a "statistical weight" $\rho(\Omega)$. We set $\rho(\emptyset) = 1$. For every finite subset $\mathbf{L} \subseteq \mathbf{K}$ and polymer functional ρ define a *partition function*

$$\mathcal{Z}(\mathbf{L}; \rho) := \sum_{\Omega: \, \Omega \in \mathcal{D}(\mathbf{L})} \rho(\Omega) \ . \tag{1.1}$$

Denote the set of all finite subsets of $\mathbf{K}$ by $\mathcal{P}_{\text{fin}}(\mathbf{K})$. Suppose that $\mathcal{Z}(\mathbf{L}; \rho) \neq 0$ for all $\mathbf{L} \in \mathcal{P}_{\text{fin}}(\mathbf{K})$. Then $\ln \mathcal{Z}(\mathbf{L}; \rho)$ exists and there exists a unique polymer function $\rho^T : \mathbf{K} \to \mathbb{R}$ such that

$$\ln \mathcal{Z}(\mathbf{L}; \rho) = \sum_{\mathbf{C}: \, \mathbf{C} \subseteq \mathbf{L}} \rho^T(\mathbf{C}) \ . \tag{1.2}$$

ρ^T is explicitly given by

$$\rho^T(\mathbf{C}) = \sum_{\mathbf{B}: \, \mathbf{B} \subseteq \mathbf{C}} (-1)^{|\mathbf{C}| - |\mathbf{B}|} \ln \mathcal{Z}(\mathbf{B}; \rho) \ . \tag{1.3}$$

ρ^T is called the Moebius transform of $\ln \mathcal{Z}$. We present in the following a general definition of the Moebius transform.

Let us define the *Moebius transform* of a function $f : X \to \mathbb{R}$ where $(X, \prec)$ is a partially ordered set such that there exists a function $d : X \to \mathbb{N}$, called grading, which obeys $d(x_1) < d(x_2)$ if $x_1 \prec x_2$. Elements $x \in X$ with

$d(x) = 0$ are called *atoms*. For atoms $x \in X$ there exist no elements $y \in X$ such that $y \prec x$. The Moebius transform $f^T : X \to \mathbb{R}$ of $f : X \to \mathbb{R}$ is uniquely defined by

$$f(x) = \sum_{y:\, y \preceq x} f^T(y) \; . \tag{1.4}$$

This is proven by induction in $n = d(x)$. For an atom $x \in X$, $d(x) = 0$, we have by (1.4)

$$f^T(x) = f(x) \; . \tag{1.5}$$

Suppose that $f^T(y)$ is known for all y with grading $d(y) < n$ for $n \in \mathbb{N}$, $n \geq 2$. Let $x \in X$ be an element of grading $d(x) = n$. Then we have by (1.4)

$$f^T(x) = f(x) - \sum_{y:\, y \prec x} f^T(y) \; . \tag{1.6}$$

Since $d(y) < d(x) = n$ for all $y \prec x$ the terms on the right hand side of (1.6) are all known by induction hypothesis. Thus $f^T(x)$ is well-defined. Since $(\mathcal{P}_{\text{fin}}(\mathbf{K}), \subset)$ is a poset with grading $d(\mathbf{L}) = |\mathbf{L}|$ we see that ρ^T is the Moebius transform of $\ln \mathcal{Z}$ and therefore uniquely defined. Since

$$\sum_{\mathbf{B}:\, \mathbf{A} \subseteq \mathbf{B} \subseteq \mathbf{C}} (-1)^{|\mathbf{C}|-|\mathbf{B}|} = \sum_{\mathbf{D}:\, \mathbf{D} \subseteq \mathbf{C}-\mathbf{A}} (-1)^{|\mathbf{C}|-|\mathbf{A}|-|\mathbf{D}|}$$

$$= (1-1)^{|\mathbf{C}-\mathbf{A}|} = 0 \tag{1.7}$$

for all $\mathbf{A} \neq \mathbf{C}$ we obtain

$$\sum_{\mathbf{B}:\, \mathbf{B} \subseteq \mathbf{C}} (-1)^{|\mathbf{C}|-|\mathbf{B}|} \ln \mathcal{Z}(\mathbf{B}; \rho) =$$

$$= \sum_{\mathbf{B}:\, \mathbf{B} \subseteq \mathbf{C}} (-1)^{|\mathbf{C}|-|\mathbf{B}|} \sum_{\mathbf{A}:\, \mathbf{A} \subseteq \mathbf{B}} \rho^T(\mathbf{A})$$

$$= \sum_{\mathbf{A}:\, \mathbf{A} \subseteq \mathbf{C}} \rho^T(\mathbf{A}) \sum_{\mathbf{B}:\, \mathbf{A} \subseteq \mathbf{B} \subseteq \mathbf{C}} (-1)^{|\mathbf{C}|-|\mathbf{B}|} = \rho^T(\mathbf{C}) \; . \tag{1.8}$$

This proves (1.3).   □

The statistical weight $\rho : \mathcal{B} \to \mathbb{R}$ satisfies the *factorization property* if

$$\rho(\mathbf{P}_1 + \mathbf{P}_2) = \rho(\mathbf{P}_1)\rho(\mathbf{P}_2) \tag{1.9}$$

for all nonempty subsets $\mathbf{P}_1$ and $\mathbf{P}_2$ such that every pair $\omega_1 \in \mathbf{P}_1$, $\omega_2 \in \mathbf{P}_2$ is compatible. In this case we see that the statistical weight factorizes

$$\rho(\mathbf{P}) = \prod_{\omega:\, \omega \in \mathbf{P}} \rho(\{\omega\}) \tag{1.10}$$

for all $\mathbf{P} \in \mathcal{D}$. We call $\rho(\omega) = \rho(\{\omega\})$ the *polymer activity* of the polymer ω.

In section 3 we will show that for all weights $\rho : \mathcal{B} \to \mathbb{R}$ there exists a unique weight $\widetilde{\rho} : \mathcal{B} \to \mathbb{R}$ obeying the factorization property and

$$\mathcal{Z}(\mathbf{L};\rho) = \sum_{P:\, P\in\mathcal{D}(\mathbf{L})} \prod_{\omega:\,\omega\in P} \widetilde{\rho}(\omega) \ . \tag{1.11}$$

In the following we may suppose without loss of generality that the factorization property holds for the statistical weights of polymer systems.

For a statistical weight ρ obeying the factorization property (1.9) we will show that

$$\rho^T(\mathbf{C}) = 0 \tag{1.12}$$

unless $\mathbf{C}$ is a cluster. For a proof suppose that $\mathbf{L}$ is not a cluster. Since $\mathbf{L}$ is not a cluster we can decompose $\mathbf{L} = \mathbf{L}_1 + \mathbf{L}_2$ into two nonempty subsets $\mathbf{L}_1$ and $\mathbf{L}_2$ such that for all $\omega_1 \in \mathbf{L}_1$ and $\omega_2 \in \mathbf{L}_2$, ω_1, ω_2 are compatible. The factorization property implies

$$\ln \mathcal{Z}(\mathbf{L};\rho) = \ln \mathcal{Z}(\mathbf{L}_1;\rho) + \ln \mathcal{Z}(\mathbf{L}_2;\rho) \ . \tag{1.13}$$

Thus

$$\ln \mathcal{Z}(\mathbf{L};\rho) = \sum_{\substack{C:\, C\subsetneq \mathbf{L} \\ C\subseteq \mathbf{L}_1 \text{ or } C\subseteq \mathbf{L}_2}} \rho^T(\mathbf{C}) \ . \tag{1.14}$$

By uniqueness of the Moebius transform this implies $\rho^T(\mathbf{C}) = 0$ if not $\mathbf{C} \subseteq \mathbf{L}_1$ or $\mathbf{C} \subseteq \mathbf{L}_2$, i. e. if $\mathbf{C}$ is not a cluster. $\square$

2   Examples

In this section we present polymer systems of various models defined on a finite lattice $\mathbb{Z}_L^D = (\mathbb{Z}/L\mathbb{Z})^D$, $L \in \{2,3,\ldots\}$. The torus $\mathbb{Z}_L^D$ consists of L^D elements $\{(i_1,\ldots,i_D)|\, i_a \in \{[0],[1],\ldots,[L-1]\},\ a \in \{1,\ldots,D\}\}$. For $z \in \mathbb{Z}$ we denote by $[z] \subset \mathbb{Z}$ the equivalence class consisting of integers $z + m \cdot L$, $m \in \mathbb{Z}$. We could also define our models on the infinite lattice $\mathbb{Z}^D$. We use here the finite lattice $\mathbb{Z}_L^D$ since the partition functions are not defined on the infinite lattice. But since the polymer systems are defined on finite sublattices, we could easily extend the definition of the polymer systems on the finite lattice to the infinite lattice. In section 8 we will see how the examples for polymer systems presented here can be used to perform the thermodynamic limit $\lim_{L\to\infty} \mathbb{Z}_L^D = \mathbb{Z}^D$ of the corresponding models.

2.1   High Temperature Expansion, Ising Model

For a spin field $s : \mathbb{Z}_L^D \to \{\pm 1\}$ write

$$e^{Js_x s_y} = \cosh J \,(1 + s_x s_y \tanh J) \tag{2.15}$$

for each $x, y \in \mathbb{Z}_L^D$ nearest neighbor pair (bond), i.e. $\|x - y\| := \sum_{\mu=1}^D |x^\mu - y^\mu| = 1$. Denote the set of all bonds $<xy>$ in $\mathbb{Z}_L^D$ by $\mathcal{E}(\mathbb{Z}_L^D)$. The partition function of the D-dimensional Ising model is given by

$$Z_L = \sum_{s:\, \mathbb{Z}_L^D \to \{\pm 1\}} \prod_{<xy> \in \mathcal{E}(\mathbb{Z}_L^D)} e^{J s_x s_y} \; . \tag{2.16}$$

The number of bonds in $\mathcal{E}(\mathbb{Z}_L^D)$ is equal to $|\mathcal{E}(\mathbb{Z}_L^D)| = D \cdot L^D$. Thus

$$\frac{Z_L}{\cosh^{D \cdot L^D} J} = \sum_{s:\, \mathbb{Z}_L^D \to \{\pm 1\}} \prod_{<xy> \in \mathcal{E}(\mathbb{Z}_L^D)} (1 + s_x s_y \tanh J)$$

$$= \sum_{S:\, \mathbb{Z}_L^D \to \{\pm 1\}} \sum_{P:\, P \subseteq \mathcal{E}(\mathbb{Z}_L^D)} (\tanh J)^{|P|}$$

$$= 2^{L^D} \sum_{P:\, P \subseteq \mathcal{E}(\mathbb{Z}_L^D)} (\tanh J)^{|P|} \; . \tag{2.17}$$

A connected subset $P \subseteq \mathcal{E}(\mathbb{Z}_L^D)$ of bonds in $\mathbb{Z}_L^D$ is called a polymer. A set of bonds P is called *connected* if for all $< x_0 x_1 >, < x_{n-1} x_n > \in P$ there exist bonds $< x_1 x_2 >, \ldots, < x_{n-2} x_{n-1} > \in P$. We call two polymers $P_1, P_2 \subseteq \mathcal{E}(\mathbb{Z}_L^D)$ incompatible if there exists a lattice site $y \in \mathbb{Z}_L^D$ such that $< xy > \in P_1$ and $< yx' > \in P_2$. For a polymer $P \subseteq \mathcal{E}(\mathbb{Z}_L^D)$ define a polymer activity by

$$\rho(P) := (\tanh J)^{|P|} \tag{2.18}$$

and for a finite set $\mathbf{P}$ consisting of polymers

$$\rho(\mathbf{P}) := \prod_{P \in \mathbf{P}} \rho(P) = (\tanh J)^{\sum_{P \in \mathbf{P}} |P|} \; . \tag{2.19}$$

We get the following polymer representation of the Ising model

$$Z_L = (2 \cosh^D J)^{L^D} \sum_{\mathbf{P}:\, \mathbf{P} \in \mathcal{D}} \prod_{P \in \mathbf{P}} \rho(P) \; . \tag{2.20}$$

2.2 Low Temperature Expansion, Contour Expansion

For each lattice site $x \in \mathbb{Z}_L^D$ consider the hypercube of unit side length and center x. The boundary of this hypercube consists of $2D$ $(D-1)$-dimensional hypersurfaces. The hypercubes of two nearest neighbor lattice sites x, y share one of these hypersurfaces. Denote the set of all these hypersurfaces by $\mathcal{P}(\mathbb{Z}_L^D)$. For each spin field $s : \mathbb{Z}_L^D \to \{\pm 1\}$ define a subset of hypersurfaces $P(s) \subseteq \mathcal{P}(\mathbb{Z}_L^D)$ in the following way. If the hypersurface $p \in \mathcal{P}(\mathbb{Z}_L^D)$ is defined by the intersection of the hypercubes of two nearest neighbor lattice sites x, y then $p \in P(s)$ iff (if and only if) $s_x s_y = -1$. Then we see that the partition function of the Ising model is

$$Z_L = 2 \sum_{P:\, P \subseteq \mathcal{P}(\mathbb{Z}_L^D)} e^{-2J|P|} \; . \tag{2.21}$$

The sum goes over all subsets $P \subseteq \mathcal{P}(\mathbb{Z}_L^D)$ of hypersurfaces which decomposes into disjoint nonempty subsets $P_1, \ldots, P_n$, $P = P_1 + \cdots + P_n$, such that each P_i encloses a connected subset $Q \subseteq \mathbb{Z}_L^D$ of lattice sites, i.e. $\partial P_i = \emptyset$ (the boundary of P_i is empty). The fact that $P(s) = P(-s)$ is responsible for the factor 2 on the right hand side of (2.21). Define a polymer as a subset of hypersurfaces $P \subseteq \mathcal{P}(\mathbb{Z}_L^D)$ such that P encloses a connected set of lattice sites. Call two polymers P_1, P_2 enclosing $Q_1, Q_2 \subseteq \mathbb{Z}_L^D$ compatible iff Q_1 and Q_2 are disjoint, $Q_1 \cap Q_2 = \emptyset$. For a polymer P define a polymer activity by

$$\rho(P) := e^{-2J|P|} . \tag{2.22}$$

Then we have the following polymer representation for the Ising model

$$Z_L = 2 \sum_{\mathbf{P}: \, \mathbf{P} \subseteq \mathcal{D}} \prod_{P \in \mathbf{P}} \rho(P) . \tag{2.23}$$

This statistical mechanical system is also called a gas of contours characterized by an exclusion interaction and activity $e^{-2J|P|}$.

2.3 N-Component Lattice Field Theory

For an N-component field $\Phi : \mathbb{Z}_L^D \to \mathbb{R}^N$ define an action

$$S(\Phi) = \frac{1}{2} \left(\Phi, (-\Delta + m^2)\Phi \right) + V(\Phi), \qquad m^2 > 0, \tag{2.24}$$

where

$$\frac{1}{2} \left(\Phi, (-\Delta + m^2)\Phi \right) = \frac{1}{2} \sum_{\substack{x, y \in \mathbb{Z}_L^D: \\ \|x-y\|=1}} (\Phi_x - \Phi_y)^2 + \frac{m^2}{2} \sum_{x \in \mathbb{Z}_L^D} \Phi_x^2 \tag{2.25}$$

is the free part of the action and

$$V(\Phi) = \sum_{x \in \mathbb{Z}_L^D} \mathcal{V}_x(\Phi_x), \qquad \mathcal{V}_x(\Phi_x) > -\frac{\mu}{2}\Phi_x^2, \qquad \mu > m^2 \tag{2.26}$$

is a local interaction. Define the partition function by

$$Z_L := \mathcal{N} \int \left[\prod_{x \in \mathbb{Z}_L^D} d^N \Phi_x \right] e^{-S(\Phi)}, \tag{2.27}$$

where $d^N \Phi_x$ is the N-dimensional Lebesgue-measure on $\mathbb{R}^N$. $\mathcal{N}$ is a normalization constant. Define the free propagator v by

$$v := (-\Delta + m^2)^{-1} . \tag{2.28}$$

The Gaussian measure with covariance v is defined by

$$d\mu_v(\Phi) := (\det 2\pi v)^{-1/2} \int \left[\prod_{x \in \mathbf{Z}_L^D} d^N \Phi_x \right] e^{-\frac{1}{2}(\Phi, v^{-1}\Phi)} . \qquad (2.29)$$

The partition function can be written as a Gaussian integral

$$Z_L = \mathcal{N}' \int d\mu_v(\Phi) \, e^{-V(\Phi)} . \qquad (2.30)$$

We introduce here two polymer representations for the N-component field theory.

For the first polymer representation let a polymer be a finite nonempty connected subset P of $\mathbf{Z}_L^D$. P is called *connected* if for all $x_0, x_n \in P$ there exist $x_1, \ldots, x_{n-1} \in \mathbf{Z}_L^D$ such that x_i, x_{i+1} are nearest neighbors for all $i = 0, \ldots n - 1$. Two polymers P_1 and P_2 are called incompatible iff they are not disjoint, $P_1 \cap P_2 \neq \emptyset$. For a subset X of $\mathbf{Z}_L^D$ define an action restricted to X by

$$S(X|\Phi) := \frac{1}{2}(\Phi_X, (-\Delta + m^2)\Phi_X) + V(\Phi_X), \qquad (2.31)$$

where

$$\Phi_X := \chi_X \cdot \Phi, \qquad (2.32)$$

and χ_X is the characteristic function of X, i. e.

$$\chi_X := \begin{cases} 1 & : \quad y \in X, \\ 0 & : \quad y \notin X . \end{cases} \qquad (2.33)$$

For a subset $X \subseteq \mathbf{Z}_L^D$ define the partition function

$$Z(X) := \mathcal{N}_X \int \left[\prod_{x \in X} d^N \Phi_x \right] e^{-S(X|\Phi)} . \qquad (2.34)$$

$X = P_1 + \cdots + P_n$ decomposes into disjoint polymers $P_1, \ldots P_n$. The normalization constant $\mathcal{N}_X$ is chosen such that

$$\mathcal{N}_X = \prod_{i=1}^{n} \mathcal{N}_{P_i}, \qquad \mathcal{N}_{\mathbf{Z}_L^D} = \mathcal{N} . \qquad (2.35)$$

Then the partition function obeys the factorization property

$$Z(X) = \prod_{i=1}^{n} Z(P_i) . \qquad (2.36)$$

Polymer activities $A(P)$ are implicitly defined for all $P \subseteq \mathbf{Z}_L^D$ by

$$Z(X) = \sum_{X = \sum P} \prod_P A(P) \qquad (2.37)$$

for all $X \subseteq \mathbb{Z}_L^D$. The sum goes over all disjoint partitions of X into nonempty subsets P. In section 3 it will be shown that these equations determines $A(P)$ uniquely for all $P \subseteq \mathbb{Z}_L^D$. By (2.36) and (2.37) we see that $A(P) = 0$ if P is not a polymer. For a polymer P define the statistical weight $\rho(P) := A(P) - \delta_{1,|P|}$. Then (2.37) implies the polymer representation

$$Z_L = \sum_{P:P \in \mathcal{D}} \prod_{P \in \mathbf{P}} \rho(P) \ . \tag{2.38}$$

The definition of the polymer activities given here is done in an implicit way. Explicit expansion of the polymer activities and insertion into the cluster expansion formula of the free energy (cf. section 5, formula (5.87)) is equivalent to a method which is called *linked cluster expansion* (LCE). This method is explained in detail in section 4.1 of the contribution by Thomas Reisz presented in this volume. For the question of convergence and generalization see also Pordt (1996), Pordt and Reisz (1997).

We will introduce a second polymer representation of Z_L. For this polymer system *all* finite subsets of $\mathbb{Z}_L^D$ are called polymers. Two not disjoint polymers P_1, P_2, $P_1 \cap P_2 \neq \emptyset$ are called incompatible. For $X \subseteq \mathbb{Z}_L^D$ define a partition function by

$$Z(X) := \mathcal{N}_X' \int d\mu_{v_X}(\Phi) \, e^{-V(X|\Phi)}, \tag{2.39}$$

where

$$\Phi_X := \chi_X \cdot \Phi, \qquad v_X = \chi_X \cdot v \cdot \chi_X, \tag{2.40}$$

and the normalization constants are chosen such that

$$\mathcal{N}_{X_1 + X_2}' = \mathcal{N}_{X_1}' \, \mathcal{N}_{X_2}' \ . \tag{2.41}$$

Then define polymer activities as in the foregoing case. Since the factorization property (2.36) no longer holds for this new definition of a polymer system we see that $\rho(P)$ can be non-vanishing also for not connected subsets P. For more details about this polymer representation of lattice field theory see Mack and Pordt (1985, 1989).

2.4   Lattice Gauge Field Theory

We present in this section a polymer representation of a pure (without matter fields) lattice gauge field theory. For the definition of a polymer system for a lattice gauge field theory with matter we refer the reader to Mack and Meyer (1982).

Let G be a compact Lie group, e. g. $SU(N)$, and χ be a character of G. A *bond* in $\mathbb{Z}_L^D$ is an oriented nearest neighbor pair $< xy >$ of lattice sites $x, y \in \mathbb{Z}_L^D$. Denote the set of all bonds in $\mathbb{Z}_L^D$ by $\mathcal{E}(\mathbb{Z}_L^D)$. A *plaquette* p is the surface between four bonds $\{< x_1 x_2 >, < x_2 x_3 >, < x_3 x_4 >, < x_4 x_1 >\} = \partial p$ (boundary of p). For a given gauge field $g : \mathcal{E}(\mathbb{Z}_L^D) \to G$ and a plaquette p define a group element $g_{\partial p} \in G$ by

$$g_{\partial p} := \prod_{b \in \partial p} g_b = g_{x_1 x_2} g_{x_2 x_3} g_{x_3 x_4} g_{x_4 x_1} \tag{2.42}$$

and a real number by

$$S_p := \frac{1}{2} \left(\chi(g_{\partial p}) + \chi^*(g_{\partial p}) \right) . \tag{2.43}$$

Denote the set of all plaquettes in $\mathbb{Z}_L^D$ by $\mathcal{P}(\mathbb{Z}_L^D)$. For a gauge field g define an action by

$$S(g) := \frac{1}{2g_0^2} \sum_{p:\, p \in \mathcal{P}(\mathbf{z}_L^D)} S_p, \tag{2.44}$$

where g_0 is the Yang-Mills coupling constant. Define the partition function for a lattice gauge field theory by

$$Z_L := \int \left[\prod_{b \in \mathcal{E}(\mathbf{z}_L^D)} dg_b \right] \prod_{p \in \mathcal{P}(\mathbf{z}_L^D)} e^{S_p}, \tag{2.45}$$

where dg_b is the Haar measure of the Lie group G. Polymers are finite connected subsets X of plaquettes in $\mathbb{Z}_L^D$. X is called connected if for all plaquettes $p_1, p_n \in X$ there exist plaquettes $p_2, \ldots p_{n-1} \in X$ such that p_i, p_{i+1} have one bond in common. Each $X \subseteq \mathcal{P}(\mathbb{Z}_L^D)$ may be decomposed into a set of mutually disjoint polymers, i. e. $X = X_1 + \cdots + X_n$, where X_i, $i = 1, \ldots, n$, are finite nonempty and connected subsets of X. Define a partition function

$$Z(X) := \int \left[\prod_{\substack{b \in \mathcal{E}(\mathbf{z}_L^D):\\ \exists p \in X:\, b \cap p \neq \emptyset}} dg_b \right] \prod_{p \in X} e^{S_p} . \tag{2.46}$$

Then we have the following factorization property

$$Z(X) = \prod_{i=1}^{n} Z(P_i) . \tag{2.47}$$

For all $P \subseteq \mathcal{P}(\mathbb{Z}_L^D)$ define implicitly a polymer activity $A(P)$ by

$$Z(X) = \sum_{X = \sum_P P} \prod_P A(P) \tag{2.48}$$

for all $X \subseteq \mathcal{P}(\mathbb{Z}_L^D)$. Then we can show that $A(P) = 0$ if P is not a polymer. Define a statistical weight by $\rho(P) := A(P) - \delta_{1,|P|}$. Then we see that the polymer representation (2.38) holds.

More details about the use of cluster expansion methods for lattice gauge field theories can be found in the book by Seiler (1982).

2.5   Random Walks

A finite number of steps

$$(\omega(0), \omega(1)), (\omega(1), \omega(2)), \ldots, (\omega(n-1), \omega(n)) \tag{2.49}$$

which are oriented pairs of lattice sites is called a *random walk* ω. The start $\omega(0)$ of the random walk ω is denoted by $b(\omega) = \omega(0)$ and the end $\omega(n)$ is denoted by $e(\omega)$. If $b(\omega) = x$ and $e(\omega) = y$ we write $\omega : x \to y$. The support of ω is defined by supp $\omega := \{\omega(0), \ldots, \omega(n)\} \subseteq \mathbb{Z}_L^D$. A statistical weight of the random walks $\omega_1, \ldots, \omega_n$ is a real number $\rho(\omega_1, \ldots, \omega_n)$. For a given statistical weight ρ define the truncated statistical weight ρ^T by

$$\rho(\omega_1, \ldots, \omega_n) = \sum_{\{1,\ldots n\}=\sum_I I} \prod_I \rho^T(\omega_i, i \in I) . \tag{2.50}$$

Suppose that ρ obeys the factorization property

$$\rho(\omega_1, \ldots, \omega_n) = \rho^T(\omega_i, i \in I_1) \, \rho^T(\omega_j, j \in I_2) \tag{2.51}$$

if $I_1 + I_2 = \{1, \ldots n\}$ and $\bigcup_{a \in I_1}$ supp $\omega_a \cap \bigcup_{b \in I_2}$ supp $\omega_b = \emptyset$. A set P of random walks is called connected if for all $\omega_1, \omega_n \in P$ there exist walks $\omega_2, \ldots, \omega_{n-1} \in P$ such that supp $\omega_i \cap$ supp $\omega_{i+1} \neq \emptyset, i = 1, \ldots, n-1$. Then the factorization property (2.51) implies $\rho^T(\omega_1, \ldots, \omega_n) = 0$ if $\{\omega_1, \ldots, \omega_n\}$ is not connected.

For an infinite sequence of walks $\omega_1, \omega_2, \ldots$ define polymers as families of walks $(\omega_i, i \in I)$ where $I \subseteq \mathbb{N}$ is a finite subset of $\mathbb{N}$ and $(\omega_i, i \in I)$ is connected. We call two families $(\omega_i, i \in I)$ and $(\omega_j, j \in J)$ incompatible if $I \cap J \neq \emptyset$. By (2.50) follows

$$\rho(\omega_1, \ldots, \omega_n) = \sum_{\{I\}: \{1,\ldots n\} \supseteq \sum_I I} \prod_I \widetilde{\rho}^T(\omega_i, i \in I) \tag{2.52}$$

where

$$\widetilde{\rho}^T(\omega_i, i \in I) = \rho^T(\omega_i, i \in I) - \delta_{1,|I|} . \tag{2.53}$$

For lattice sites $x_1, \ldots, x_{2n}$ define the Green function

$$G(x_1, \ldots, x_{2n}) := \sum_\pi \sum_{\substack{\omega_1 : x_{\pi(1)} \to x_{\pi(2)} \\ \vdots \\ \omega_n : x_{\pi(2n-1)} \to x_{\pi(2n)}}} \rho(\omega_1, \ldots, \omega_n) \tag{2.54}$$

where the sum is over all pairings π of the set $\{1, \ldots, 2n\}$. For m odd we define $G(x_1, \ldots, x_m) = 0$. Define truncated Green functions by

$$G(x_1, \ldots, x_{2n}) = \sum_{\{I\}: \{1,\ldots n\}=\sum_I I} \prod_I G^T(\omega_i, i \in I) . \tag{2.55}$$

By definition we see that $G^T(\omega_i, i \in I) = 0$ if $|I|$ is odd. More about random walks can be found in the book by Fernando, Fröhlich, and Sokal (1992) and the talks of this volume.

2.6 Monomer-Dimer System, Polymer Systems

We will define a monomer-dimer system (see Heilmann and Lieb (1970)) and the general polymer system on the lattice $\mathbb{Z}_L^D$ in this section.

A dimer consists of two nearest neighbor lattice sites $x, y \in \mathbb{Z}_L^D$, $\|x-y\| = \sum_{\mu=1}^D |x^\mu - y^\mu| = 1$. Denote by D_k the number of ways to put k dimers on the lattice such they do not overlap. For two real numbers $m, d \in \mathbb{R}$ define the partition function of a monomer-dimer system by

$$Z_L = \sum_{k \geq 0} D_k d^k m^{L^D - 2k} . \qquad (2.56)$$

This partition function is a special case of a polymer system defined by real numbers $A(P)$ for all connected subsets $P \subseteq \mathbb{Z}_L^D$. Define a partition function by

$$Z_L = \sum_{\mathbb{Z}_L^D = \sum P} \prod_P A(P), \qquad (2.57)$$

where the sum goes over all disjoint partitions of the lattice $\mathbb{Z}_L^D$ into connected subsets P. Supposing that $A(P) = 0$ if $|P| > 2$ and $A(P) = d$ if $|P| = 2$ and $A(P) = m$ if $|P| = 1$ we recover the monomer-dimer system defined by (2.56).

As in subsection 2.3 polymers are nonempty finite and connected subsets of $\mathbb{Z}_L^D$. Two polymers P_1, P_2 are incompatible, $P_1 \iota P_2$ iff $P_1 \cap P_2 \neq \emptyset$. Then we have the polymer representation

$$Z_L = \sum_{\mathbf{P}: \mathbf{P} \in \mathcal{D}} \prod_{P \in \mathbf{P}} M(P) \qquad (2.58)$$

where $M(P) := -\delta_{1,|P|} + A(P)$. As defined in section 1 $\mathcal{D}$ is the set of all finite sets consisting of mutually compatible polymers.

3 Exponential Function and Polymer Representation

In this section we define a $\star$-product and an exponential function $\exp_\star$ based on this product (see also Fernando, Fröhlich, and Sokal (1992), section 12.7 and Ruelle (1969)).

For a denumerable set Λ define the set of all finite subsets of Λ by $\mathcal{P}_{\text{fin}}(\Lambda) := \{X \subseteq \Lambda \mid |X| < \infty\}$ and the set of functionals $\mathcal{A} := \{f : \mathcal{P}_{\text{fin}}(\Lambda) \to \mathbb{C}\}$. Define an algebra on $\mathcal{A}$ by

$$(f + g)(P) = f(P) + g(P), \qquad (3.59)$$

$$(f \star g)(P) = \sum_{Y: Y \subseteq P} f(Y) g(P - Y), \qquad (3.60)$$

$$(\lambda \cdot f)(P) = \lambda f(P), \qquad (3.61)$$

for all $f, g \in \mathcal{A}$, $P \in \mathcal{P}_{\text{fin}}(\Lambda)$, and $\lambda \in \mathbb{C}$. Define a subset $\mathcal{A}_c \subset \mathcal{A}$ for $c \in \mathbb{C}$ by $\mathcal{A}_c := \{f : \mathcal{P}_{\text{fin}}(\Lambda) \to \mathbb{C} | f(\emptyset) = c\}$. The unit element of $(\mathcal{A}, \star)$ is defined by

$$\mathbb{1}(P) := \begin{cases} 1 & : \quad P = \emptyset, \\ 0 & : \quad P \neq \emptyset . \end{cases} \tag{3.62}$$

For $f \in \mathcal{A}$ define the nth power of F recursively by

$$f^{\star n} := \begin{cases} \mathbb{1} & : \quad n = 0, \\ f \star f^{\star n-1} & : \quad n > 0 . \end{cases} \tag{3.63}$$

The important relation is

$$f^{\star n}(X) = \sum_{\substack{X_1, \ldots, X_n \subseteq X: \\ X_1 + \cdots X_n = X, \ X_i \neq \emptyset}} \prod_{i=1}^{n} f(X_i) \tag{3.64}$$

and $f^{\star n}(X) = 0$ if $n > |X|$.

Define the exponential function $\exp_\star : \mathcal{A}_0 \to \mathcal{A}_1$ by

$$\exp_\star f := \sum_{n \geq 0} \frac{1}{n!} f^{\star n} . \tag{3.65}$$

For $f \in \mathcal{A}_0$ and $X \in \mathcal{P}_{\text{fin}}(\Lambda)$ we see that the series expansion of $\exp_\star f(X)$ is finite and

$$\exp_\star f(X) = \sum_{X = \sum_P P} \prod_P f(P) . \tag{3.66}$$

It is easy to show (see next section) that $\exp_\star : \mathcal{A}_0 \to \mathcal{A}_1$ is a bijective mapping. The inverse of $\exp_\star$ is given by the function $\ln_\star : \mathcal{A}_1 \to \mathcal{A}_0$ defined by

$$\ln_\star(\mathbb{1} + f) := \sum_{n \geq 1} \frac{(-1)^{n-1}}{n} f^{\star n} \tag{3.67}$$

for all $f \in \mathcal{A}_0$. Explicitly

$$\ln_\star(\mathbb{1} + f)(X) = \sum_{n \geq 1} (-1)^{n-1} (n-1)! \sum_{X = \sum_{i=1}^{n} P_i} \prod_{i=1}^{n} f(P_i) . \tag{3.68}$$

Define $\square \in \mathcal{A}_0$ by

$$\square(P) := \delta_{1, |P|} := \begin{cases} 1 & : \quad |P| = 1, \\ 0 & : \quad |P| \neq 1 \end{cases} \tag{3.69}$$

for all $P \in \mathcal{P}_{\text{fin}}(\Lambda)$. Then we get for all $X \in \mathcal{P}_{\text{fin}}(\Lambda)$

$$\exp_\star(\square + f)(X) = \sum_{\{P\}: X \supseteq \sum_P P} \prod_P f(P) . \tag{3.70}$$

By the definition of the exponential function $\exp_*$ we see that

$$\exp_* f(X_1 + X_2) = \exp_* f(X_1) * \exp_* f(X_2) \tag{3.71}$$

is equivalent to

$$f(P) = 0 \quad \text{for all } P \cap X_1 \neq \emptyset \text{ and } P \cap X_2 \neq \emptyset . \tag{3.72}$$

4 Kirkwood-Salsburg and Mayer-Montroll Equations

Kirkwood-Salsburg equations are the basic relations to obtain polymer expansions of expectation values (Ruelle (1969), Gallavotti and Miracle-Sole (1968), Brascamp (1975)). The Mayer-Montroll equations express the reduced correlation function in terms of Mayer activites.

For $A \in \mathcal{A}_0$, $Z = \exp_*(A)$, and for all $X, Q \in \mathcal{P}_{\text{fin}}(\Lambda)$ where $X \supseteq Q$, we have the *Kirkwood-Salsburg* equation

$$Z(X) = \sum_{Y:\, X \supseteq Y \supseteq Q} \sum_{\substack{Y = \sum P \\ P \cap Q \neq \emptyset}} \left[\prod_P A(P) \right] Z(X - Y) . \tag{4.73}$$

The Kirkwood-Salsburg equation (4.73) reads for the special case $Q = \{x\}$, $x \in \Lambda$

$$Z(X) = \sum_{P:\, x \in P \subseteq X} A(P) Z(X - P) . \tag{4.74}$$

For $A \in \mathcal{A}_0$ there exists a unique solution $Z \in \mathcal{A}_1$ of (4.74). This is shown by induction in the numbers of elements of X. For $|X| = 1$ (4.74) implies $Z(X) = A(X)$. Suppose that we know $Z(P)$ for all $|P| < N$ and $|X| = N$. Then we know the right hand side of (4.74). This determines $Z(X)$. Using the equation

$$A(X) = Z(X) - \sum_{\substack{P:\, x \in P \subset X \\ P \neq X}} A(P) Z(X - P) . \tag{4.75}$$

we see by induction that for a given $Z \in \mathcal{A}_1$ there exists a unique solution $A \in \mathcal{A}_0$ of (4.74). This proves that $\exp_*$ is bijective.

We call A the *polymer activity* and Z the *partition function*. Define the *Mayer activity* by

$$M = -\square + A. \tag{4.76}$$

For $X, Y \in \mathcal{P}_{\text{fin}}(\Lambda)$ define *reduced correlation functions* by

$$\rho_X(Y) := \frac{Z(X - Y)}{Z(X)} = \frac{\partial}{\partial A(Y)} \ln Z(X) . \tag{4.77}$$

The Mayer-Montroll equations express the reduced correlation functions in terms of Mayer activities. For the formulation of these equations we need some further notations and definitions. We write $P_1 \sim P_2$ for disjoint pairs $P_1, P_2 \in \mathcal{P}_{\text{fin}}(\Lambda) - \{\emptyset\}$, $P_1 \cap P_2 = \emptyset$. A finite set $\mathcal{P} = \{P_1, \ldots, P_n\}$ is called *admissible* iff $P \sim P'$ for all $P, P' \in \mathcal{P}$, $P \neq P'$. $K(X)$ denotes the set of all admissible $\mathcal{P}$ which consists of subsets $P \subseteq X$

$$K(X) := \{\mathcal{P} = \{P_1, \ldots, P_n\} | \mathcal{P} \text{ admissible}, P_i \subseteq X\}. \tag{4.78}$$

Denote by $\Pi(Y)$ the set of all partitions of Y into nonempty disjoint subsets. Define $\Pi(\emptyset) := \emptyset$. Then we have

$$K(X) = \bigcup_{Y : Y \subseteq X} \Pi(Y). \tag{4.79}$$

Two admissible sets $\mathcal{P}^{(1)} = \{P_1^{(1)}, \ldots, P_n^{(1)}\}$ and $\mathcal{P}^{(2)} = \{P_1^{(2)}, \ldots, P_n^{(2)}\}$ are called *compatible*, $\mathcal{P}^{(1)} \sim \mathcal{P}^{(2)}$ iff $P \sim P'$ for all $P \in \mathcal{P}^{(1)}$, $P' \in \mathcal{P}^{(2)}$. For a set $P \in \mathcal{P}_{\text{fin}}(\Lambda)$ we write $P \sim \mathcal{P}$ if $\{P\} \sim \mathcal{P}$.

Denote by $\text{Conn}_X(\mathcal{P})$ the set which consists of admissible sets $\mathcal{P}' \in K(X)$ which contain sets in $\mathcal{P}_{\text{fin}}(\Lambda) - \{\emptyset\}$ that are incompatible with at least one P in $\mathcal{P}$

$$\text{Conn}_X(\mathcal{P}) := \{\mathcal{P}' \in K(X) | P' \nsim \mathcal{P} \; \forall P' \in \mathcal{P}'\}. \tag{4.80}$$

We will use the notation $\text{Conn}(\mathcal{P}) \equiv \text{Conn}_\Lambda(\mathcal{P})$. Define

$$M^{\mathcal{P}} := \prod_{P \in \mathcal{P}} M(P), \qquad \text{for } \mathcal{P} = \{P_1, P_2, \ldots\} \; . \tag{4.81}$$

$Z = \exp_*(\Box + M)$ and (4.76) imply

$$Z(X) = \sum_{\mathcal{P} : \mathcal{P} \in K(X)} M^{\mathcal{P}} \; . \tag{4.82}$$

We are ready to state the Mayer-Montroll equations of the reduced correlation functions

$$\rho_\Lambda(X) = \sum_{n \geq 0} \sum_{\substack{\mathcal{P}_1, \ldots, \mathcal{P}_n \in K(X): \\ \emptyset \neq P_i \in \text{Conn}(\mathcal{P}_{i-1}), i=1,\ldots,n}} (-M)^{\mathcal{P}_1 + \cdots \mathcal{P}_n} \; . \tag{4.83}$$

For a proof of the Mayer-Montroll equations see Mack and Pordt (1989).

5   Cluster Expansion Formula

The thermodynamic limit of the partition function does not exist in general. But the thermodynamic limit of the logarithm of the partition function divided by the volume (free energy density) exists. Therefore it is useful to have a polymer expansion of the free energy.

The representation of the logarithm of a partition function $\ln Z(X)$ in terms of Mayer activities is called *cluster expansion formula* of the free energy.

We start with some definitions and notations. Consider a finite family of sets $\mathbf{P} = (P_1, \ldots, P_n)$, $P_i \in \mathcal{P}_{\text{fin}}(\Lambda)$. A *Venn-diagram* of $\mathbf{P}$, denoted by $\gamma(\mathbf{P})$ is a graph defined by n vertices $P_1, \ldots, P_n$ and lines $(P_a P_b)$ if $P_a \cap P_b \neq \emptyset$. $\mathbf{P}$ is called a *cluster* if $\gamma(\mathbf{P})$ is connected. Denote by $\mathcal{C}(X)$ the set of all clusters consisting of polymers $P \subseteq X$. For a cluster $\mathbf{P}$ define a combinatorial factor

$$a(\mathbf{P}) := \sideset{}{'}\sum_{G: G \subseteq \gamma(\mathbf{P})} (-1)^{|G|}, \tag{5.84}$$

where the sum $\sum'$ goes over all connected subgraphs of $\gamma(\mathbf{P})$ containing all vertices of the graph $\gamma(\mathbf{P})$. $|G|$ denotes the lines in G. For $\mathbf{P} = (P_1, \ldots, P_n)$ and $P \in \mathcal{P}_{\text{fin}}(\Lambda)$ define

$$\mathbf{P}(P) := |\{i \in \{1, \ldots n\} \,|\, P_i = P\}| \tag{5.85}$$

and

$$\mathbf{P}! := \prod_P \mathbf{P}(P)! \ . \tag{5.86}$$

With these definitions we have for all $X \in \mathcal{P}_{\text{fin}}(\Lambda)$

$$\ln\left(1 + \sum_{\{P\}: \sum P \subseteq X} \prod_P M(P)\right) = \sum_{\mathbf{P}: \mathbf{P} \in \mathcal{C}(X)} \frac{a(\mathbf{P})}{\mathbf{P}!} \prod_{P \in \mathbf{P}} M(P) \ . \tag{5.87}$$

For a proof of this relation see Glimm and Jaffe (1987), Brydges (1984), or Pordt (1990).

6 Convergence Criteria

We have presented various polymer expansions for expectation values, correlation functions and free energies. It remains to find the conditions under which these series expansions are convergent. We present here two criteria.

The first of these criteria is due to Kotecký and Preiss (1986). Let a polymer representation be given by

$$\mathcal{Z}(\mathbf{L}; \rho) = \sum_{\mathbf{P}: \mathbf{P} \in \mathcal{D}(\mathbf{L})} \prod_{\omega \in \mathbf{P}} \rho(\omega) \tag{6.88}$$

for $\mathbf{L} \subseteq \mathbf{K}$, and

$$\ln \mathcal{Z}(\mathbf{L}; \rho) = \sum_{\mathbf{C}: \mathbf{C} \subseteq \mathbf{L}} \rho^T(\mathbf{C}) \ . \tag{6.89}$$

Suppose that there are functions $a : \mathbf{K} \to [0, \infty)$, and $d : \mathbf{K} \to [0, \infty)$ such that

$$\sum_{\omega': \omega' \iota \omega} e^{a(\omega') + d(\omega')} |\rho(\omega')| \leq a(\omega) \tag{6.90}$$

for all $\omega \in \mathbf{K}$. Then

$$\sum_{\substack{C:\, C \in B \\ \omega \perp C}} |\rho^T(\mathbf{C})|\, e^{d(\mathbf{C})} \leq a(\omega) \tag{6.91}$$

for all $\omega \in \mathbf{K}$. For a proof see Kotecký and Preiss (1986).

The second criterion is due to Cammarota (1982). For $Z : \mathcal{P}_{\text{fin}}(\Lambda) \to \mathbb{C}$, $Z(\emptyset) = 1$ and $Z = \exp_*(A)$ define a *reduced activity* $\overline{A}$ by

$$\overline{A}(X) := \frac{A(X)}{\prod_{x \in X} A(\{x\})} \; . \tag{6.92}$$

Then we have

$$\frac{Z(X)}{\prod_{x \in X} Z(\{x\})} = \sum_{X = \sum P} \prod_{P} \overline{A}(P) \tag{6.93}$$

and

$$\overline{A}(P) = 1, \qquad \text{if } |P| = 1 \; . \tag{6.94}$$

Suppose in the following that $A(P) = 1$ if $|P| = 1$. Furthermore, suppose that for each integer $n \geq 2$

$$\sup_{x \in \Lambda} \sum_{\substack{P:\, x \in P \\ |P| = n}} |A(P)| \leq \epsilon^n \tag{6.95}$$

and

$$\frac{\epsilon}{1 - \epsilon} < \frac{1}{2e} \; . \tag{6.96}$$

Then

$$\sum_{\substack{\mathbf{P}:\, P \in \mathcal{C}(\Lambda) \\ Y \in P}} |\frac{a(\mathbf{P})}{\mathbf{P}!} \prod_{P \in \mathbf{P}} M(P)| \leq$$

$$|M(Y)| \left(1 + |Y|e^{|Y|}\frac{1}{2}\ln(1 - 2e\frac{\epsilon}{1 - \epsilon})^{-1} \right) \; . \tag{6.97}$$

For a proof see Cammarota (1982). The bound (6.97) implies immediately a convergence criteria for the right hand side of the cluster expansion formula and the series expansion (5.87) converges absolutely for all $X \in \mathcal{P}_{\text{fin}}(\Lambda)$.

7   Polymer Expansions for Green Functions and Expectation Values

By the above definitions we are now in a position to formulate the polymer expansions of Green functions of Euclidean lattice (gauge) field theories. At two examples we discuss the polymer expansions for an N-component field theory (see Mack and Pordt (1985,1990)) and a lattice gauge field theory (see Seiler (1982)).

Consider the generating functional of an N-component lattice field theory

$$Z[J] = \mathcal{N} \int d\mu_v(\varPhi)\, e^{-V(\varPhi)+(J,\varPhi)}, \qquad (7.98)$$

where

$$\mathcal{N} = [\int d\mu_v(\varPhi)\, e^{-V(\varPhi)}]^{-1}, \qquad (7.99)$$

and

$$(J,\varPhi) := \sum_{x\in\mathbb{Z}_L^D} \sum_{i=1}^N J_x^i\, \varPhi_x^i \ . \qquad (7.100)$$

The Gaussian measure $d\mu_v(\varPhi)$ is defined by (2.29). The n-point Green function is defined by

$$G(x_1,\ldots,x_n) := \frac{\partial^n}{\partial J_{x_1}\cdots\partial J_{x_n}} Z[J]|_{J=0} \qquad (7.101)$$

and the connected n-point Green function by

$$G^c(x_1,\ldots,x_n) := \frac{\partial^n}{\partial J_{x_1}\cdots\partial J_{x_n}} \ln Z[J]|_{J=0} \ . \qquad (7.102)$$

According to the example of a polymer system of section 2.3 for an N-component field theory define for all finite nonempty subset $X \subseteq \mathbb{Z}_L^D$ a partition function

$$Z[X|J] := \mathcal{N}_X \int d\mu_{v_X}(\varPhi)\, e^{-V(\varPhi_X)+(J,\varPhi_X)}, \qquad (7.103)$$

where

$$\mathcal{N}_X = [\int d\mu_{v_X}(\varPhi)\, e^{-V(\varPhi_X)}]^{-1} \ . \qquad (7.104)$$

The above defined partition functions Z obey the *localization property*

$$\frac{\partial}{\partial J_y} Z[X|J] = 0, \qquad \text{if } y \notin X \qquad (7.105)$$

and the normalization condition

$$Z[X|J=0] = 1 \ . \qquad (7.106)$$

Define a polymer activity A for the partition function Z by $A = \ln_* Z$. By (7.105) follows the localization property of the polymer activity

$$\frac{\partial}{\partial J_y} A[X|J] = 0, \qquad \text{if } y \notin X \qquad (7.107)$$

and

$$A[X|J=0] = \delta_{1,|X|} \ . \qquad (7.108)$$

Using the Kirkwood-Salsburg equation (4.73) with $Q = \{x_1, \ldots, x_n\}$ we obtain

$$G(x_1, \ldots, x_n) = \sum_{k \geq 1} \frac{1}{k!}$$

$$\sum_{\substack{P_1, \ldots, P_k \in \mathcal{P}_{\text{fin}}(\Lambda): \\ P_a \cap P_b = \emptyset, \ a \neq b}} \sum_{\substack{I_1, \ldots, I_k: \{x_i, i \in I_a\} \subseteq P_a \\ \sum_{i=1}^{k} I_i = \{1, \ldots n\}}} \prod_{a=1}^{k} \left(\prod_{i \in I_a} \frac{\partial}{\partial J_{x_i}} \right) A[P_a|J]|_{J=0} \quad (7.109)$$

If Z does not obey the normalization condition (7.106) we obtain instead

$$G(x_1, \ldots, x_n) = \sum_{k \geq 1} \frac{1}{k!} \sum_{\substack{P_1, \ldots, P_k \in \mathcal{P}_{\text{fin}}(\Lambda): \\ P_a \cap P_b = \emptyset, \ a \neq b}}$$

$$\sum_{\substack{I_1, \ldots, I_k: \{x_i, i \in I_a\} \subseteq P_a \\ \sum_{i=1}^{k} I_i = \{1, \ldots n\}}} \prod_{a=1}^{k} \left(\prod_{i \in I_a} \frac{\partial}{\partial J_{x_i}} \right) A[P_a|J]|_{J=0} \, \rho_\Lambda(\sum_{a=1}^{k} P_a) \ . \quad (7.110)$$

For $\mathbb{Z}_2$-symmetric models, i. e. models with even interactions $V(-\Phi) = V(\Phi)$ the partition function is even, $Z[X| - J] = Z[X|J]$. In this case the polymer activity is also an even function and $\left(\prod_{i \in I} \frac{\partial}{\partial J_{x_i}} \right) A[P_a|J]|_{J=0} = 0$ if $|I|$ is odd. For example, the polymer representation of the 2-point Green function in the $\mathbb{Z}_2$-symmetric case is given by

$$G(x_1, x_2) = \sum_{P} \frac{\partial^2}{\partial J_{x_1} \partial J_{x_2}} A[P|J]|_{J=0} \cdot (\rho_\Lambda(P)) \quad (7.111)$$

if the normalization conditions (7.106) do (not) hold.

An *observable* is a functional $\mathcal{O} : \{\Phi : \Lambda \to \mathbb{R}^N\} \to \mathbb{C}$. The *expectation value* of an observable $\mathcal{O}$ is defined by

$$< \mathcal{O} >:= \frac{\int d\mu_v(\Phi) \, \mathcal{O} \, e^{-V(\Phi)}}{\int d\mu_v(\Phi) \, e^{-V(\Phi)}} \ . \quad (7.112)$$

For example, the expectation value of the monomial $\Phi_{x_1} \cdots \Phi_{x_n}$ is the n-point Green function

$$< \Phi_{x_1} \cdots \Phi_{x_n} > = G(x_1, \ldots, x_n) \ . \quad (7.113)$$

Define *connected Gaussian measures* $d\mu_{vX}^c$ by

$$d\mu_{vX}(\Phi) = \sum_{X = \sum P} \bigotimes_{P} d\mu_{vP}^c(\Phi) \ . \quad (7.114)$$

An explicit expression of the connected Gaussian measures $d\mu_{vX}^c$ is given by tree formulas for pair interactions. They can be found in Brydges (1984), Glimm and Jaffe (1987), or Abdesselam and Rivasseau (1994).

The normalized polymer activities are in terms of connected Gaussian measures given by

$$A[P|J] = \int d\mu^c_{v_P}(\varPhi)\, e^{-V(\varPhi_P)+(J,\varPhi_P)} \ . \tag{7.115}$$

The *support* supp $\mathcal{O}$ of the observable $\mathcal{O}$ is the largest set Q such that $\frac{\partial}{\partial \varPhi_y}\mathcal{O}(\varPhi) = 0$ for all $y \notin Q$. Using the Kirkwood-Salsburg equation (4.73) with $Q = \text{supp } \mathcal{O}$ we obtain for the expectation value

$$< \mathcal{O} > = \sum_{k \geq 1} \sum_{\substack{P_1,\ldots P_k \in \mathcal{P}_{\text{fin}}(A):\, P_a \cap \text{supp } \mathcal{O} \neq \emptyset \\ P_a \cap P_b = \emptyset,\, a \neq b}}$$

$$\left[\int \prod_{a=1}^k d\mu^c_{v_{P_a}}(\varPhi)\, e^{-V(\varPhi_{P_a})}\, \mathcal{O}(\varPhi) \right] \rho_A(\sum_{i=1}^k P_i) \ . \tag{7.116}$$

In the rest of this section we study polymer representations of lattice gauge field theories. An observable $\mathcal{O}(g_b, b \in \mathcal{E}(\mathbb{Z}^D_L))$ is a complex function of the gauge group elements g_b. The support of $\mathcal{O}$ is defined by

$$\text{supp } \mathcal{O} := \{ b \in \mathcal{E}(\mathbb{Z}^D_L) |\, \mathcal{O} \text{ depends on } b \} \ . \tag{7.117}$$

For a subset of bonds $B \subseteq \mathcal{E}(\mathbb{Z}^D_L)$ define a subset of plaquettes $\overline{B} \subseteq \mathcal{P}(\mathbb{Z}^D_L)$ by

$$\overline{B} := \{ p \in \mathcal{P}(\mathbb{Z}^D_L) |\, \exists b \in B : b \in \partial p \} \ . \tag{7.118}$$

Define the expectation value of the observable $\mathcal{O}$ by

$$< \mathcal{O} > := \frac{\int \left(\prod_{b \in \mathcal{E}(\mathbf{z}^D_L)} dg_b \right) \mathcal{O}(g_b, b \in \mathcal{E}(\mathbb{Z}^D_L)) \prod_{p \in \mathcal{P}(\mathbf{z}^D_L)} e^{S_p}}{\int \left(\prod_{b \in \mathcal{E}(\mathbf{z}^D_L)} dg_b \right) \prod_{p \in \mathcal{P}(\mathbf{z}^D_L)} e^{S_p}} \ . \tag{7.119}$$

For a subset $X \subseteq \mathcal{P}(\mathbb{Z}^D_L)$ of plaquettes define polymer activities $\mathcal{A}$ of the Boltzmann factor by

$$\prod_{p \in X} e^{S_p} = \sum_{X = \sum Q} \prod_Q \mathcal{A}(Q) \ , \tag{7.120}$$

where

$$\mathcal{A}(Q) := \begin{cases} 1 + \prod_{p \in Q}(e^{S_p} - 1) & : \text{ if } Q \text{ is connected,} \\ 0 & : \text{ if } Q \text{ is not connected .} \end{cases} \tag{7.121}$$

The polymer activities defined by (2.37) are explicitly given by

$$A(X) = \int \left[\prod_{\substack{b \in \mathcal{E}(\mathbb{Z}^D_L) \\ \exists p \in X :\, b \in \partial p}} dg_b \right] \mathcal{A}(X) \ . \tag{7.122}$$

From the Kirkwood-Salsburg equation (4.73) follows the polymer representation of the expectation value

$$< \mathcal{O} >= \sum_{k \geq 1} \sum_{\substack{Q_1, \ldots, Q_n \subseteq \mathcal{P}(\mathbf{Z}_L^D): \, Q_a \text{ connected} \\ Q_a \cap \mathrm{supp}\ \overline{\mathcal{O}} \neq \emptyset}}$$

$$\int \left[\prod_{\substack{b \in \mathcal{E}(\mathbf{Z}_L^D) \\ \exists a: \, b \in \partial p, \, p \in P_a}} dg_b \right] \prod_{a=1}^k \mathcal{A}(Q_a) \, \mathcal{O}(\Phi) \, \rho_\Lambda(\overline{\mathrm{supp}\ \mathcal{O}}) \ . \quad (7.123)$$

This representation was used to prove the existence of the thermodynamic limit of expectation values for sufficiently large coupling constants g_0.

8   Thermodynamic Limit

In statistical mechanics and field theory one is interested in the existence and behavior of physical objects in an infinite volume Λ, $|\Lambda| = \infty$. For a definition of quantities in the infinite volume Λ one introduces an approximation in a finite volume $X \subset \Lambda$ and takes the *thermodynamic limit* $X \nearrow \Lambda$. For example we have defined expectation values $< \mathcal{O} >_{\mathbf{Z}_L^D}$ on a finite torus $\mathbf{Z}_L^D$ and define the thermodynamic limit of the expectation value $< \mathcal{O} >_{\mathbf{Z}^D}$ on the infinite lattice $\mathbf{Z}^D$ by

$$< \mathcal{O} >_{\mathbf{Z}^D} = \lim_{L \to \infty} < \mathcal{O} >_{\mathbf{Z}_L^D} \ . \quad (8.124)$$

For systems defined by polymer representations on an infinite volume Λ the thermodynamic limit can be performed in the following way. Let $\Phi : \mathcal{P}_{\mathrm{fin}}(\Lambda) \to \mathbb{R}$ be a real-valued functional on $\mathcal{P}_{\mathrm{fin}}(\Lambda) = \{P \subset \Lambda | \, |P| < \infty\}$, i. e. $\Phi(P) \in \mathbb{R}$ is defined for all finite subset P of Λ. We want to define the thermodynamic limit

$$\lim_{P \nearrow \Lambda} \Phi(P) = \Phi(\Lambda) \ . \quad (8.125)$$

Define the *Moebius transform* $\tilde{\Phi} : \mathcal{P}_{\mathrm{fin}}(\Lambda) \to \mathbb{R}$ of Φ by

$$\Phi(X) = \sum_{P: \, P \subseteq X} \tilde{\Phi}(P) \quad (8.126)$$

for all $X \in \mathcal{P}_{\mathrm{fin}}(\Lambda)$. Then the thermodynamic limit exists if the following norm is finite

$$\|\tilde{\Phi}\| := \sum_{P: \, P \in \mathcal{P}_{\mathrm{fin}}(\Lambda)} |\tilde{\Phi}(P)| < \infty \ . \quad (8.127)$$

If $\|\tilde{\Phi}\| < \infty$ the thermodynamic limit of $\Phi(X)$ is defined by an absolutely convergent series

$$\Phi(\Lambda) := \sum_{P:\, P \in \mathcal{P}_{\text{fin}}(\Lambda)} \widetilde{\Phi}(P) \ . \tag{8.128}$$

Suppose that $\widetilde{\Phi}(P)$ are analytic functions for all $P \in \mathcal{P}_{\text{fin}}(\Lambda)$ and $\|\widetilde{\Phi}\| < \infty$. Then it follows by the dominated convergence theorem that the thermodynamic limit $\Phi(\Lambda)$ is also an analytic function.

For example let us consider the thermodynamic limit of the free energy density

$$f := \lim_{X \nearrow \mathbf{Z}^D} \frac{1}{|\Lambda|} \ln Z(X) \tag{8.129}$$

where the partition functions $Z(X)$, for all finite subsets X of Λ define a polymer system on the infinite lattice $\mathbb{Z}^D$. The free energy F and the Moebius transform $\widetilde{F}$ are defined by

$$F(X) := \ln Z(X) = \sum_{P:\, P \subseteq X} \widetilde{F}(P) \ . \tag{8.130}$$

We cannot simply take the thermodynamic limit of the free energy F since F is an extensive quantity, i. e. $F(X) \propto O(|X|)$ for $|X| \to \infty$. Before taking the thermodynamic limit one has to divide $F(X)$ by the volume $|X|$. For that let us rewrite (8.130)

$$F(X) = \sum_{x:\, x \in X} \sum_{P:\, x \in P \subseteq X} \frac{\widetilde{F}(P)}{|P|} \ . \tag{8.131}$$

Supposing translation-invariance (8.131) simplifies to

$$F(X) = |X| \sum_{P:\, x \in P \subseteq X} \frac{\widetilde{F}(P)}{|P|} \tag{8.132}$$

for an arbitrary element $x \in X$. Define for all $x \in \Lambda$ a functional $\widetilde{\Phi}_x : \mathcal{P}_{\text{fin}}(\Lambda) \to \mathbb{R}$ by

$$\widetilde{\Phi}_x(P) := \begin{cases} \frac{\widetilde{F}(P)}{|P|} & : \text{ if } x \in P, \\ 0 & : \text{ if } x \notin P \ . \end{cases} \tag{8.133}$$

Then we have in the general case

$$F(X) = \sum_{x \in X} \sum_{\substack{P:\, P \in \mathcal{P}_{\text{fin}}(\Lambda) \\ P \subseteq X}} \widetilde{\Phi}_x(P) \tag{8.134}$$

and for the translation-invariant case

$$F(X) = |X| \sum_{\substack{P:\, P \in \mathcal{P}_{\text{fin}}(\Lambda) \\ P \subseteq X}} \widetilde{\Phi}(P), \tag{8.135}$$

where we have set $\widetilde{\Phi} = \widetilde{\Phi}_x$.

If $\|\widetilde{\Phi}_x\| < \infty$ for all $x \in \Lambda$ we see that the thermodynamic limit (8.129) of the free energy density f exists and

$$|f| \leq \sup_{x \in \Lambda} \|\widetilde{\Phi}_x\| \tag{8.136}$$

respectively for the translation-invariant case

$$|f| \leq \|\widetilde{\Phi}\| \ . \tag{8.137}$$

In the case $\|\widetilde{\Phi}\| < \infty$ we may resum the absolutely convergent series

$$\Phi(\Lambda) = \sum_{P:\, P \in \mathcal{P}_{\text{fin}}(\Lambda)} \widetilde{\Phi}(P) \tag{8.138}$$

and write

$$\Phi(\Lambda) = \sum_{n \geq 1} \frac{1}{n!} \sum_{\substack{x_1,\ldots,x_n \in \Lambda \\ x_a \neq x_b \text{ if } a \neq b}} \widetilde{\Phi}(\{x_1,\ldots,x_n\}) \ . \tag{8.139}$$

Supposing that there exist positive real constants c_1, c_2 such that

$$\sum_{\substack{x_1,\ldots,x_n \in \Lambda \\ x_a \neq x_b \text{ if } a \neq b}} |\widetilde{\Phi}(\{x_1,\ldots,x_n\})| \leq c_1\, c_2^n \tag{8.140}$$

we obtain an upper bound for the thermodynamic limit of Φ

$$|\Phi(\Lambda)| \leq \|\widetilde{\Phi}\| \leq c_1\, (e^{c_2} - 1) < \infty \ . \tag{8.141}$$

Thus (8.140) yields a criterion for the finiteness of the norm $\|\widetilde{\Phi}\|$ and therefore the existence of the thermodynamic limit of Φ.

References

Abdesselam, A., Rivasseau, V. (1994) : Trees, Forests and Jungles: A Botanical Garden for Cluster Expansions, in: Constructive Physics, Proceedings of the Conference held at Ecole Polytechnique, Palaiseau, France 25-27 July 1994, Rivasseau, V. (Ed.), Lecture Notes in Physics, Springer

Brascamp, H .J. (1975) : The Kirkwood-Salsburg Equations : Solutions and Spectral Properties, Commun. Math. Phys. **40**, 235

Brydges, D. (1984) : A Short Course on Cluster Expansions, *In Critical Phenomena, Random Systems*, Gauge Theories 1984

Cammarota, C. (1982) : Decay of Correlations for Infinite Range Interactions in Unbounded Spin Systems, Commun. Math. Phys. **85**, 517

Fernando, R., Fröhlich, J., Sokal, A. D. (1992) : Random Walks, Critical Phenomena, and Triviality in Quantum Field Theory, Springer

Gallavotti, G., Miracle-Sole, S. (1968) : Correlation Functions of a Lattice System, Commun. Math. Phys. **7**, 274

Glimm, J., Jaffe, A. (1987) : Quantum Physics, Second Edition, Springer

Groeneveld, J. (1962) : Two Theorems on Classical Many-Particle Systems, Phys. Lett. Vol. **3**, Nr. 1, 50

Gruber, C., Kunz, H. (1971) : General Properties of Polymer Systems, Commun. Math. Phys. **22**, 133

Heilmann, O. J., Lieb, E. (1970) : Monomers and Dimers, Phys. Rev. Lett. <u>24</u>, Nr. **25**, 1412

Itzykson C., Drouffe, J.-M. (1989) : Statistical Filed Theory, Volume I and II, Cambridge University Press

Kotecký, R., Preiss, D. (1986) : Cluster Expansion for abstract Polymer Models, Commun. Math. Phys. **103**, 491

Mack, G., Meyer, H. (1982) : A Disorder Parameter that Tests for Confinement in Gauge Theories with Quark Fields, Nucl. Phys. **B200[FS4]**, 249

Mack, G., Pordt, A. (1985) : Convergent Perturbation Expansions for Euclidean Quantum Field Theory, Commun. Math. Phys. **97**, 267

Mack, G., Pordt, A. (1989) : Convergent Weak Coupling Expansions for Lattice Field Theories that Look Like Perturbation Series, Rev. Math. Phys. **1**, 47

Pordt, A. (1990) : Convergent Multigrid Polymer Expansions and Renormalization for Euclidean Field Theory, DESY 90-020

Pordt, A. (1994) : On the Renormalization Group Flows and Polymer Algebras, in: Constructive Physics, Proceedings of the Conference held at Ecole Polytechnique, Palaiseau, France 25-27 July 1994, Rivasseau, V. (Ed.), Lecture Notes in Physics, Springer

Pordt, A. (1996) : A Convergence Proof for Linked Cluster Expansions, Uni. Münster preprint MS-TPI-96-05, and e-print archive: hep-lat@xxx.lanl.gov 9604010

Pordt, A. and Reisz, T. (1997) : Linked Cluster Expansions Beyond Nearest Neighbour Interactions: Convergence and Graph Classes, Int. Journ. of Mod. Phys. A, Vol. 12, No. 21, 3739

Ruelle, D. (1969) : Statistical Mechanics, W. A. Benjamin Inc.

Seiler, E. (1982) : Gauge Theories as Problem of Constructive Quantum Field Theory and Statistical Mechanics, Lecture Notes in Physics, Vol. **159**, Berlin, Heidelberg, New York , Springer

Polymers, Spin Models and Field Theory

Bo Söderberg

Lund University, Dept. of Theoretical Physics 2, Sölvegatan 14, S-223 63 Lund, Sweden

Abstract. The generic relation between continuous polymers and zero-component Euclidean field-theories is reviewed, and exemplified by polymers with contact and Coulomb interactions. An analogous relation on the lattice is also discussed, relating the statistics of self-avoiding walks to a zero-component spin-model.

1 Introduction

Relations between polymer statistics and Euclidean field theories have been known for a long time (Symanzik (1969)), and are in fact very general (see e.g. des Cloizeaux and Jannink (1990)). They derive from the formal similarity between the statistical mechanics of polymers and the path-integral formulation of quantum mechanics. Such relations are quite useful, providing access to the rich toolbox of field theory to polymer theory.

For the case of a continuous polymer with a contact interaction (representing a generic short-range interaction), the relation to a zero-component ϕ^4 field theory was first demonstrated by de Gennes (de Gennes (1972)), and has been exploited by several people (see e.g. de Gennes (1979), Parisi (1988), Itzykson and Drouffe (1989), and references therein).

There also exists an analogous relation between the statistics of self-avoiding walks on the lattice and a certain zero-component spin-model (see e.g. Zinn-Justin (1989)).

Another interesting case is that of a polyelectrolyte, i.e. a polymer with a Coulomb pair interaction. It is related to a certain simple field theory with two interacting fields (Pfeuty et. al. (1977)).

In this talk, I will discuss these relations, both the generic one and the various special cases, after an initial survey of polymer representations.

2 Idealized Representations of Free Polymers

2.1 "Realistic" Model

An ideal non-interacting polymer can be described by a set of $N+1$ pointlike *monomers*, connected in a sequence by N spring-like bond forces. Thus, the Hamiltonian (neglecting the kinetic term) takes the form

$$H[\mathbf{x}] = \frac{1}{2} \sum_i (\mathbf{x}_{i+1} - \mathbf{x}_i)^2 \ , \qquad (2.1)$$

where x_i is the position of monomer i.

The thermodynamics is defined by the Boltzmann distribution, and the corresponding partition function is given by the multiple integral

$$Z_N = \int d[\mathbf{x}] \exp(-H[\mathbf{x}]) \; , \tag{2.2}$$

where the temperature for simplicity has been scaled away. Fixing the endpoints $x_0 = 0, x_N = \mathbf{R}$, we obtain the restricted partition function

$$Z_N(\mathbf{R}) \propto \exp\left(-\frac{\mathbf{R}^2}{2N}\right) \; . \tag{2.3}$$

This representation of a polymer, albeit strongly idealized, still has some resemblance to a real chemical molecule. For analytical or numerical investigations, however, one often prefers to further simplify the model.

2.2 Continuous Model

For general theoretical considerations, a fully continuous representation, where the discrete monomers are replaced by a continuum, is preferable. Thus, we denote by $\mathbf{x}(\tau)$ the position of the point labelled by the continuous "monomer index" τ in some interval $0 < \tau < \mu$, with μ replacing N as a measure of the length of the chain. Consequently, instead of (2.1) we use as a Hamiltonian for a free polymer

$$S_0[\mathbf{x}(\tau)] = \frac{1}{2} \int_0^\mu \left(\frac{d\mathbf{x}}{d\tau}\right)^2 d\tau \; . \tag{2.4}$$

The corresponding restricted partition function, with the endpoints fixed at $\mathbf{x}(0) = 0, \mathbf{x}(\mu) = \mathbf{R}$, amounts to

$$Z_0(\mu, \mathbf{R}) \propto \exp\left(-\frac{\mathbf{R}^2}{2\mu}\right) \; , \tag{2.5}$$

in analogy to (2.3). Dividing this by $Z_0(\mu) \equiv \int d\mathbf{R} Z_0(\mu, \mathbf{R})$ results in the normalized Gaussian $\mathbf{R}$-distribution

$$z_0(\mu, \mathbf{R}) = (2\pi\mu)^{-D/2} \exp\left(-\frac{\mathbf{R}^2}{2\mu}\right) \; , \tag{2.6}$$

describing the distribution of the endpoints, if these are let free. We get the typical Brownian result

$$\langle \mathbf{R}^2 \rangle = D\mu \; , \tag{2.7}$$

with the endpoint distance being proportional to the squareroot of the chainlength, due to $\mathbf{R}$ being the sum of many small independent moves.

2.3   Lattice Model

For numerical investigations, a lattice approach is often preferred. On the lattice, a polymer can be modelled by a *walk*. Given a lattice $L \in \mathbb{R}^D$ and a starting point $x_0 \in L$, a walk of length N is defined as a sequence of N steps, where in each step the walk proceeds to one of the neighboring sites.[1]

In particular, a free polymer of length N is modelled by a *random walk*, where all possible distinct walks of length N are given the same statistical weight, leading to the usual Brownian behaviour of the endpoint distance $\mathbf{R}$,

$$\langle \mathbf{R}^2 \rangle = N \ . \tag{2.8}$$

3   Interacting Walks and Spin Models

3.1   Modified Walk Ensemble

Now let us turn to the less trivial case of interacting polymers.

On a D-dimensional lattice, a polymer with a short-range interaction can be modelled by a self-repelling walk, where self-intersections are penalized. Consider the extreme case of *self-avoiding walks* (SAW:s), where self-intersecting walks are discarded altogether, i.e. one requires for a valid walk

$$\mathbf{x}_i \neq \mathbf{x}_j, \text{ if } i \neq j \ . \tag{3.9}$$

The geometrical scaling properties will change as compared to the free case. Thus, the squared endpoint distance will scale, for large N, as

$$\langle \mathbf{R}^2 \rangle \propto N^{2\nu} \ , \tag{3.10}$$

with a scaling index ν larger than the free value $\nu_0 = \frac{1}{2}$, at least for $D < 4$ – in four dimensions and higher, a typical random walk self-intersects too seldom for the interaction to make any fundamental difference. For the physical case of $D = 3$, the index has been measured with Monte Carlo methods, yielding $\nu \approx 0.588$, whereas for $D = 2$, it is known to be $3/4$ (see e.g. Itzykson and Drouffe (1989)). For $D \geq 4$, $\nu = \nu_0$.

3.2   Relation to a Spin Model

Consider a ferromagnetic $O(n)$-symmetric nonlinear sigma model on the lattice. At every site i, an n-component spin $\mathbf{S}_i$ is defined, restricted to the sphere $\mathbf{S}_i^2 = n$, with a spherically symmetric measure $D[\mathbf{S}] = \prod_i d^n \mathbf{S}_i \delta(\mathbf{S}_i^2 - n)$. A ferromagnetic interaction is introduced by using a Boltzmann distribution $D[\mathbf{S}]e^{-H[\mathbf{S}]}$, with the Hamiltonian

---

[1] See also the contribution by A. Pordt to these proceedings, on the subject of random walks in Euclidean field theory.

$$H[\mathbf{S}] = -\beta \sum_{<i,j>} \mathbf{S}_i \cdot \mathbf{S}_j \; , \tag{3.11}$$

where the sum is restricted to neighbouring sites $< i, j >$.

We will be interested in the two-point function

$$G(\mathbf{R}) = \frac{1}{n} \langle \mathbf{S}_0 \cdot \mathbf{S}_{\mathbf{R}} \rangle = \frac{1}{n} \frac{\langle \mathbf{S}_0 \cdot \mathbf{S}_{\mathbf{R}} e^{-H[\mathbf{S}]} \rangle_0}{\langle e^{-H[\mathbf{S}]} \rangle_0} \; , \tag{3.12}$$

where $\langle \rangle_0$ stands for an expectation value based on the measure D[S] alone. In a strong-coupling expansion, $G(\mathbf{R})$ is expanded in a power series in β, obtained by expanding the exponentials e^{-H}, and collecting together terms of equal power in β. Both in the denominator and in the numerator, only terms with an even number of spin-factors from each site will contribute. Since every term in H contains the product of two neighbouring spins, the expansion of the denominator yields a sum over closed walks, while the numerator leads to walks between 0 and $\mathbf{R}$, with or without extra closed walks (loops).

For the special (and somewhat unnatural) case of the spins having *zero components*, $n = 0$, only terms with at most two spin-factors from each site will contribute, due to factors of n. For the same reason, terms with loops will vanish. As a result, the denominator has only the trivial term 1, while the numerator becomes a sum over self-avoiding walks between 0 and $\mathbf{R}$, each walk contributing β^N, with N the length of the walk. We obtain

$$G(\mathbf{R}) = \sum_N m_N(\mathbf{R}) \beta^N \; , \tag{3.13}$$

where $m_N(\mathbf{R})$ is the number of SAWs of length N from 0 to $\mathbf{R}$, establishing the relation between an $n = 0$ spin model and SAW statistics.

4   Interacting Continuous Polymers and Field Theories

4.1   Generic Interacting Polymer

For the case of a continuous polymer of length μ in D dimensions, a quite generic interaction is described by the Hamiltonian (or action)

$$S[\mathbf{x}(\tau)] = \frac{1}{2} \int_0^\mu d\tau \left(\frac{d\mathbf{x}}{d\tau} \right)^2 + \int_0^\mu d\tau \, V_1(\mathbf{x}(\tau)) + \frac{1}{2} \int_0^\mu \int_0^\mu d\tau d\tau' \, V_2(\mathbf{x}(\tau), \mathbf{x}(\tau'))$$

$$+ \frac{1}{6} \int_0^\mu \int_0^\mu \int_0^\mu d\tau d\tau' d\tau'' \, V_3(\mathbf{x}(\tau), \mathbf{x}(\tau'), \mathbf{x}(\tau'')) + \dots \; , \tag{4.14}$$

where $V_1(\mathbf{x})$ is a generic one-particle potential, $V_2(\mathbf{x}, \mathbf{y})$ a ditto pair potential, etc. The interaction is the most general one that has no explicit dependence on the chain parameter τ: V_n depends on $(\tau, \tau', \dots)$ only via $(\mathbf{x}(\tau), \mathbf{x}(\tau'), \dots)$.

To begin with, we will assume fixed B.C., with the endpoints of the chain locked at $\mathbf{x}(0) = \mathbf{a}$, $\mathbf{x}(\mu) = \mathbf{b}$. The main object of interest is the *partition function*, which is given by the functional integral

$$Z(\mu, \mathbf{a}, \mathbf{b}) = \int_{\mathbf{x}(0)=\mathbf{a}}^{\mathbf{x}(\mu)=\mathbf{b}} d[\mathbf{x}(\tau)] \exp\left(-S[\mathbf{x}(\tau)]\right) . \qquad (4.15)$$

The proper partition function for free boundary conditions is obtained simply by integrating $Z(\mu, \mathbf{a}, \mathbf{b})$ over the endpoints $\mathbf{a}, \mathbf{b}$.

4.2   The Relation to a Field Theory

The absence of an explicit τ-dependence ensures that the interaction part of the action, S_{int}, is a functional entirely of the *monomer density*, $\rho(\mathbf{y}) = \int_0^\mu d\tau \delta\left(\mathbf{y} - \mathbf{x}(\tau)\right)$, in terms of which it is given by

$$S_{\text{int}}[\rho] = \int d\mathbf{x}\, \rho(\mathbf{x})\, V_1(\mathbf{x}) + \frac{1}{2} \iint d\mathbf{x} d\mathbf{y}\, \rho(\mathbf{x})\rho(\mathbf{y})\, V_2(\mathbf{x}, \mathbf{y})$$
$$+ \frac{1}{6} \iiint d\mathbf{x} d\mathbf{y} d\mathbf{z}\, \rho(\mathbf{x})\rho(\mathbf{y})\rho(\mathbf{z})\, V_3(\mathbf{x}, \mathbf{y}, \mathbf{z}) + \dots \qquad (4.16)$$

as a functional Taylor expansion. Now, we assume that the corresponding Boltzmann factor, $\exp(-S_{\text{int}})$, can be functionally Laplace transformed with respect to $\rho(\mathbf{x})$,

$$F[\xi(\mathbf{x})] = \int d[\rho(\mathbf{x})] \exp\left(-\int d\mathbf{x}\xi(\mathbf{x})\rho(\mathbf{x}) - S_{\text{int}}[\rho]\right) . \qquad (4.17)$$

This is well-defined for any combination of arbitrary one-particle potentials and non-negative many-particle potentials. The functional Laplace transform can be inverted to yield, formally,

$$\exp\left(-S_{\text{int}}[\rho]\right) \propto \int d[\xi(\mathbf{x})]F[\xi(\mathbf{x})] \exp\left(\int d\mathbf{x}\xi(\mathbf{x})\rho(\mathbf{x})\right) , \qquad (4.18)$$

where each $\xi(\mathbf{x})$ is to be integrated along a complex contour to the right of all singularities. We then obtain for the partition function

$$Z(\mu, \mathbf{a}, \mathbf{b}) \propto \int d[\xi(\mathbf{x})]F[\xi(\mathbf{x})] \times$$
$$\int_{\mathbf{x}(0)=\mathbf{a}}^{\mathbf{x}(\mu)=\mathbf{b}} d[\mathbf{x}(\tau)] \exp\left(-\frac{1}{2}\int_0^\mu d\tau \left(\frac{d\mathbf{x}}{d\tau}\right)^2 + \int_0^\mu d\tau \xi(\mathbf{x}(\tau))\right) , \qquad (4.19)$$

and $-\xi(\mathbf{x})$ appears as an effective one-particle potential for the polymer. Now the $\mathbf{x}$ integral is recognized as the Euclidean quantum-mechanical path integral for the amplitude $\langle \mathbf{a}| \exp(-\mu\hat{H})|\mathbf{b}\rangle$ to go from $\mathbf{a}$ to $\mathbf{b}$ in imaginary time μ, with the quantum-mechanical Hamiltonian

$$\hat{H} = -\frac{1}{2}\nabla^2 - \xi(\mathbf{x}) + const. \, . \tag{4.20}$$

We thus have

$$Z(\mu, \mathbf{a}, \mathbf{b}) \propto \int d[\xi(\mathbf{x})] F[\xi(\mathbf{x})] \left\langle \mathbf{a} \left| \exp\left(-\mu\hat{H}\right) \right| \mathbf{b} \right\rangle \, . \tag{4.21}$$

Now we can make an ordinary Laplace transform in μ,

$$W(s, \mathbf{a}, \mathbf{b}) = \int_0^\infty d\mu \exp(-s\mu) Z(\mu, \mathbf{a}, \mathbf{b}) \, , \tag{4.22}$$

yielding

$$W(s, \mathbf{a}, \mathbf{b}) \propto \int d[\xi(\mathbf{x})] F[\xi(\mathbf{x})] \left\langle \mathbf{a} \left| \left(\hat{H} + s\right)^{-1} \right| \mathbf{b} \right\rangle \, . \tag{4.23}$$

The matrix-element can be written as a functional integral over a field $\phi(\mathbf{x})$ having $n = 0$ components,

$$\left\langle \mathbf{a} \left| \left(\hat{H} + s\right)^{-1} \right| \mathbf{b} \right\rangle =$$
$$2 \int d[\phi(\mathbf{x})] \phi_1(\mathbf{a}) \phi_1(\mathbf{b}) \exp\left(-\int d\mathbf{x} \phi \left(\hat{H} + s\right) \phi\right) \, , \tag{4.24}$$

and we obtain

$$W(s, \mathbf{a}, \mathbf{b}) \propto \int d[\xi(\mathbf{x})] F[\xi(\mathbf{x})] \int d[\phi(\mathbf{x})] \phi_1(\mathbf{a}) \phi_1(\mathbf{b}) \times$$
$$\exp\left(-\int d\mathbf{x} \left\{ \frac{1}{2}\left(\nabla\phi\right)^2 + s\phi^2 - \xi(\mathbf{x})\phi^2 \right\}\right) \, . \tag{4.25}$$

Now we can do the ξ integral to recover $\exp(-S_{\text{int}}[\rho(\mathbf{x})])$, however with the monomer density replaced by the *field density*, $\rho(\mathbf{x}) \to \phi(\mathbf{x})^2$. Thus, we finally arrive at the following result:

$$W(s, \mathbf{a}, \mathbf{b}) \propto \int d[\phi(\mathbf{x})] \phi_1(\mathbf{a}) \phi_1(\mathbf{b}) \exp\left(-\hat{S}[\phi(\mathbf{x})]\right) \, . \tag{4.26}$$

The RHS defines the two-point function $G(s, \mathbf{a}, \mathbf{b})$ of an interacting field theory, with an action $\hat{S}$ given by

$$\hat{S}[\phi(\mathbf{x})] = \int d\mathbf{x} \left\{ \frac{1}{2}\left(\nabla\phi\right)^2 + s\phi^2(\mathbf{x}) + V_1(\mathbf{x})\,\phi^2(\mathbf{x}) \right\}$$
$$+ \frac{1}{2}\iint d\mathbf{x}d\mathbf{y}\, V_2(\mathbf{x}, \mathbf{y})\,\phi^2(\mathbf{x})\phi^2(\mathbf{y})$$
$$+ \frac{1}{6}\iiint d\mathbf{x}d\mathbf{y}d\mathbf{z}\, V_3(\mathbf{x}, \mathbf{y}, \mathbf{z})\,\phi^2(\mathbf{x})\phi^2(\mathbf{y})\phi^2(\mathbf{z}) + \ldots \tag{4.27}$$

This shows that the partition function for a generic interacting polymer, when Laplace transformed with respect to the chain-length μ, yields the propagator of a zero-component field theory, with a bare mass determined by the corresponding conjugate variable s, and where the interaction appears in terms of the square of the field. Upon collecting the neglected constants of proportionality accumulated in the tranformations, the exact relation is given by

$$2G(s, \mathbf{a}, \mathbf{b}) = \int_0^\infty d\mu \exp(-s\mu) z(\mu, \mathbf{a}, \mathbf{b}) , \qquad (4.28)$$

where $z(\mu, \mathbf{a}, \mathbf{b})$ stands for $Z(\mu, \mathbf{a}, \mathbf{b})/Z_0(\mu)$, with $Z_0(\mu)$ denoting the partition function of the non-interacting polymer, integrated over one endpoint. The correctness can be verified by considering the case of no interactions, since the constants of proportionality involved are not affected by the interaction.

Likewise, the higher multi-point functions of the field theory are related to systems of several interacting polymers.

4.3 Example 1: Contact Interaction

The simplest non-trivial polymer interaction is an excluded-volume pair potential,

$$V_2(\mathbf{x}, \mathbf{y}) = g\delta(\mathbf{x} - \mathbf{y}) , \qquad (4.29)$$

which should be seen as representing a generic repulsive short-range interaction. The related field theory is a zero-component ϕ^4 one,

$$\hat{S}[\phi] = \int d\mathbf{x} \left\{ \frac{1}{2} \left(\nabla \phi \right)^2 + s\phi^2(\mathbf{x}) + \frac{g}{2} \left(\phi^2(\mathbf{x}) \right)^2 \right\} . \qquad (4.30)$$

The critical indices of the ϕ^4 theory are closely related to the scaling properties of polymers with a short-range interaction (de Gennes (1972)). Note the analogy to the relation between SAW statistics and an $n = 0$ spin-model, which are the corresponding lattice models.

4.4 Example 2: The Polyelectrolyte

In the physically interesting case of a polyelectrolyte, one has a pair interaction given by a (possibly screened) Coulomb potential $V_2(\mathbf{x} - \mathbf{y})$ satisfying $\left(-\nabla^2 + \kappa^2 \right) V_2(\mathbf{R}) = g\delta(\mathbf{R})$, and in three dimensions given by $V_2(\mathbf{R}) = g \exp(-\kappa r)/4\pi r$. The related field-theory is defined by the *non-local* action

$$\hat{S}[\phi] = \int d\mathbf{x} \left\{ \frac{1}{2} \left(\nabla \phi \right)^2 + s\phi^2(\mathbf{x}) \right\} + \frac{1}{2} \iint d\mathbf{x} d\mathbf{y} \, V_2(\mathbf{x}, \mathbf{y}) \, \phi^2(\mathbf{x}) \phi^2(\mathbf{y}) .$$
$$(4.31)$$

Noting that V_2 is the integral kernel of the inverse modified Helmholtz operator,

$$V_2(\mathbf{x}, \mathbf{y}) = g \left(-\nabla^2 + \kappa^2\right)^{-1} (\mathbf{x}, \mathbf{y}) \ , \qquad (4.32)$$

we can make a further simplification by introducing an auxiliary scalar field ψ, the exchange of which mediates the interaction. The resulting combined action $\tilde{S}$ is *local* and given by

$$\tilde{S}[\phi, \psi] = \int d\mathbf{x} \frac{1}{2} \left\{ (\nabla\phi)^2 + 2s\phi^2 + (\nabla\psi)^2 + \kappa^2\psi^2 + 2i\sqrt{g}\phi^2\psi \right\} \ . \qquad (4.33)$$

The variable s, conjugate to the chain length μ, appears only in the (bare) ϕ mass, while the inverse range κ of the interaction potential appears as the ψ mass; for the case of an unscreened potential, the ψ-field will be massless.

Due to translational and rotational invariance, the partition function $z(\mu, \mathbf{a}, \mathbf{b})$ can depend only on $R \equiv |\mathbf{a} - \mathbf{b}|$. Integrating over $\mathbf{R}$ would yield the corresponding partition function for free boundary conditions. However, it adds no complication to include a factor $\exp(i\mathbf{P} \cdot \mathbf{R})$ in the integral, yielding $\tilde{z}(\mu, \mathbf{P})$, which upon a Laplace transform in μ gives

$$\int_0^\infty d\mu \exp(-s\mu)\tilde{z}(\mu, \mathbf{P}) = \int d\mathbf{R} \int_0^\infty d\mu \ \exp(-s\mu + i\mathbf{P} \cdot \mathbf{R})z(\mu, 0, \mathbf{R})$$

$$= 2 \int d\mathbf{R} \exp(i\mathbf{P} \cdot \mathbf{R})G(0, \mathbf{R}) \equiv 2\tilde{G}(\mathbf{P}) \ . \qquad (4.34)$$

This yields twice the two-point function in momentum space, at momentum $\mathbf{P}$, for the ϕ field. It contains all relevant information on the distribution of the end-to-end distance $\mathbf{R}$ of the polymer. Thus, e.g., the average end-to-end distance squared, $\langle R^2 \rangle$, can be obtained from

$$\frac{\tilde{z}(\mu, \mathbf{P})}{\tilde{z}(\mu, 0)} = 1 - \frac{P^2 \langle R^2 \rangle}{2D} + \mathcal{O}(P^4) \ . \qquad (4.35)$$

5   Relation in Terms of Perturbation Theory

A perturbative expansion for a self-interacting polymer is obtained by applying an inverse Laplace transform to the loop expansion for the resulting (not necessarily local) field theory.

In the examples above, the interaction is in the form of a translation-invariant pair-potential; furthermore, the resulting field-theory is local. Thus, the loop-expansion of the two-point function $\tilde{G}(\mathbf{P})$ in momentum space amounts to the well-known computation of Feynman graphs.[2]

For the *polyelectrolyte* case, the Feynman rules for computing $\tilde{G}(\mathbf{P})$ are simple: On a single ϕ line, add lines of emitted and re-absorbed ψ lines joining the ϕ line in vertices. The external ϕ momentum is given by $\mathbf{P}$, and momentum is conserved at every vertex. A ϕ propagator carries a factor

---

[2] See also the contribution by E. Eisenriegler to these proceedings, on the subject of random walks in polymer physics.

$1/(q^2 + 2s)$ and a ψ propagator a factor $1/(q^2 + \kappa^2)$. Every vertex represents a factor $-2i\sqrt{g}$. For every loop, add a momentum integral $\int dq/(2\pi)^D$. It is easy to see from dimensional counting, that below six dimensions, the result will be ultra-violet finite for all graphs. (See Liverpool and Stapper (1997) for a recent RG analysis of the polyelectrolyte.)

Note that no ϕ-loops are allowed (due to $n = 0$); this can be interpreted as the ϕ-field corresponding to a *quantum-mechanical particle* that is not second-quantized; thus, there can be no creation or annihilation of the corresponding particles.

To lowest order, $\tilde{G}(\mathbf{P})$ is given by the bare propagator

$$\tilde{G}^{(0)}(\mathbf{P}) = \frac{1}{P^2 + 2s} , \qquad (5.36)$$

which is indeed half of the Laplace transform of $\exp(-\mu P^2/2)$, reflecting the distribution of $\mathbf{R}$ for a free polymer being proportional to $\exp(-\mathbf{R}^2/2\mu)$.

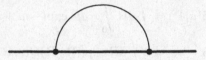

Fig. 1. The only graph contributing to first order in the perturbation expansion for $\tilde{G}(\mathbf{P})$.

To first order in g, corresponding to one loop, there is only one graph contributing, displayed in Fig. 1, yielding

$$\tilde{G}^{(1)}(\mathbf{P}) = \frac{-4g}{(P^2 + 2s)^2} \int \frac{d^3\mathbf{q}}{(2\pi)^3} \frac{1}{(q^2 + \kappa^2)((\mathbf{P} - \mathbf{q})^2 + 2s)}$$

$$= \frac{-4g}{(P^2 + 2s)^2} \frac{1}{4\pi P} \arctan\left(\frac{P}{\sqrt{2s} + \kappa}\right) . \qquad (5.37)$$

To second order there are three distinct graphs, displayed in Fig. 2.

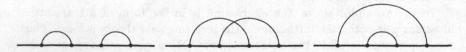

Fig. 2. The three graphs contributing to second order in the perturbation expansion for $\tilde{G}(\mathbf{P})$.

To compute the partition function Z, and related entities like $\langle S_{\text{int}} \rangle$, it is enough to consider $\mathbf{P} = 0$, while for $\langle R^2 \rangle$, the leading small P^2 corrections must be taken into account.

In a similar way, any pair-potential that only depends on the relative position, $V_2(\mathbf{x}, \mathbf{x}') = U(\mathbf{x} - \mathbf{x}')$, can be perturbatively treated; provided the ψ propagator is replaced by the Fourier transform of the potential, $\tilde{U}(\mathbf{q}) = \int d\mathbf{x} \exp(i\mathbf{q} \cdot \mathbf{x}) U(\mathbf{x})$.

Thus, for the *contact interaction*, the momentum space Feynman rules are even simpler: The "interaction lines" carry a unit factor, while the interaction vertices carry a factor of $-2ig^{1/2}$. Upon contracting the trivial interaction lines to points, we obtain four-vertices as expected for ϕ^4, carrying a factor of $-4g$. The critical dimension in this case is four, as verified by dimensional counting.

In all cases, the only role of $n = 0$ is to prevent ϕ loops. Thus the field ϕ is not second-quantized, and corresponds rather to a quantum-mechanical particle of mass $m^2 = 2s$.

6 Conclusions

The relation between self-interacting polymers and zero-component field theories has been reviewed. Loosely, it can be understood as follows: The polymer partition-function is essentially a Euclidean quantum-mechanical path integral. Quantum-mechanics corresponds to $n = 0$ field theory – no creation or annihilation.

Thus, scaling properties of polymer systems can be related to critical indices of the field theory, which can be computed using e.g. the well-developed machinery of renormalization group analysis.

References

des Cloizeaux, J., Jannink, G. (1990): *Polymers in Solution* (Clarendon Press, Oxford), chapter 11.

de Gennes, P.G. (1972): Phys. Lett. **38 A**, 339.

de Gennes, P.G. (1979): *Scaling Concepts in Polymer Physics* (Cornell Univ. Press, Ithaca), chapter X.

Itzykson, C., Drouffe, J.-M. (1989): *Statistical Field Theory* (Cambridge Univ. Press, New York), volume 1, chapter 1.

Liverpool, T.B., Stapper, M. (1997): cond-mat/9707036.

Parisi, G. (1988): *Statistical Field Theory* (Addison-Wesley, Urbana), chapter 16.

Pfeuty, P., Velasco, R.M., de Gennes, P.G. (1977): J. de Physique **38**, L-5.

Symanzik, K. (1969): Euclidean Quantum Field Theory, in *Local Quantum Theory*, *Proc. International School of Physics "Enrico Fermi"* XLV, Ed. L. Jost (Academic Press, New York), 152.

Zinn-Justin, J. (1989): *Quantum Field Theory and Critical Phenomena* (Clarendon Press, Oxford), chapter A25.

Reaction-Diffusion Mechanisms and Quantum Spin Systems

Gunter M. Schütz[1]

Institut für Festkörperforschung, Forschungszentrum Jülich, D-52425 Jülich, Germany

Abstract. We present a brief tutorial introduction into the quantum Hamiltonian formalism for stochastic many-body systems defined in terms of a master equation for their time evolution. These models describe interacting classical particle systems where particles hop on a lattice and may undergo reactions such as $A + A \to 0$. The quantum Hamiltonian formalism for the master equation provides a convenient general framework for the treatment of such models which, by various mappings, are capable of describing a wide variety of phenomena in non-equilibrium physics and in random media. The formalism is particularly useful if the quantum Hamiltonian has continuous global symmetries or if it is integrable, i.e. has an infinite set of conservation laws. This is demonstrated in the case of the exclusion process and for a toy model of tumor growth. Experimental applications of other integrable reaction-diffusion models in various areas of polymer physics (gel electrophoresis of DNA, exciton dynamics on polymers and the kinetics of biopolymerization on RNA) are pointed out.

1 Classical Stochastic Many-Body Dynamics in the Quantum Hamiltonian Formalism

The time evolution of many systems encountered in nature is most appropriately described by *stochastic laws* rather than by deterministic equations as in Newtonian mechanics. This randomness may be intrinsic in the underlying theory (e.g. radioactive decay as understood in terms of quantum mechanics) or be due to the inability to cope with the required amount of data for an accurate microscopical description, as in chaotic systems or in the study of thermodynamical properties of macroscopic objects. Whatever its origin might be, the randomness of the time evolution leads to the description of the process in terms of *random variables* and to the connection of the theory with measurements in terms of *expectation values*. Rather than predicting the actual value of some observable in a given measurement, one calculates e.g. the mean value of the observable as obtained by repeating the same measurement many times, starting from the same initial state. The classical example for such a random process is Brownian motion. As time passes on, it is not possible to calculate the position of the particle with any fixed accuracy. Instead one calculates quantities such as the probability of finding the particle in a given region of space, or simply the mean position of the particle and its fluctuations around this mean value. The purpose of these lecture notes is to study some reaction-diffusion models which are stochastic systems of *many interacting particles*. The obvious field of application of reaction-diffusion systems is the theory of diffusion-limited chemical reactions. However, as we shall see below, many phenomena in many-body physics and in biological physics are equivalent to reaction-diffusion mechanisms.

There are various approaches to such a stochastic description (van Kampen (1981)) such as the Langevin or Fokker-Plank equations. Here we adopt the strategy of describing the system by a probability distribution the time evolution of which is governed by a *master equation*. Solving the master equation, which is a first-order linear differential equation in the time variable, yields the probability of finding any given state the system may take given that it started from some initial state. From this distribution one can then calculate expectation values and hence make contact with actual measurements. This program may appear rather ambitious at first sight. However, as shown below, technically it is of the same difficulty as quantum mechanics and, indeed, the point of using this description is to make use of the tools developed so successfully for the treatment of quantum mechanical problems.

The crucial idea in this program is the formulation of the master equation in terms of a many-body quantum Hamiltonian, see Kadanoff and Swift (1968), Doi (1976) and Grassberger and Scheunert (1980) for the "bosonic" form of such a description of stochastic processes. In this framework the renormalization group has proved to be a powerful tool (Cardy (1997)). More recently, the mapping to *quantum spin Hamiltonians* (Sandow and Trimper (1993), Alcaraz et al. (1994), Schütz (1995a)) has produced a remarkable series of *exact* results. This mapping is the leitmotif of this lecture. Many new insights in different areas of physics have been obtained through the application of quantum mechanical methods for quantum spin systems to classical stochastic dynamics. It is our aim to review some of these recent methods and results.

1.1   The Master Equation

Discrete-Time Dynamics Before discussing many-particle systems we illustrate the master equation approach to stochastic processes on a simple example. The simplest stochastic model system that can be described in this way is a two-state system such as a coin which can show either heads or tails, or a spin which can either point up or down. We imagine some random mechanism (such as tossing the coin or a thermally activated spin-flip process) which may alter the state of the system after a time lapse Δt and we assume that the random updating is independent of the previous history of the system. Such processes are Markov processes. The probability $P_\eta(t + \Delta t)$ of finding the system at time $t + \Delta t$ after one updating event in the state η depends only on the state of the system at time t. The dynamical evolution can be represented by a set of rules which state the respective probabilities of moving from some state η to a state η' in one update. If spin up ($\uparrow$) flips to spin down ($\downarrow$) with probability p and spin down flips to spin up with probability q, then the probabilities $P_{\uparrow,\downarrow}(t + \Delta t)$ of finding the system in either of the two states evidently satisfy the *master equation*

$$P_\uparrow(t + \Delta t) = (1 - p)P_\uparrow(t) + qP_\downarrow(t) \tag{1.1}$$

$$P_\downarrow(t + \Delta t) = pP_\uparrow(t) + (1 - q)P_\downarrow(t). \tag{1.2}$$

From this description it is also obvious that $P_\eta(t + \Delta t)$ is related to $P_\eta(t)$ by a linear map. It is then convenient to write the master equation as a vector equation for the probability vector $|P(t)\rangle$ which has as its components the probabilities $P_\eta(t)$. The master equation then reads

$$|P(t + \Delta t)\rangle = T|P(t)\rangle \tag{1.3}$$

with the *transfer matrix* or transition matrix T. For the two-state spin model defined above T reads

$$T = \begin{pmatrix} 1 - p & q \\ p & 1 - q \end{pmatrix} \tag{1.4}$$

and the probability vector $|P(t)\rangle = P_\uparrow(t)|\uparrow\rangle + P_\downarrow(t)|\downarrow\rangle$ is given by

$$|P(t)\rangle = \begin{pmatrix} P_\uparrow(t) \\ P_\downarrow(t) \end{pmatrix}. \tag{1.5}$$

Generally, the strategy of writing the master equation can be summarized as follows: To each state $\eta \in X$ one assigns a basis vector $|\eta\rangle$ of the vectorspace $\mathbf{X} = \mathbb{C}^n$. Here n is the cardinality of X, i.e. the number of distinct states of the system. Together with their transposed vectors $\langle\eta|$ they form an orthonormal basis, $\langle\eta|\eta'\rangle = \delta_{\eta,\eta'}$. The probability vector is $|P(t)\rangle = \sum_\eta P_\eta(t)|\eta\rangle$ and the transfer matrix has as matrix elements $T_{\eta,\eta'} \equiv \langle\eta|T|\eta'\rangle$ the transition probabilities from state η' to state η. If the transition probabilities are time-independent (as we assume throughout this paper), then the solution to the master equation (1.3) at time $t = n\Delta t$ with an initial distribution $|P(0)\rangle$ can formally be written

$$|P(t)\rangle = T^n|P(0)\rangle. \tag{1.6}$$

For later reference it is useful to introduce the row vector $\langle s| = \sum_\eta \langle\eta|$ all components of which are equal to one. Conservation of probability, i.e. $1 = \sum_\eta P_\eta(t) = \langle s|P(t)\rangle$ implies

$$\langle s|T = \langle s|. \tag{1.7}$$

This means that in each column η of T all matrix elements, i.e., transition probabilities $p_{\eta\to\eta'}$ add up to 1 (see (1.4) for illustration). This is nothing but a technical way of expressing completeness of the set X: the system always moves to some state $\eta \in X$.[1] A matrix with the property (1.7) and in which all matrix elements are real and satisfy $0 \leq T_{\eta,\eta'} \leq 1$ is called a stochastic transfer matrix. The action of the transfer matrix has a simple interpretation in terms of the history of a given realization of the random process: In any given realization the system starts at some initial state η_0 and proceeds through a series of n states to a final state η_n at time $t = n\Delta t$. This realization of the stochastic time evolution happens with probability $p_{\eta_0 \to \eta_1} p_{\eta_1 \to \eta_2} \cdots p_{\eta_{n-1} \to \eta_n}$. The matrix element $\langle \eta' | T^n | \eta_0 \rangle$ is just the sum of all probabilities of histories which lead from η_0 to some $\eta' = \eta_n$ in n steps.

Expectation values of an observable F are by definition the quantities $\langle F \rangle = \sum_\eta F(\eta) P_\eta(t)$. Here $F(\eta)$ is some function of the random variables η, e.g. the spin $F(\uparrow) = 1$, $F(\downarrow) = -1$. In a series of measurements the system may be found in states η of the system with probabilities $P_\eta(t)$. Hence the expression $\langle F \rangle$ is the average value of what one measures in a series of many identical experiments, using the same initial state. If the initial states are not always the same fixed state, but some collection of different states, given by an initial distribution $P_\eta(0)$, then the expression $\langle F \rangle$ involves not only averaging over many realizations of the same process, but also averaging over the initial states. In vector notation one finds

$$\langle F \rangle = \langle s | F T^n | P(0) \rangle \tag{1.8}$$

with the observable represented by the diagonal matrix $F = \sum_\eta F(\eta) | \eta \rangle \langle \eta |$. If we want to specify time and initial condition, we write $\langle F(t) \rangle_{P_0}$. We mention in passing that for many systems of interest, e.g. random walks in disordered media, Eq. (1.8) represents a convenient way of a numerically exact calculation of expectation values by iterating on a computer the action of the transfer matrix on the starting vector $| P(0) \rangle$. It is not necessary to take averages over histories and initial states as in a Monte Carlo simulation. Our main interest is, however, in the analytical treatment of the expectation value for interacting many-body systems.

Continuous-Time Dynamics The dynamics in the previous subsection was defined in terms of discrete-time updates as performed e.g. on a computer. Now we pass on to the continuous-time formulation of Markov processes. This will eventually lead to the quantum Hamiltonian formalism for interacting particle systems.

One obtains a continuous-time representation of the processes described above by defining the off-diagonal transition probabilities in terms of rates $w_{\eta \to \eta'} = p_{\eta \to \eta'} / \Delta t$ which are the transition probabilities per time unit. This

---

[1] One may restrict the description of a stochastic process to its motion on a subset of X. In this case (which we shall not consider here) the system has a finite probability of being outside this subset and hence (1.7) is violated for such a transfer matrix T'.

allows one to write the transfer matrix in the form $T = 1 - H\Delta t$. The off-diagonal matrix elements of H are the (negative) transition rates, $H_{\eta,\eta'} = -w_{\eta'\to\eta}$. The diagonal elements $H_{\eta,\eta}$ are the (positive) sum of all the rates in each column η, i.e. the sum of all outgoing rates $w_{\eta\to\eta'}$. Taking the limits $\Delta t \to 0$, $n \to \infty$ in such a way that $t = n\Delta t$ remains fixed the time evolution may be written $T^n = T^{t/(\Delta t)} \to e^{-Ht}$ with a 'quantum Hamiltonian' H. The term 'quantum Hamiltonian' originates from the observation that by expanding the master equation (1.3) up to first order in Δt one finds

$$\frac{d}{dt}|P(t)\rangle = -H|P(t)\rangle \tag{1.9}$$

with the formal solution $|P(t)\rangle = e^{-Ht}|P(0)\rangle$. This equation has the form of a quantum mechanical Schrödinger equation in imaginary time. The standard notation $d/dt P_\eta(t) = \sum_{\eta'} [w_{\eta'\to\eta} P_{\eta'}(t) - w_{\eta\to\eta'} P_\eta(t)]$ of the master equation can be recovered from (1.9) by inserting the definition of $|P(t)\rangle$ and taking the scalar product with $\langle\eta|$.

For the two-state spin model with $p = w\Delta t$, $q = v\Delta t$ the corresponding quantum Hamiltonian reads

$$H = \begin{pmatrix} w & -v \\ -w & v \end{pmatrix}. \tag{1.10}$$

In terms of Pauli matrices $H = (w + v)/2 + (w - v)\sigma^z/2 - ws^- - vs^+$. Here $s^\pm = (\sigma^x \pm i\sigma^y)/2$. The off-diagonal part of H represents the flip events with their respective rates, while the diagonal elements ensure conservation of probability: In each column all matrix elements add up to zero, as required by probability conservation $\langle s|e^{-Ht} = \langle s|$, or, equivalently,

$$\langle s|H = 0. \tag{1.11}$$

Expectation values are calculated analogously to (1.8). Introducing the time-dependent operator in the Heisenberg representation $F(t) = e^{Ht}Fe^{-Ht}$ one can write, using (1.11),

$$\langle F \rangle = \langle s|F(t)|P(0)\rangle = \langle s|Fe^{-Ht}|P(0)\rangle. \tag{1.12}$$

Note that the expression 'quantum Hamiltonian' is misleading in three respects: Firstly, a stochastic quantum Hamiltonian may be, but need not be Hermitian. Instead, the physical meaning of the expectation value (1.12) is guaranteed by the probability conservation (1.11) and positivity of the transition rates $H_{\eta,\eta'} = -w_{\eta'\to\eta} \le 0$. Secondly, expectation values $\langle F \rangle$ are not of the usual quantum mechanical form $\langle\Psi|F|\Psi\rangle$. In fact, the probability vector $|P(t)\rangle$ represents the probability itself, rather than a complex probability amplitude. Finally, time is euclidean. However, the name quantum Hamiltonian formalism has become fairly standard and is justified not only by the form of the master equation (1.9), but also by the fact that for numerous interacting particle systems of interest H is indeed the Hamiltonian of some well-known quantum many-body system.[2] We want to stress though that whether the stochastic Hamiltonian is hermitian or not is not important for our purposes. Neither the physical interpretation as a stochastic system nor the methods applied below for the solution of the master equation (1.9) - i.e. symmetries, similarity transformation and the Bethe ansatz for integrable systems - depend on hermiticity.

Since the complete time evolution of the system is encoded in the master equation (1.9) and since the transition rates are not constrained by the condition of detailed balance the formalism applies not only to equilibrium dynamics, but equally to the stochastic description of systems far from thermal equilibrium.

1.2   Many-Body Systems

The Tensor Basis In the two-state spin model discussed above we had just a single spin flipping up and down. In many-body physics and in the study of interacting particle systems one is interested in the behaviour of many coupled spins sitting on some lattice. By identifying a spin up with a vacancy and a spin down with the presence of a particle on the lattice site, spin models can be seen as particle systems where each lattice site may be occupied by at most one particle. This correspondence can be generalized: Allowing for different species of particles, or site occupation by more than one particle, one obtains models where each lattice site can be in one of n distinct states. Such a model can be viewed as spin-$(n-1)/2$ system.[3]

The natural extension of the vector description of a single spin to many spins on a lattice is by taking a tensor basis as basis of the state space. Using the convention of considering spin down as a particle, a many particle configuration η is represented by the basis vector $|\eta\rangle = |\eta(1)\rangle \otimes \ldots \otimes |\eta(L)\rangle$. These vectors form a basis of the tensor space $(\mathbb{C}^n)^{\otimes L}$. We shall from now on use mainly particle language rather than spin language. States η are denoted by a set of occupation numbers $\eta = \{\eta(1), \ldots, \eta(L)\}$ for a lattice of L sites. For two-states models one has $\eta(x) = 0, 1$.

Construction of the Quantum Hamiltonian As in the single-site two-states model above, the flip events are represented by offdiagonal matrices. A matrix s_i^+ annihilates a particle at site i and s_i^- represents a creation event. To be precise, they represent attempts rather than actual events: Acting on an already occupied site with s^- yields zero, i.e. no change in the probability

---

[2] An example is the symmetric exclusion process discussed below. In other cases one may obtain a hermitian Hamiltonian after a (non-unitary) similarity transformation of the stochastic generator.

[3] Here we restrict ourselves to n-states systems, and mostly to the simplest case $n = 2$. For integrable 3-states systems see e.g. Alcaraz et al. (1994), Dahmen (1995), Schulz and Trimper (1996) and Sec. 4.

vector. This reflects the rejection of any attempt at creating a second particle on a given site. The reason is the exclusion of double occupancy which is encoded in the properties of the Pauli matrices. Simultaneous events are represented by products of Pauli matrices. E.g. hopping of a particle from site i to site j is equivalent to annihilating a particle at site i and at the same time creating one at site j. Thus it is given by the matrix $s_i^+ s_j^-$. The hopping attempt is successful only if site i is occupied and site j is empty. Otherwise acting with $s_i^+ s_j^-$ on the state gives zero and hence no change. The rate of hopping (or of any other possible stochastic event) is the numerical prefactor of each hopping matrix (or other attempt matrix). Of course, in principle the rate may depend on the configuration of the complete system. Suppose the hopping rate is given by a function $w(\eta)$ where η is the configuration prior to hopping. In this case the hopping matrix is given by $s_i^+ s_j^- w(\eta)$ where in $w(\eta)$ one replaces any $\eta(i)$ by the projector $n_i = (1 - \sigma_i^z)/2$. This is the projector on states with a particle at site i. If e.g. for some reason hopping from site i to site j should occur only if a third site k is empty, then the hopping matrix would be given by $s_i^+ s_j^- (1 - n_k)$. The construction of the attempt matrices for other processes or for n-states model is analogous. The formulation of the master equation in terms of a many-body quantum-spin operator does not only allow for a very convenient derivation of the equations of motion for expectation values (see e.g. Peschel et al. (1994), Schulz and Trimper (1996)), but suggests also analysis by specifically quantum mechanical methods.

The summation vector $\langle s |$ for a many-particle system is a tensor product of the single-site summation vectors. For two-states models one notes

$$\langle s | s_i^+ = \langle s | n_i , \quad \langle s | s_i^- = \langle s | (1 - n_i). \tag{1.13}$$

With these relations it is easy to construct the diagonal part of the quantum Hamiltonian in order ensure conservation of probability. To each off-diagonal attempt matrix one constructs a diagonal matrix by replacing all s_i^+ by n_i and by replacing all s_i^- by $1 - n_i$. E.g. to hopping from i to j with rate w represented by $-w s_i^+ s_j^-$ one adds $w n_i (1 - n_j)$. The (negative) sum of all attempt matrices minus their diagonal counterparts is then the full quantum Hamiltonian. In the same way one constructs the diagonal parts of n-states models by using the analogues of Eqs. (1.13). Conservation of probability (1.11) is then automatically satisfied.

Expectation Values The expectation values one is usually interested in are k-point correlation functions

$$\langle n_{i_1}(t_1) \ldots n_{i_k}(t_k) \rangle = \langle s | n_{i_1} e^{-H(t_1 - t_2)} n_{i_2} \ldots n_{i_k} e^{-H t_k} | P(0) \rangle. \tag{1.14}$$

These are the joint probabilities of finding particles at sites i_l at times $t_l \geq t_{l+1}$. Since we are dealing with many-body systems we shall from now on denote a vector $| \eta \rangle$ which represents a state with particles on sites $k_1, \ldots, k_N$ by $| k_1, \ldots, k_N \rangle$. The empty lattice is represented by the vector $| 0 \rangle$, the summation vector is always denoted $\langle s |$.

1.3 Stationary States

One of the most basic questions to ask is the behaviour of the system at late times of the stochastic evolution. For ergodic systems the probability distribution in the limit $t \to \infty$ is independent of the initial state and one would like to know quantities like the mean density, density fluctuations, or the spatial structure of the density distribution and its correlations. For transition rates that are constant in time this asymptotic distribution is invariant under time translations and hence called stationary. We shall denote such a distribution by $|P^*\rangle$. From the considerations of the previous subsections it is clear that $|P^*\rangle$ is an eigenvector of H with eigenvalue zero,

$$H|P^*\rangle = 0. \tag{1.15}$$

One may wonder whether such an eigenvector exists and how many linearly independent stationary states there are. Existence in a system with finite state space is easy to prove: By construction there is at least one *left* eigenvector with vanishing eigenvalue, see (1.11). This guarantees the existence of at least one *right* eigenvector (1.15) with vanishing eigenvalue. However, there is no equally simple general argument which gives the number of different stationary states. Instead, one has to study the possibilities of moving from one given state η to some other state η' after a finite time. If one manages to identify a subset of states such that one can go from each of these states to any other state within this subset with non-zero probability after some finite time, then there is exactly one stationary distribution in this subset. Furthermore, the support of the distribution is identical to the subset. To illustrate this somewhat abstract statement, consider a lattice gas on a finite lattice with particle number conservation. If the dynamics are such that for fixed particle number each possible state can be reached from any initial state after finite time with finite probability then there is exactly one stationary distribution for each subset of states with fixed total particle number. If one also allows for particle creation and annihilation processes, then one can move from any initial state to any other state, irrespective of particle number. In this case there is only one stationary distribution for the whole system.

Finally, note that, by construction, zero is the eigenvalue of H with the lowest real part. This follows from a theorem by Gershgorin (Gradshteyn and Ryzhik (1980)). Therefore in quantum mechanical language the stationary vector corresponds to the ground state of H. However, if H is not hermitian $|P^*\rangle$ is *not* the transposed vector $(\langle s|)^T$, but a more complicated object.

2 Diffusion of Hard-Core Particles

Before studying systems with particle reactions we first consider purely diffusive models. A classical example where the quantum Hamiltonian formalism has turned out to be fruitful is the exclusion process. In its simplest form this is a one-species process where each particle hops between nearest neighbour sites with constant rate D (Spitzer (1970), Ligget (1985)). The particles have a hard-core exclusion interaction: hopping attempts which would lead to a double occupancy of a site are rejected. Hence forward and backward hopping across a bond between neighbouring sites is given by the matrix $h_{ij} = D(n_i(1-n_j)+(1-n_i)n_j-s_i^+ s_j^- - s_i^- s_j^+) = D(1-\sigma_i^x\sigma_j^x-\sigma_i^y\sigma_j^y-\sigma_i^z\sigma_j^z)/2$. Hence the Hamiltonian for the full symmetric exclusion process is given by the Heisenberg quantum ferromagnet (Alexander and Holstein (1978))

$$H^{SEP} = -\frac{D}{2} \sum_{<i,j>} \left(\sigma_i^x\sigma_j^x + \sigma_i^y\sigma_j^y + \sigma_i^z\sigma_j^z - 1\right). \tag{2.16}$$

This is the first instance where the stochastic time evolution operator is indeed identical to the quantum Hamiltonian of a non-trivial many-body system which has a long history in condensed matter physics. In particular, from the form of H^{SEP} one reads off the important and well-known property that H^{SEP} commutes with the generators $S^{x,y,z} = \sum_i \sigma_i^{x,y,z}$ of the angular momentum algebra $SU(2)$.[4]

This process can be visualized by representing particles by the symbol A and vacancies by the symbol $\emptyset$ and writing

$$\emptyset A \rightleftharpoons A \emptyset \quad \text{with rate 1}$$

for the hopping events between sites i, j. The symmetric exclusion process was first introduced and studied in detail by Spitzer (1970), the main result being duality relations which express time-dependent k-point correlation functions $\langle n_{i_1}(t) \dots n_{i_k}(t) \rangle$ for an arbitrary many-particle initial state in terms of correlators of the same system, but containing only k particles. In particular, the density expectation value $\langle n_k(t) \rangle$ is completely given in terms of the dynamics of just a single particle. A single particle in this system is a random walker hopping with constant rate between neighbouring sites. This result was later rederived and extended using the global $SU(2)$ symmetry of H (Schütz and Sandow (1994)). The $SU(2)$ symmetry allows for the derivation of similarly strong results for multi-time correlation functions $\langle n_{i_1}(t_1) \dots n_{i_k}(t_k) \rangle$ and also for their extension to the partial exclusion process. The partial exclusion process is the spin-s version of this model where each lattice site can be occupied by at most $2s$ particles which hop with rates $n_i(2s - n_j)$ from site i to site j. The underlying mathematical structure which generates these reductions are the selection rules of $SU(2)$ (Stinchcombe et al. (1993)).

Generally, any k-point correlator can be reduced to a k-particle transition probabability. Spin-wave theory (Mattis (1965)) and, for the one-dimensional model, Bethe ansatz (Bethe (1931), Thacker (1981)) can then be employed

---

[4] For the remainder of this section we shall drop the superscript SEP (for symmetric exclusion process).

to analyze the resulting simplified expressions (see below). The point of using the continuous $SU(2)$ symmetry becomes already clear by considering the density expectation value $\langle n_k(t) \rangle_{P_0}$ for some arbitrary many-particle initial distribution. This quantity describes the evolution of the spatial density structure as the system approaches equilibrium where all configurations are equally likely and the density is constant. Using the factorization of $\langle s |$ into a tensor product one first notes $\langle s | = \langle 0 | e^{S^+}$ where $\langle 0 |$ represents the empty lattice (all spins up) and $S^+ = \sum_i s_i^+$. Also, because of factorization, $e^{S^+} n_i = (n_i + s_i^+) e^{S^+}$. Next we observe that because of the $SU(2)$-symmetry the time-evolution matrix e^{-Ht} commutes with e^{S^+}. Therefore, $\langle s | n_i = \langle i | e^{S^+}$ where we have denoted the state $\langle s | s_i^+$ with a single particle on site i by $\langle i |$. Finally, using particle number conservation and inserting a complete set of one-particle states $\sum_l | l \rangle \langle l |$ between e^{-Ht} and s^{S^+} yields

$$\langle n_k(t) \rangle = \langle 0 | (s_k^+ + n_k) e^{-Ht} e^{S^+} | P(0) \rangle$$

$$= \sum_l \langle k | e^{-Ht} | l \rangle \langle n_l(0) \rangle. \tag{2.17}$$

The quantity $\langle k | e^{-Ht} | l \rangle$ is nothing but the conditional probability $P(k; t | l; 0)$ for a single random walker to be found at point k at time t given that it started at site l at time $t = 0$. Hence the dynamics of the density expectation value is completely determined by the single-particle conditional probabilities and the density profile in the initial state. This relation is valid on any lattice. On an infinite one-dimensional lattice with nearest neighbour hopping $P(k; t | l; 0)$ satisfies the differential-difference equation $d/dt P(k; t | l; 0) = D(P(k+1; t | l; 0) + P(k-1; t | l; 0) - 2P(k; t | l; 0))$ with initial value $P(k; 0 | l; 0) = \delta_{k,l}$. This equation is readily solved by Fourier transformation and one finds

$$P(k; t | l; 0) = e^{-2Dt} I_{k-l}(2Dt) \tag{2.18}$$

with the modified Bessel function $I_n(2Dt)$. From the diffusive behaviour of the system one reads off the dynamical exponent $z = 2$ which relates the scaling behaviour in spatial direction to the temporal scaling on large scales. Rescaling spatial coordinates by a factor λ and at the same time rescaling time by λ^2 leaves correlation functions invariant up to an overall amplitude.

Now we consider the two-point correlator $\langle n_i(t_w + t) n_j(t_w) \rangle$ measured in a system with particles initially distributed in some as yet unspecified way. This quantity measures the correlation between finding a particle at site j after a "waiting time" t_w and then finding a particle at site i after an additional time lapse t. Going through analogous steps as above one finds after inserting a complete set of two-particle states $\sum_{m,l} | l, m \rangle \langle l, m |$

$$\langle n_i(t_w + t) n_j(t_w) \rangle = \langle i | e^{-Ht} (s_j^+ + n_j) e^{-Ht_w} e^{S^+} | P(0) \rangle$$

$$= \sum_{k \neq j} \langle i | e^{-Ht} | k \rangle \sum_{m,l} \langle n_l(0) n_m(0) \rangle \langle k, j | e^{-Ht_w} | l, m \rangle$$

$$+\langle i | e^{-Ht} | j \rangle \sum_l \langle n_l(0) \rangle \langle j | e^{-Ht_w} | l \rangle. \qquad (2.19)$$

This relation expresses the correlator in terms of one and two-particle transition probabilities and the value of the correlator in the initial state. For the equal-time correlator $(t = 0)$ one gets

$$\langle n_i(t_w) n_j(t_w) \rangle = \sum_{m,l} \langle n_l(0) n_m(0) \rangle \langle i, j | e^{-Ht_w} | l, m \rangle. \qquad (2.20)$$

If the initial distribution is uncorrelated with density ρ the system is in thermal equilibrium and stationary. Then the correlator does not depend on t_w and becomes a function of time difference t only: $\langle n_i(t) n_j(0) \rangle = \rho^2 + \rho(1 - \rho) \langle i | e^{-Ht} | j \rangle$ since in this equilibrium initial state $\langle n_j(0) n_l(0) \rangle = \rho^2 + \rho(1 - \rho) \delta_{j,l}$. On the other hand, if the system is started in some non-equilibrium state and evolves freely until the waiting time t_w, analysis of the correlator for i close to j shows that the system exhibits a simple form of *aging*: The density at a given site i at some later observation time $t_w + t$ decorrelates with the density at site j at t_w on a time scale which grows with the waiting time t_w. The longer one waits, the more time it takes for the system to decorrelate. This is not the case in thermal equilibrium. Measuring this relaxation time gives knowledge about the "age" t_w of the system.

To obtain this result one has to solve for the two-particle conditional probability. Since the particles have no long-range interaction, but only on-site repulsion one might wonder to which extent these conditional probabilities deviate from those obtained for completely non-interacting particles. We address this problem for the one-dimensional system where one expects the hard-core constraint to be most relevant for the dynamics of the system. The system with nearest neighbour hopping can now be solved using the Bethe ansatz (Bethe (1931)). The idea behind the Bethe ansatz is first to turn the master equation for the N-particle conditional probability $P(x_1, \ldots, x_N; t) = \langle x_1, \ldots, x_N | e^{-Ht} | y_1, \ldots, y_N \rangle$ into an eigenvalue equation by the ansatz $P(x_1, \ldots, x_N; t) = e^{-\epsilon t} P_\epsilon(x_1, \ldots, x_N)$ and then to write $P_\epsilon(x_1, \ldots, x_N)$ as a superposition of plane waves with pseudo momenta p_i conjugate to the particle positions x_l. Since all particles are identical this superposition is a sum over permutations of the momenta in the plane waves $e^{i \sum_l p_{j(l)} x_l}$. The magic of the ansatz (which originates in the underlying integrability of the system) consists in the fact that the amplitude of each permutation in the sum factorizes into a product of corresponding permutations of two-particle amplitudes. We do not go here into any detail, but merely demonstrate how the Bethe ansatz works for the two-particle problem. Adapted to the problem at hand, the Bethe ansatz for the conditional probability reads

$$P(x_1, x_2; t | y_1, y_2; 0) = \frac{1}{(2\pi)^2} \int dp_1 \int dp_2 e^{-(\epsilon_1 + \epsilon_2)t - ip_1 y_1 - ip_2 y_2} \Psi_{p_1, p_2}(x_1, x_2)$$

$$(2.21)$$

with the Bethe wave function

$$\Psi_{p_1,p_2}(x_1,x_2) = e^{ip_1x_1+ip_2x_2} + S(p_1,p_2)e^{ip_2x_1+ip_1x_2}, \qquad (2.22)$$

the "energies"

$$\epsilon_i = 2D(1 - \cos p_i) \qquad (2.23)$$

and the two-particle scattering amplitude

$$S(p_1,p_2) = -\frac{1 + e^{ip_1+ip_2} - 2e^{ip_2}}{1 + e^{ip_1+ip_2} - 2e^{ip_1}}. \qquad (2.24)$$

The energy expression arises from the diffusive motion of the particles: The time evolution operator acts on the conditional probability like a lattice Laplacian if the difference between the coordinates is larger than 1, i.e. if the two particles do not "feel" the presence of the other. The scattering amplitude arises from the need to satisfy the master equation with this energy term also if the difference in coordinates is equal to one. This requires to define $\Psi_{p_1,p_2}(x_1,x_2)$ in the unphysical domain $x_1 = x_2$ by $\Psi_{p_1,p_2}(x,x) + \Psi_{p_1,p_2}(x+1,x+1) = 2\Psi_{p_1,p_2}(x,x+1)$ which fixes the relative amplitude S of the two plane waves. The contour of integration is determined by the initial condition $P(x_1,x_2;0|y_1,y_2;0) = \delta_{x_1,y_1}\delta_{x_2,y_2}$ and we consider the coordinates to be ordered, $x_1 < x_2$ and $y_1 < y_2$.[5]

To analyze (2.21) we note that at late times the main contribution to the integral arises from small values of p_1, p_2. So we can expand the cosine in the energy term to first non-vanishing order and make a substitution of variables $p_i \to \tilde{p}_i = p_i\sqrt{t}$, $x_i, y_i \to \tilde{x}_i, \tilde{y}_i = x_i/\sqrt{t}, y_i/\sqrt{t}$. Expanding S for small arguments $\tilde{p}_i/\sqrt{t}$ leads to $S = 1 + O(t^{-1/2})$. Thus we arrive at the somewhat surprising conclusion that the leading contribution to the conditional probability comes simply from $S = 1$, corresponding to non-interacting particles. Furthermore, because of the factorization of the plane wave amplitudes for n-particle conditional probabilities, the same result holds true in this case. Thus, *all n-point correlation functions of the symmetric exclusion process are, to leading order in time, identical to the same n-point correlators of non-interacting particles.* Corrections are of order $n/\sqrt{t}$.

Before concluding this section we want to emphasize that this simple asymptotic equivalence to non-interacting particles does not hold for the asymmetric exclusion process where particles hop with different rates to the right and left respectively. This system can also be solved with the Bethe ansatz (Gwa and Spohn (1992), Kim (1995), Schütz (1997)). Here the interaction leads to strong non-linear behaviour. It results in the evolution of shocks (Derrida et al. (1993)) and to a collective diffusion coefficient which diverges with the square root of the number of particles in the system (Derrida and Evans (1994)). The dynamical exponent for states with finite density turns

---

[5] See Schütz (1997) where the general case of biased hopping is treated. For an earlier full solution of the symmetric case see Dieterich and Peschel (1983).

out to be $z = 3/2$ (Gwa and Spohn (1992), Kim (1995)), consistent with the superdiffusive behaviour indicated by the divergent collective diffusion coefficient. A special case is this process with closed (reflecting) boundaries. In this case one can use the symmetry under the quantum group $SU(2)_q$ to obtain further results on the dynamics of the system (Sandow and Schütz (1994)). For a review of many properties of this rather interesting system, see Ligget (1985), Spohn (1991) and Privman (1997).

Another important limitation of the equivalence of the unbiased exclusion process to non-interacting particles is the tagged particle process. If one wants to follow the motion of a specific tagged particle rather than studying the probabilities of finding any (unspecified) particles on a given set of sites, then clearly the hard-core repulsion becomes relevant. The effect is particularly strong in one dimension, where the motion of a tagged particle is bounded by the positions of its nearest neighbours. Even though also the tagged particle problem is integrable, exact results are much harder to obtain as they require a full solution of the many-particle problem. There is no simple reduction of interesting quantities to a few-particle problem.

3   Coarsening and Domain Growth

3.1   Glauber Dynamics for the Ising Model and the Voter Model

A phenomenon of wide interest in physics and chemistry is the growth of domains in non-equilibrium two-phase systems. The best-known example is perhaps the Ising model with domains of up- and down spins, separated by domain walls. The energy of the Ising model is given by the nearest neighbour sum $E = -J \sum s_i s_j$. Since the creation of a local domain wall costs an energy J the system tries to organize itself at low temperature into large domains of uniform magnetization. Starting from a high-temperature equilibrium state with many domain walls and quenching to low temperatures leads to a coarsening process: Small domains of uniform magnetization merge to form larger domains since then the total length of the domain walls and thus the energy decreases. Glauber (1963) introduced spin-flip dynamics which ensure that the system reaches the equilibrium distribution at temperature $T = 1/\beta$ of the one-dimensional Ising model. In this model a spin within a domain of equal magnetization is flipped with a rate $\mu = 1 - \tanh \beta J$, whereas a spin in a region of opposite magnetization is flipped with a rate $\lambda = 1 + \tanh \beta J$. At domain boundaries spins are flipped with unit rate, since no change in energy is involved. This process can be visualized in the following way:

$$\uparrow\uparrow\uparrow \rightarrow \uparrow\downarrow\uparrow \quad \text{and} \quad \downarrow\downarrow\downarrow \rightarrow \downarrow\uparrow\downarrow \quad \text{with rate } \mu$$
$$\uparrow\downarrow\uparrow \rightarrow \uparrow\uparrow\uparrow \quad \text{and} \quad \downarrow\uparrow\downarrow \rightarrow \downarrow\downarrow\downarrow \quad \text{with rate } \lambda$$
$$\uparrow\uparrow\downarrow \rightleftharpoons \uparrow\downarrow\downarrow \quad \text{and} \quad \downarrow\downarrow\uparrow \rightleftharpoons \downarrow\uparrow\uparrow \quad \text{with rate } 1$$

In one dimension this process can be mapped to a reaction-diffusion system. By identifying a domain wall ($\uparrow \downarrow$ or $\downarrow \uparrow$) with a particle of type A on the dual lattice and no domain wall with a vacancy $\emptyset$ this process becomes a reaction-diffusion system with pair annihilation and pair creation of exclusion particles (Rácz (1985), Family and Amar (1991), Santos (1997))

$$\emptyset\,\emptyset \;\rightarrow\; A\,A \quad \text{with rate } \mu$$
$$A\,A \;\rightarrow\; \emptyset\,\emptyset \quad \text{with rate } \lambda$$
$$\emptyset\,A \;\rightleftharpoons\; A\,\emptyset \quad \text{with rate } 1.$$

At zero temperature there is no creation ($\mu = 0$, $\lambda = 2$). Thus the system evolves into the single absorbing state with no particles at all. In spin language this is the totally ferromagnetic state with all spins up or all spins down. This process of diffusion-limited pair-annihilation (DLPA) is of interest not only for the study of spin relaxation and coarsening, but also for the understanding of the dynamics of laser-induced excitons on polymers as seen in experiments (see next section, for an extensive review, see Privman (1997)). The quantum Hamiltonian describing the stochastic time-evolution of the domain walls is given by (Siggia (1977))

$$H^{DW} = -\sum_i \left[\mu s_i^- s_{i+1}^- + \lambda s_i^+ s_{i+1}^+ + s_i^+ s_{i+1}^- + s_i^- s_{i+1}^+ + (\mu - 1)\sigma_i^z - 1 \right]$$

$$(3.25)$$

It is related to the quantum Hamiltonian of the anisotropic transverse XY model in a magnetic field by a simple diagonal similarity transformation $B = q^{S^z}$ with $q = \sqrt{\mu/\lambda}$. There are various ways of calculating correlation functions for this system. The Hamiltonian can be turned by a Jordan-Wigner transformation (Jordan and Wigner (1928), Lieb et al. (1961)) into an integrable free fermion system (Siggia (1977), Lushnikov (1987)). This provides a convenient framework for the calculation of various expectation values (Lushnikov (1987), Alcaraz et al. (1994), Grynberg et al., Schütz (1995b), Santos et al. (1996)). The zero temperature limit $\mu \rightarrow 0$ leads to the quantum Hamiltonian H^{DLPA} for diffusion-limited pair annihilation, treated already earlier by Bramson and Griffeath (1980), Torney and McConnell (1983), Spouge (1988) and more recently in a discrete-time description by Privman (1994). In the version (3.25) of the process the hopping time scale is determined by the spin-flip time scale of the Glauber process. In the general pair-annihilation/creation process it is an independent constant.

Glauber dynamics can also be seen as a reaction-diffusion system without reference to domain walls. One simply identifies an up-spin with a vacancy and a down-spin with a particle. In one dimension at zero temperature the process can then be described as follows:

$$A\,\emptyset \text{ or } \emptyset\,A \;\rightarrow\; A\,A \quad \text{with rate } 1$$
$$A\,\emptyset \text{ or } \emptyset\,A \;\rightarrow\; \emptyset\,\emptyset \quad \text{with rate } 1$$

This can obtained by a translational rearrangement of the three-site interaction terms in terms of two-site processes. The quantum Hamiltonian for this process may be written (Felderhof (1971))

$$H^{VM} = -\frac{1}{2} \sum_{<i,j>} (\sigma_i^x + \sigma_j^x - 2)(1 - \sigma_i^z \sigma_j^z). \qquad (3.26)$$

The generalization of this nearest neighbour process to higher dimensions is generally known as the voter model. The flip rate for a given spin at some lattice site is equal to the number of nearest neighbour spins of opposite value. Thus an atom (= human being) changes his magnetization (= opinion yes or no) at a rate proportional to the opinions of his neighbours! Note that there is no simple mapping to DLPA via the domain wall picture in higher dimensions. In the presence of a weak magnetic field one finds the same process, but with different rates for ending up in the states AA or $\emptyset\emptyset$ respectively.

3.2   Order/Disorder Competition: A Simple Biological Model

One can also observe domain growth from a single nucleus by a particle branching process. In this case there is no accompanying coarsening process. One can imagine particles which diffuse with unit rate and occasionally create offsprings with rate p. If one assumes site exclusion, i.e. offsprings are created only on empty neighbouring sites because of spatial constraints, this process is a simple toy model of growing tissue cell populations (Drasdo (1996)). The branching process describes cell division, while the particle hopping corresponds to the diffusive motion of cells in their environment.[6] In addition to that we allow for a death process with rate q which kills both the original cell and its offspring during the branching (cell mitosis). This leads to the following reaction diffusion system:

$$\emptyset A \rightleftharpoons A \emptyset \quad \text{with rate } 1$$
$$\emptyset A, \ A \emptyset \rightarrow \emptyset \emptyset \quad \text{with rate } p$$
$$\emptyset A, \ A \emptyset \rightarrow A A \quad \text{with rate } q$$

It is intuitively clear that for $p > q$ the cell population will grow until all space is covered, while for $p < q$ the population will eventually die out. Therefore it is of more interest to study the case $p = q$ when creation of offsprings and the death process balance each other. This process is also of interest from a different angle: By inspection one realizes that this process is a combination

---

[6] Of course this is only a very simplified description of what actually happens in a real system. One neglects, to name just one of many approximations, completely any attractive interaction between cells due to their surface structure. The incorporation of such an interaction is conceptually easy, but not the point of these lecture notes.

of the exclusion process (pure diffusion) and the voter model. The diffusion process tries to disorder the system, while the branching/death process tries to create an ordered system of uniform structure, as discussed above in the context of the voter model. Hence the questions arise, which process wins, and how is the stationary state reached.

3.3   Diffusion-Limited Annihilation Revisited

Before answering the questions raised at the end of the last subsection we note that the density expectation value $\langle n_k \rangle$ satisfies a continuity equation

$$\frac{d}{dt}\langle n_k \rangle = \sum_{j=n.n.} (\langle n_j \rangle - \langle n_k \rangle) \tag{3.27}$$

where the sum runs over all nearest neighbour sites. Hence the total number of particles $\langle N \rangle$ is on average conserved. This does not reflect a symmetry as for processes with actual particle number conservation. Rather it is a very weak result because it gives no information on whether the system orders or remains disordered. If the system orders, i.e. reaches a state where all lattice sites are empty or full, then conservation of average particle number simply means that the probability of reaching the full state is equal to N/L while the probability of reaching the empty state equals $1 - N/L$. Hence, in order to understand the dynamical properties of the process one has to understand the behaviour of the particle correlations. If the system reaches a disordered state particle correlations will become weak. Otherwise, in the ordered state the correlations are strong and independent of the distance in the lattice.

In one dimension this problem was addressed by studying the dynamical particle-particle correlations for a translationally invariant initial state [Droz et al. (1989)], using the fact that the equations of motion for correlators decouple into closed subsets. Here we take a different strategy which not only explains *why* this happens, but also that such a decoupling occurs in any dimension. In a second step, we use the Bethe ansatz to study the behaviour of all dynamical k-point correlation functions in one dimension.

To examine the two-point correlation function we introduce a new tool, viz. a change of basis, which is borrowed like the Bethe ansatz from quantum mechanics. The definition of a stochastic matrix introduced above is manifestly basis-dependent. A change of basis, i.e. a similarity transformation of the stochastic time evolution operator, will generically not lead to a stochastic quantum Hamiltonian. However, in special cases either such a transformed Hamiltonian or its transposed does define a new stochastic process and thus one can relate results obtained for one process to quantities of the transformed process. If two processes are related by a similarity transformation, $\tilde{H} = \mathcal{B}H\mathcal{B}^{-1}$, we call these two processes *equivalent*. On the other hand, if the transposed and transformed matrix $\tilde{H}^T = \mathcal{B}H\mathcal{B}^{-1}$ describes some stochastic process, then these two processes are called *enantiodromic*.

For $p = q$ the quantum Hamiltonian for the biological growth process may be written $H = H^{SEP} + pH^{VM}$ where we set the hopping time scale of the exclusion process $D = 1$. Now we introduce the matrix $\mathcal{B} = B^{\otimes L}$ where $B = 1 + i\sigma^y$. It is then easy to show that $H^{DLPA} = (\mathcal{B}H\mathcal{B}^{-1})^T$ where the hopping time scale of DLPA is equal to $1 + p$ (instead of 1 as above) and the annihilation rate $\lambda = 2p$. Now we are in a position to relate the density two-point correlation function between points k, l on an arbitrary lattice to quantities in DLPA, defined on the same lattice. Since (i) $\langle s | n_k n_l e^{-Ht} | P_0 \rangle = \langle P_0 | e^{-H^T t} n_k n_l | s \rangle$, (ii) $B^T = 2B^{-1}$ and (iii) by using the factorization of $\mathcal{B}$ into a tensor product one finds

$$\langle n_k(t) n_l(t) \rangle_{P_0} = \langle P_0 | \mathcal{B}^{-1} e^{-H^{DLPA} t} \mathcal{B} n_k n_l | s \rangle$$
$$= \frac{1}{4} \langle s | Q e^{-H^{DLPA} t} (|0\rangle + |k\rangle + |l\rangle + |k,l\rangle) \quad (3.28)$$

where the observable Q of DLPA is defined by the relation $\langle s | Q = \langle P_0 | \mathcal{B}^{-1}$ and has to be calculated from the initial condition of the growth model. The more important result is the enantiodromy relation of the observable $(n_k n_l)$ of the growth process to the initial state of the DLPA process: We find that the calculation of the two-point correlator for an arbitrary initial state reduces to the calculation of a two-particle problem in DLPA. To measure correlations we define $C_{k,l}(t) = \langle n_k(t) n_l(t) \rangle_{P_0} - \langle n_k(t) \rangle_{P_0} \langle n_l(t) \rangle_{P_0}$ for which one finds in the same way

$$C_{k,l}(t) = \frac{1}{4} \left(\langle s | Q e^{-H^{DLPA} t} | k,l \rangle - \langle s | Q e^{-H^{DLPA} t} | k \rangle \langle s | Q e^{-H^{DLPA} t} | l \rangle \right).$$
$$(3.29)$$

In order to answer the question whether the growth system orders we assume a worst case initial scenario and take as initial state random initial conditions where each possible configuration is found with equal probability $1/2^L$. This corresponds to the initial state $| P_0 \rangle = |s\rangle / 2^L$ and thus leads to $\langle P_0 | \mathcal{B}^{-1} = \langle 0 |$, i.e., $Q = \prod_k (1 - n_k)$. Thus the two-point correlator of the growth model is determined by the particle survival probability in DLPA. A single particle in DLPA never decays, hence $\langle s | Q e^{-H^{DLPA} t} | l \rangle = 0$. Two particles will eventually disappear, $\lim_{t \to \infty} \langle 0 | e^{-H^{DLPA} t} | k,l \rangle = 1$. Therefore $C_{k,l}(\infty) = 1/4$. Even though we started with a maximally disordered state the system orders completely and can be found with equal probability in the empty or full state respectively.

The approach to this final distribution depends on the dimensionality of the system. In one dimension the correlation function can be calculated exactly using again the Bethe ansatz for the two-particle master equation. One finds by expanding the two-body scattering matrix $S(p_i, p_j)$ around the free fermion value $S = -1$ that (to leading order in time) the ratio between the annihilation rate and the hopping rate drops out of the correlation function. This generalizes to higher order correlators and therefore all calculations involving correlators of finite order can be done in the free fermion limit corresponding to $S = -1$. This is just the DLPA process (3.25) derived from Glauber dynamics. Adapting the result by Torney and McConnell (1983) to the present problem one finds $C_{k,l}(t) = (1 - \sum_{n=-|k-l|+1}^{|k-l|} P(n; 2t|0; 0))/4$. This expresses the correlator in terms of the conditional probability (2.18) with $D = 1 + p$ for a random walker on a one-dimensional lattice. Hence the correlator approaches its constant stationary value $1/4$ with a power law $t^{-1/2}$. In two dimensions (growth on a substrate) and three dimensions one can use Smoluchovsky theory (von Smoluchovsky (1917)) or more sophisticated correlation truncation schemes (Lindenberg et al. (1995)) to show that the approach is faster: $\propto \ln t/t$ in $d = 2$ and $\propto 1/t$ in three dimensions respectively.

4   Experimental Realizations of Integrable Reaction-Diffusion Systems

4.1   Gel-Electrophoresis of DNA

A widely used and simple method for the separation of DNA fragments of different length is DC gel-electrophoresis. The DNA mixture to be separated is introduced into a gel matrix. Since the DNA is charged, it will move in a constant electric field E with velocity $v(E, N)$ where N is the length of the fragment. After some time fragments of different length will have travelled a distance in the gel depending on their length and can therefore be separated. Clearly it would be desirable to have a quantitative understanding of the motion of DNA in gels.

Based on the earlier concepts of the confining tube (Edwards (1967)) and of reptation de Gennes (1971), Rubinstein (1987) and Duke (1989) introduced a lattice gas model for the motion of a polymer in a gel matrix. In this model the gel is idealized by a cubic lattice where the cells are the pores of the gel through which the polymer reptates. The polymer itself is represented by a string of N reptons, where N is the length of the polymer divided by its persistence length. These reptons hop stochastically from pore to pore according to rules derived from reptation and assuming local detailed balance. Since in electrophoresis only the average velocity of the center of mass in field direction is of interest, one can project the motion of the reptons onto this direction. Some mappings that we will not describe here lead finally to a lattice gas model representing the relative motion of all reptons in field direction. The motion perpendicular to the field is diffusive with a diffusion constant (Widom et al. (1991), Prähofer and Spohn (1997)) $D = 1/3N^2$ entering the drift velocity $v(E, N)$ for small E through the Nernst-Einstein relation $v = DNE$ (Widom et al. (1991), van Leeuwen (1991)).

The lattice gas dynamics for the reptation-dynamics is as follows: There are two kinds of particles, A and B, moving on a lattice of $L = N - 1$ sites and each site can be occupied by at most one particle, A or B. A-particles hop to right (left) with rate q (q^{-1}) if the site is unoccupied. Here $\ln(q)$ is the energy gain when a repton moves into a pore in field direction. On site 1 of the chain A-particles are created (annihilated) with rate q (q^{-1}), while on site L they are annihilated (created) with rate q (q^{-1}). For the B-particles the same rules hold, but with q and q^{-1} interchanged (Barkema et al. (1994)). The average drift velocity $v(E, N)$ is given by the difference $j_A(E, N) - j_B(E, N)$ between the stationary currents of A particles and B particles. The stationary distribution of the system is not known except in the periodic system (van Leeuwen and Kooiman (1992)) which does not have a direct interpretation in terms of polymers moving through a gel. However, extensive Monte-Carlo studies (Barkema et al. (1994)) have provided a good and reliable knowledge of v in the framework of the model. The surprise is that these results are in excellent agreement with experimental data (Barkema et al. (1996), Barkema and Schütz (1996)). This gives confidence that despite all its simplifications, the Rubinstein-Duke model captures the essential physical processes involved and allows for reliable predictions in real gel-electrophoresis.

In order to make contact with quantum spin chains we write the master equation of the process in the quantum Hamiltonian formalism. Following the construction outlined in the introduction one finds that the stochastic time evolution of the system is given by the Hamiltonian of a three-states quantum chain (Barkema and Schütz (1996))

$$H(\alpha, q) = b_1(\alpha, q) + b_L(\alpha, q^{-1}) + \sum_{i=1}^{L-1} u_i(q) \qquad (4.30)$$

where $b_i(\alpha, q) = \alpha q(1 - n_i^A - a_i^+ - b_i) + \alpha q^{-1}(1 - n_i^B - a_i - b_i^+)$ and $u_i(q) = q(n_i^A n_{i+1}^0 + n_i^0 n_{i+1}^B - a_i a_{i+1}^+ - b_i^+ b_{i+1}) + q^{-1}(n_i^0 n_{i+1}^A + n_i^B n_{i+1}^0 - a_i^+ a_{i+1} - b_i b_{i+1}^+)$. Here $n_i^A \equiv E_i^{11}$, $n_i^B \equiv E_i^{33}$ and $n_i^0 = 1 - n_i^A - n_i^B \equiv E_i^{22}$ are projection operators on states with an A-particle, vacancy and B-particle resp. on site i. The operators $a_i \equiv E_i^{21}$, $a_i^+ \equiv E_i^{12}$, $b_i \equiv E_i^{23}$, and $b_i^+ \equiv E_i^{32}$ are annihilation and creation operators for A- and B particles. E_i^{jk} is the 3×3 matrix with matrix elements $(E_i^{jk})_{\alpha,\beta} = \delta_{j,\alpha}\delta_{k,\beta}$ acting on site i. The factor α takes into account the possibility of a different mobility of the end-reptons compared to those in the bulk.

Nothing is known about the integrability of the model in non-zero field. However, if no field is applied ($q = 1$) and if the ends of the polymer are fixed in the gel ($\alpha = 0$, e.g. by making a chemical bond with an immobile particle), then H is integrable. In this case the model describes the internal random fluctuations of the polymer within the gel. The integrability of $H(0, 1)$ can be seen by using an algebraic property of the Hamiltonian that has turned out to be useful also for other reaction-diffusion systems. For $\alpha = E = 0$ the Hamiltonian for both the isotropic spin-1/2 Heisenberg chain (2.16) with reflecting boundary conditions and the Rubinstein-Duke model have the form

$H = \sum_{i=1}^{L-1} u_i$ where the hopping matrices u_i satisfy the same Temperley-Lieb algebra $u_i^2 = 2u_i$, $u_i u_{i\pm1} u_i = u_i$, $[u_i, u_j] = 0$ for $|i-j| \geq 2$. This property gurantees integrability and also gives information about the spectrum of H (Alcaraz and Rittenberg (1993)). Using the Bethe ansatz one can compute the relaxation of the DNA to equilibrium (where each configuration is equally probable). If the ends of the polymer are not kept fixed, then the model has at least an integrable subspace with a spectrum which is identical to that of the isotropic Heisenberg chain with non-diagonal, symmetry breaking boundary fields. This can be shown by using a similarity transformation on H and projecting on one of its invariant subspaces. In this case one can use the integrability to obtain the dynamics of the distribution of vacancies (Stinchcombe and Schütz (1995)), i.e. the relaxation of stored length in a freely diffusing polymer.

4.2 Kinetics of Biopolymerization

Back in 1968 MacDonald et al. (1968), MacDonald and Gibbs (1969) studied the kinetics of biopolymerization on nucleic acid templates. The mechanism they try to describe is (in a very simplified manner) the following: Ribosomes attach to one end of a messenger-RNA chain and "read" the genetic information which is encoded in triplets of base pairs by moving along the m-RNA.[7] At the same time the ribosome adds monomers to a biopolymer attached to it: Each time a unit of information is being read a monomer is added to a biopolymer attached to the ribosom and which is in this way synthesized by the ribosom. After having added the monomer the ribosom moves one triplet further and reads again. So in each reading step the biopolymer grows in length by one monomer. Which monomer is added depends on the genetic information read by the ribosom. The ribosoms are much bigger than the triplets on the m-RNA, they cover 20-30 of such triplets. Therefore different ribosomes hopping stochastically on the m-RNA cannot overtake each other. When a ribosome has reached the other end of the m-RNA the biopolymer is fully synthesized and the ribosome is released.

In order to describe the kinetics of this process MacDonald et al. introduced the following simple model. The m-RNA is represented by one-dimensional lattice of L sites where each lattice site represents one triplet of base pairs. The ribosom is a hard-core particle covering r neighbouring sites (for real systems $r = 20 \ldots 30$) but moving forward by only one lattice site with constant rate p. At the beginning of the chain particles are added with rate αp and at the end of the chain they are removed with rate βp. One can also allow for back-hopping with rate q. In the idealized case $r = 1$ this model became later known as the asymmetric exclusion process with open boundary conditions (see Sec. 2). Its steady state was first studied using a mean-field approach by MacDonald et al. (1968). Then in a following paper (MacDonald

---

[7] The m-RNA is a long molecule made up of such consecutive triplets.

and Gibbs (1969)) the generalized case $r > 1$ was studied numerically and compared to experimental data on the stationary density distribution of ribosomes along the chain. These were found to be consistent with the results obtained from the model with $q = 0$ and $\alpha = \beta < p/2$. Furthermore it turned out that the phase diagram for general r is similar to the much simpler case $r = 1$ in the sense that there are three distinct phases, a low density phase, a high density phase and a maximal current phase (see below).

These observations allow for a physical understanding of certain dynamical aspects of this biological system. The experimentally relevant case is the phase transition line from the low-density phase to the high-density phase. Both mean field and numerical calculations predict a region of low density of ribosomes from the beginning of the chain up to some point where the density suddenly jumps (over a few lattice sites) to a high density value, comparable to a jam in traffic flow.[8] These predictions make an exact solution of at least the simple case $r = 1$ desirable. The stochastic dynamics of this model are given by the integrable Hamiltonian of the anisotropic spin-1/2 Heisenberg chain with non-diagonal boundary fields (de Vega and Gonzalez-Ruiz (1994))

$$H = -\alpha p \left[s_1^- - (1 - n_1) \right] - \beta p \left[s_1^+ - n_1 \right]$$
$$- \sum_{i=1}^{L} \left[p \left(s_i^+ s_{i+1}^- - n_i(1 - n_{i+1}) \right) + q \left(s_i^- s_{i+1}^+ - (1 - n_i)n_{i+1} \right) \right] \tag{4.31}$$

For $\alpha = \beta = 0$ this reduces to the $SU(2)_q$ symmetric quantum chain with diagonal boundary fields which can be solved by the coordinate or algebraic Bethe ansatz. However, the boundary fields given here break the $U(1)$ symmetry of the model and other approaches are necessary to find at least the steady state of the system, i.e. the ground state of H. In what follows we will consider only $q = 0$. We set $p = 1$ which is no loss in generality since it sets only the time scale of the process.

The breakthrough to the exact solution came only more than 20 years after the work on biopolymerization and independently of it. It turned out that the stationary distribution of a system of L sites can be expressed recursively in terms of the solution for $L - 1$ sites (Derrida et al. (1992)). The exact solution obtained from the solution of these recursion relations (Schütz and Domany (1993), Derrida et al. (1993)) reproduces the three phases predicted by mean field, but also reveals an intricate interplay between two correlation lengths which determine the phase diagram and the nature of the phase transitions. In particular, it turns out that the correlation length on the phase transition line between the low-density phase and the high-density phase is

---

[8] This description of the stationary mean-field density profile describes correctly the situation for $r = 1$, but disregards a more complicated sublattice structure for $r > 1$. However the figures provided by MacDonald and Gibbs (1969) suggest that the description remains qualitatively correct if one averages over this sublattice structure.

infinite, which is incompatible with the mean field result. The exact solution gives a linearly increasing density profile rather than the sharp shock predicted by mean field. This can be explained by assuming that a sharp shock exists, but, due to current fluctuations, performs a random walk along the lattice. Therefore, if one waits long enough, the shock will have been at each lattice site with equal probability. This picture discussed in (Schütz and Domany (1993)) yields a linearly increasing density and is confirmed by an exact solution of dynamical properties of a related exclusion process with deterministic bulk dynamics (Schütz (1993)). What one therefore expects for an experimental sample is indeed a region of low density of ribosoms followed by a sharp transition to a region of high density of ribosoms as found experimentally. This rapid increase can be anywhere on the m-RNA, but with a probability distribution given by the effective initialization and release rates α, β. If $\alpha = \beta$ the distribution of shock position would be constant over the lattice, otherwise exponential on a length scale $\xi = 1/(\ln{[\alpha(1 - \alpha)/\beta(1 - \beta)]})$.[9]

4.3   Exciton Dynamics on Polymer Chains

Finally, we discuss briefly an experiment in which excitons on polymer chains are created by laser excitations and then hop on the chain and coagulate when they meet. The carrier substance is $(CH_3)_4NMnCl_3$ (TMMC). The particles are excitons of the Mn^{2+} ion and move along the widely separated $MnCl_3$ chains. A single exciton has a decay time of about $0.7ms$. The on-chain hopping rate is $10^{11} - 10^{12}s^{-1}$. If two excitons arrive on the same Mn^{2+} ion, they undergo a coagulation reaction $A + A \rightarrow A$ with a reaction time $\approx 100fs$ (Kroon et al. (1993), Privman (1997)).

On a one-dimensional lattice diffusion-limited coagulation is equivalent to DLPA by a similarity transformation (Krebs et al. (1995)). The annihilation rate is equal to the coagulation rate of the original process. Since the experimental data suggest that the coagulation is approximately instantaneous, one finds an annihilation rate which equals twice the hopping rate. In the quantum Hamiltonian formalism the stochastic time evolution of this transformed process is then given by the free fermion Hamiltonian H^{DLPA} (3.25) with hopping rate D and annihilation rate $2D$. Here D sets the time scale for the diffusion. The finite life time τ of the excitons is much larger than D^{-1}, thus a decay term $\tau^{-1} \sum (s_i^+ - n_i)$ can be neglected. One then finds that the average density of excitons decays algebraically in time with an exponent $x = 1/2$ (Bramson and Griffeath (1980)) in good agreement with the experimental result $x = 0.48(2)$ Kroon et al. (1993).

---

[9] For completeness we add that in the maximal current phase $\alpha, \beta > 1/2$, i.e. when polymerization determines the dynamics, the exact solution predicts an algebraic decay of the density to its bulk value $1/2$ with exponent $b = 1/2$ rather than $b = 1$ predicted by the mean field solution.

5   Conclusions

It has been realized in recent years that the stochastic time evolution of many one-dimensional reaction-diffusion processes can be mapped to quantum spin systems, and in special cases to integrable quantum chains. This insight has made available the tool box of quantum mechanics and particularly of integrable models for these interacting particle systems far from equilibrium. With these methods many new exact results for their dynamical and stationary properties have been derived. It is also amusing to note that the Hamiltonians for such systems are mostly not hermitian and therefore from a quantum mechanical point of view not interesting. The interpretation as time evolution operators for stochastic dynamics thus extends the physical relevance of integrable systems to non-hermitian models. Reaction-diffusion mechanisms which can be described in this way are not only actual chemical systems, but comprise a large variety of phenomena in physics and beyond.

Acknowledgments

It is a pleasure to acknowledge fruitful collaborations and numerous useful discussions with many active workers in the various fields of non-equilibrium physics covered in these lecture notes.

References

Alcaraz, F. C., Rittenberg, V. (1993), Phys. Lett. B **314**, 377.
Alcaraz, F. C., Droz, M., M. Henkel and V. Rittenberg (1994), Ann. Phys. (USA) **230**, 250.
Alexander, S., Holstein, T. (1978), Phys. Rev. B **18**, 301.
Barkema, G. T., Marko, J., Widom, B. (1994), Phys. Rev. E **49**, 5303.
Barkema, G. T., Caron, C., Marko, J.F.(1996), Biopolymers **38** 665.
Barkema, G. T., Schütz, G. M. (1996), Europhys. Lett. **35** 139.
Bethe, H. (1931), Z. Phys. **71**, 205.
Bramson, M., Griffeath, D. (1980), Z. Wahrsch. Verw. Geb. **53**, 183.
Cardy, J. (1997), in *Proceedings of Mathematical Beauty of Physics*, J.-B. Zuber, ed., Advanced Series in Mathematical Physics **24**, 113.
de Gennes, P.G. (1971), J. Chem. Phys. **55**, 572.
Dahmen, S.R. (1995), J. Phys. A **28**, 905.
Derrida, B., Domany, E., Mukamel, D. (1992), J. Stat. Phys. **69**, 667.
Derrida, B., Evans, M.R., Hakim, V.,Pasquier, V. (1993a), J. Phys. A **26**, 1493.
Derrida, B., Janowsky, S. A., Lebowitz, J. L., Speer, E. R. (1993b), Europhys. Lett. **22**, 651.
Derrida, B., Evans, M. R. (1994), in *Probability and Phase Transition*, G. Grimmet, ed. (Kluwer Academic, Dordrecht).
de Vega, H.-J., Gonzalez-Ruiz, A. (1994), J. Phys. A **27**, 6129.
Dieterich, W., Peschel, I. (1983), J. Phys. C **16**, 3841.

Doi, M. (1976), J. Phys. A **9**, 1465 (1976).

Drasdo, D. (1996) in *Self-organization of Complex Structures: From Individual to Collective Dynamics*, F. Schweitzer (ed.) (Gordon and Breach, London).

Duke, T. A. J. (1989), Phys. Rev. Lett. **62**, 2877.

Edwards, S.F. (1967), Proc. Phys. Soc., London **92**, 9.

Family, F., Amar, J. G. (1991), J. Stat. Phys. **65**, 1235.

Felderhof, B. U. (1971), Rep. Math. Phys. **1**, 215.

Glauber, R. J. (1963), J. Math. Phys. **4**, 294.

Gradshteyn, I. S., Ryzhik, I.M. (1980), *Tables of Integrals, Series and Products*, fourth edition (Academic Press, London), Ch. 15

Grassberger, P., Scheunert, M. (1980), Fortschr. Phys. **28**, 547 (1980).

Grynberg, M. D., Newman, T. J., Stinchcombe, R.B. (1994), Phys. Rev. E **50**, 957.

Gwa, L.-H., Spohn, H. (1992), Phys. Rev. A **46**, 844.

Jordan, P., Wigner, E. (1928), Z. Phys. **47**, 631.

Kadanoff, L. P., Swift, J. (1968), Phys. Rev. **165**, 310 (1968).

Kim, D. (1995), Phys. Rev. E **52**, 3512.

Krebs, K., Pfannmüller, M. P., Wehefritz, B., Hinrichsen, H. (1995), J. Stat. Phys. **78**, 1429.

Kroon, R., Fleurent, H., Sprik, R. (1993), Phys. Rev. E **47**, 2462.

Lieb, E., Schultz, T., Mattis, D. (1961), Ann. Phys. (N.Y.) **16**, 407.

Ligget, T. (1985), *Interacting Particle Systems*, (Springer, New York).

Lindenberg, K., Agyrakis, P., Kopelman, P. (1995), J. Phys. Chem. **99**, 7542.

Lushnikov, A. A. (1987), Phys. Lett. A **120**, 135.

MacDonald, J. T., Gibbs, J. H., Pipkin, A. C. (1968), Biopolymers **6**, 1.

MacDonald, J. T., Gibbs, J. H. (1969), Biopolymers **7**, 707.

Mattis, D. C., (1965), *The Theory of Magnetism*, (Harper and Row, New York).

Peschel, I., Rittenberg, V., Schulze, U. (1994), Nucl. Phys. B **430**, 633.

Prähofer, M., Spohn, H. (1997), Physica A **233**, 191.

Privman, V. (1994), Phys. Rev. E **50**, 50.

Privman, V. (ed.) (1997), *Nonequilibrium Statistical Mechanics in One Dimension* (Cambridge University Press, Cambridge, U.K.).

Rácz, Z. (1985), Phys. Rev. Lett. **55**, 1707.

Rubinstein, M. (1987), Phys. Rev. Lett. **59**, 1946.

Sandow, S., Trimper, S. (1993), Europhys. Lett. **21**, 799.

Sandow, S., Schütz, G. (1994), Europhys. Lett. **26**, 7.

Santos, J. E., Schütz, G. M., Stinchcombe, R. B. (1996), J. Chem. Phys. **105**, 2399.

Santos, J. E. (1997), J. Phys. A **30**, 3249.

Schmittmann, B., Zia, R. K .P., (1996) in *Phase Transitions and Critical Phenomena*, eds. C. Domb and J. Lebowitz (Academic Press, London).

Schulz, M., Trimper, S. (1996), Phys. Lett. A **216**, 235.

Schütz, G. (1993), Phys. Rev. E **47**, 4265.

Schütz, G., Domany, E. (1993), J. Stat. Phys. **72**, 277.

Schütz, G., Sandow, S. (1994), Phys. Rev. E **49**, 2726.

Schütz, G. M. (1995a), J. Stat. Phys. **79**, 243.

Schütz, G. M. (1995b), J. Phys. A **28**, 3405.

Schütz, G. M. (1997), J. Stat. Phys. **88**, 427.

Siggia, E. (1977), Phys. Rev. B **16**, 2319.

Spitzer, F. (1970), Adv. Math. **5**, 246.

Spohn, H. (1991), *Large Scale Dynamics of Interacting Particle Systems* (Springer, New York).

Spouge, J.L. (1988), Phys. Rev. Lett. **60**, 871.

Stinchcombe, R. B., Grynberg, M. D., Barma, M. (1993), Phys. Rev. E **47**, 4018.

Stinchcombe, R. B., Schütz, G. M. Phys. (1995), Phys. Rev. Lett. **75**, 140.

Thacker, H. B. (1981), Rev. Mod. Phys. **53**, 253.

Torney, D. C., McConnell, H. M. (1983), J. Phys. Chem. **87**, 1941.

van Leeuwen, J. M. J. (1991), J. Phys. I **1**, 1675.

van Leeuwen, J. M. J., Kooiman, A. (1992), Physica A **184**, 79.

van Kampen, N. G. (1981) *Stochastic methods in Physics and Chemistry* (North Holland, Amsterdam).

von Smoluchovsky, M. (1917), Z. Phys. Chem. **92**, 129.

Widom, B., Viovy, J. L. Défontaines A. D. (1991), J.Phys. I France **1**, 1759.

Droz, M., Rácz, Z., Schmidt, J. (1989), Phys. Rev. A **39**, 2141.

Bosonization in Particle Physics

Dietmar Ebert

Institut für Physik, Humboldt-Universität zu Berlin,
Invalidenstrasse 110, D-10115, Berlin, Germany

Abstract. Path integral techniques in collective fields are shown to be a useful analytical tool to reformulate a field theory defined in terms of microscopic quark (gluon) degrees of freedom as an effective theory of collective boson (meson) fields. For illustrations, the path integral bosonization approach is applied to derive a (non)linear σ model from a Nambu-Jona-Lasinio (NJL) quark model. The method can be extended to include higher order derivative terms in meson fields or heavy-quark symmetries. It is also approximately applicable to QCD.

1 Introduction

In this lecture, I want to demonstrate the powerfulness of the path integral approach in collective fields for the bosonization of quark models containing (effective) 4-quark interactions [1]. For illustrations, let me consider Nambu-Jona-Lasinio (NJL) type of models [2-4] with *local* quark interactions $\sim G\left(\bar{q}\Gamma q\right)^2$ representing a relativistic version of the superconductor BCS theory [5,6]. These models lead to a gap equation for a dynamical quark mass signalling the spontaneous breakdown of chiral symmetry (SBχS). Furthermore, the collective field of Cooper pairs of the superconductor is now replaced by collective meson fields of $(q\tilde{q})$-bound states.

To be more explicit, let me consider the generating functional $\mathcal{Z}$ of the NJL model defined by a path integral of the exponential of the corresponding action over quark fields q as the underlying microscopic degrees of freedom. Path integral bosonization then means to transform this generating functional into an integral of the exponential of an effective meson action where the new collective integration variables $\sigma, \pi, \rho, ...$ denote the observable meson fields,

$$ \mathcal{Z} = \int D\mu(q)\mathrm{e}^{\mathrm{i}\int \mathcal{L}_{\mathrm{NJL}}\mathrm{d}^4x} \overset{I}{\Longrightarrow} \int D\mu(\sigma,\pi,\rho,...)\,\mathrm{e}^{\mathrm{i}\int \mathcal{L}_{\mathrm{Eff}}.\mathrm{d}^4x}, \qquad (1.1) $$

with $D\mu(q) = DqD\bar{q}$, $D\mu(\sigma,\pi,\rho,...) = D\sigma D\pi D\rho...$ being the respective integration measures of fields.

The basic ingredient of the path integral bosonization (1) is the use of the Hubbard-Stratonovich transformation [7,8] which replaces the (effective) 4-quark interactions of NJL models by a Yukawa-type coupling of quarks with collective meson fields $\phi_i = (\sigma, \pi, \rho_\mu, ...)$. After this the primary path integral over quark fields on the L.H.S. of (1) becomes Gaussian resulting in a quark determinant containing meson fields. Further important steps are:

1. the use of the loop expansion of the quark determinant in powers of meson fields
2. the evaluation of the resulting Feynman diagrams in the low-momentum region
3. the limit of large numbers of colours, $N_c \to \infty$, GN_c fixed (G being the 4-quark coupling constant) in order to apply a saddle point approximation to the integration over meson fields.

Quark loop diagrams emitting two ϕ-fields then generate in the low-momentum (low-derivative) approximation kinetic and mass terms of mesons. Finally, quark loops emitting $n > 2$ meson fields lead to meson interactions with effective small coupling constants $g_n \sim O\left(\left(\frac{1}{\sqrt{N_c}}\right)^{n-2}\right)$ allowing for a modified perturbation theory in terms of meson degrees of freedom.

Notice that the NJL quark Lagrangian incorporates the global chiral flavour symmetry of Quantum Chromodynamics (QCD) as well as its explicit and spontaneous breaking. The equivalent effective meson theory on the R.H.S. of (1) just reproduces this symmetry breaking pattern at the meson level. As mentioned above, masses and coupling constants of collective mesons are now calculable from quark loop diagrams and expressed by the parameters of the underlying quark theory (current quark masses, four-fermion coupling G and a loop momentum cut-off Λ).

The final aim is, of course, bosonization of QCD, i.e. the transformation

$$\mathcal{Z} = \int D\mu\left(q, G_\mu\right) e^{i \int \mathcal{L}_{QCD} d^4 x} \overset{II}{\Longrightarrow} \int D\mu\left(\sigma, \pi, \rho, ...\right) e^{i \int \mathcal{L}_{Eff.} d^4 x}, \qquad (1.2)$$

where q, G_μ denote quark and gluon fields (ghost fields are not shown explicitly). In order to begin with an effective 4-quark interaction as intermediate step, one first has to integrate away the gluon (ghost) fields on the L.H.S. of (2). This can only be done exactly for space-time dimensions $D = 2$ in the light-cone gauge, $G_- = \frac{1}{2}(G_0 - G_1) = 0$, where all self-interactions of gluon fields vanish. The resulting expression contains an effective *nonlocal* current×current quark interaction with a known gluon propagator which then can be bosonized by introducing *bilocal* meson fields $\phi(x, y) \sim \bar{q}(x)\Gamma q(y)$, employing the limit of large numbers of colours N_c [9]. Clearly, the analogous bosonization of four-dimensional QCD is more complicated. It requires some additional approximations: first a truncation of the arising multilocal quark current interactions retaining only the bilocal two-current term and secondly a modelling of the unknown nonperturbative gluon propagator (Cf. Fig. 1a) (see [1,10-13]). As a final result one derives an effective chiral Lagrangian describing the low-energy interactions of observable mesons including nontrivial meson form factors.

Nevertheless, as anticipated in (1), also simpler *local* NJL type of models are expected to yield a reasonable low-energy description of the chiral sector of QCD_4. Indeed, discarding the complicated question of the unknown

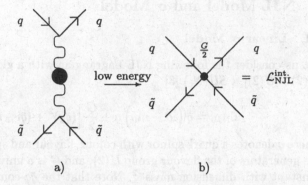

Fig. 1. a-b) Low-energy approximation of a nonlocal current × current interaction with nonperturbative gluon propagator (wavy line) (a), by a local Nambu-Jona-Lasinio type interaction (b).

structure of nonperturbative gluon and quark propagators (related to confinement), one can try to approximate the nonperturbative gluon propagator in the region of low energies by a universal constant G leading to a *local* NJL type of current × current interaction (Cf. Fig. 1). (For other interesting applications of the above path integral approach the reader is further referred to the nuclear many-body problem [10,14] and the Hubbard model [15,16].)

In conclusion, let us notice that in the very special case of two space-time dimensions there exists a different realization of the bosonization idea based on an explicit construction of fermionic fields in terms of bosonic fields due to Mandelstam [17]. This then allows one to replace, at the operator level, quark bilinears by bosonic fields, e.g.

$$\bar{\psi}\gamma_\mu\psi \sim -\pi^{-\frac{1}{2}}\varepsilon_{\mu\nu}\partial^\nu\phi, \quad \bar{\psi}\psi \sim M\cos\left(2\sqrt{\pi}\phi\right),$$

and to prove the equivalence of fermion models with four-fermion interactions (Thirring model) with a corresponding boson model (Sine-Gordon model). Subsequently this bosonization scheme was generalized to nonabelian symmetries. Using in particular Witten's non-Abelian bosonization rules [18] one has derived in the strong coupling limit (which is contrary to the weak coupling limit, GN_c fixed, for $N_c \to \infty$, of the above described path integral approach) a low-energy effective action from QCD$_2$ [19]. The lecture of T.Giamarchi at this Workshop discussed just the application of the operator bosonization in Condensed Matter Physics. Unfortunately, the elegant and powerful techniques of operator bosonization cannot be extended in a straightforward way to field-theoretical models in four dimensions.

In the next section I shall now describe the path integral bosonization of the NJL model along line I closely following the original papers [3,4].

2 NJL Model and σ Model

2.1 Linear σ Model

Let us consider the following NJL Lagrangian with a global symmetry $[U(2) \times U(2)] \times SU(N_c)$ [3]

$$\mathcal{L}_{\text{NJL}} = \bar{q}\left(\text{i}\hat{\partial} - m_0\right)q + \frac{G}{2}\left[(\bar{q}q)^2 + (\bar{q}\text{i}\gamma_5\tau q)^2\right], \qquad (2.3)$$

where q denotes a quark spinor with colour, flavour and spinor indices, τ_i's are the generators of the flavour group $U(2)$, and G is a universal quark coupling constant with dimension mass^{-2}. Note that the $\bar{q}q$-combinations in (3) are colour singlets, and we have admitted an explicit chiral symmetry-breaking current quark mass term $-m_0\bar{q}q$. The integration over the quark fields in the generating functional $\mathcal{Z}$ of the NJL model (given by the L.H.S. of (1)) can easily be done after bi-linearizing the four-quark interaction terms with the help of colour-singlet collective meson fields σ, π. To this end, we use the Hubbard-Stratonovich transformation [7,8]

$$\exp\left\{\text{i}\int\frac{G}{2}\left[(\bar{q}q)^2 + (\bar{q}\text{i}\gamma_5\tau q)^2\right]\text{d}^4x\right\} =$$

$$= \mathcal{N}\int D\sigma D\pi_i \exp\left\{\text{i}\int\left[-\frac{1}{2G}\left(\sigma^2 + \pi^2\right) - \bar{q}\left(\sigma + \text{i}\gamma_5\tau\cdot\pi\right)q\right]\text{d}^4x\right\}, \quad (2.4)$$

which replaces the 4-quark interaction by a Yukawa coupling with collective fields σ, π, Cf. Fig.2.

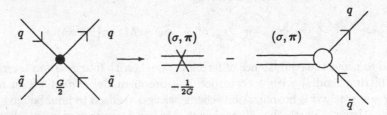

Fig. 2. Graphical representation of the integral identity (4).

Inserting (4) into the L.H.S. of (1) leads to the intermediate result

$$\mathcal{Z} = \int D\sigma D\pi_i \int DqD\bar{q}\,\text{e}^{\text{i}\int\mathcal{L}_{\text{NJL}}^{\text{qM}}\text{d}^4x} \qquad (2.5)$$

with the semi-bosonized meson-quark Lagrangian

$$\mathcal{L}_{\text{NJL}}^{\text{qM}} = -\frac{1}{2G}\left((\sigma - m_0)^2 + \boldsymbol{\pi}^2\right) + \bar{q}\left(i\hat{\partial} - \sigma - i\gamma_5\boldsymbol{\tau}\cdot\boldsymbol{\pi}\right)q, \qquad (2.6)$$

where we have absorbed the current quark mass in the σ field by a shift $\sigma \to \sigma - m_0$ of the integration variable. The Gaussian integral over the Grassman variable q in (5) can easily be performed leading to the fermion determinant

$$\det S^{-1} = \exp N_c \text{Tr} \ln S^{-1} = \exp iN_c \int -i \, \text{tr} \left(\ln S^{-1}\right)_{(x,x)} d^4x \qquad (2.7)$$

where

$$S^{-1}(x,y) = \left[i\hat{\partial}_x - \sigma(x) - i\gamma_5\boldsymbol{\tau}\cdot\boldsymbol{\pi}(x)\right]\delta^4(x-y) \qquad (2.8)$$

is the inverse quark propagator in the presence of collective fields $\sigma, \boldsymbol{\pi}$ and the trace tr in (7) runs over Dirac and isospin indices. Note that the factor N_c results from the colour trace. Combining (5)-(7) we obtain

$$\mathcal{Z} = \int D\sigma D\pi_i e^{i\int \mathcal{L}_{\text{Eff.}}(\sigma,\pi)d^4x}, \qquad (2.9)$$

$$\mathcal{L}_{\text{Eff.}}(\sigma,\pi) = -\frac{1}{2G}\left((\sigma - m_0)^2 + \boldsymbol{\pi}^2\right) - i\,N_c \text{tr} \ln\left(i\hat{\partial} - \sigma - i\gamma_5\boldsymbol{\tau}\cdot\boldsymbol{\pi}\right)_{(x,x)}. \qquad (2.10)$$

Let us analyze (9) in the limit of large N_c with GN_c fixed where one can apply the saddle point approximation. The stationary point $\sigma = \langle\sigma\rangle_0 \equiv m, \pi = 0$ satisfies the condition

$$\left.\frac{\delta\mathcal{L}_{\text{Eff.}}}{\delta\sigma}\right|_{\langle\sigma\rangle_0, \pi=0} = 0, \qquad (2.11)$$

which takes the form of the well-known Hartree-Fock gap equation determining the dynamical quark mass m

$$m = m_0 + 8mGN_cI_1(m), \qquad (2.12)$$

$$I_1(m) = i\int^{\Lambda} \frac{d^4k}{(2\pi)^4} \frac{1}{k^2 - m^2} \qquad (2.13)$$

with Λ being a momentum cut-off which has to be held finite (see Fig.3).

Note that in the chiral limit, $m_0 \to 0$, (12) admits two solutions $m = 0$ or $m \neq 0$ in dependence on $GN_c \lessgtr (GN_c)_{\text{crit}}$. Thus, for large enough couplings we find a nonvanishing quark condensate $\langle\bar{q}q\rangle \sim \text{tr}\,S(x,x)$ signalling spontaneous breakdown of chiral symmetry.

It is further convenient to perform a shift in the integration variables

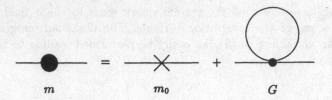

$$m \qquad\qquad m_0 \qquad\qquad G$$

Fig. 3. Graphical representation of the gap equation (12). The quark loop-diagram corresponds to the integral I_1.

$$\sigma \to m + \frac{\sigma'}{\sqrt{N_c}},$$

$$\pi \to \frac{\pi}{\sqrt{N_c}}.$$

In order to obtain from the nonlocal expression (10) a local effective meson Lagrangian we have to apply the following recipes:

1. loop expansion of the determinant in the fields σ', π, i.e.

$$N_c \operatorname{tr} \ln \left\{ \left(i\hat{\partial} - m \right) \left[1 - \frac{1}{i\hat{\partial} - m} \left(\frac{\sigma'}{\sqrt{N_c}} + i\gamma_5 \tau \frac{\pi}{\sqrt{N_c}} \right) \right] \right\} =$$

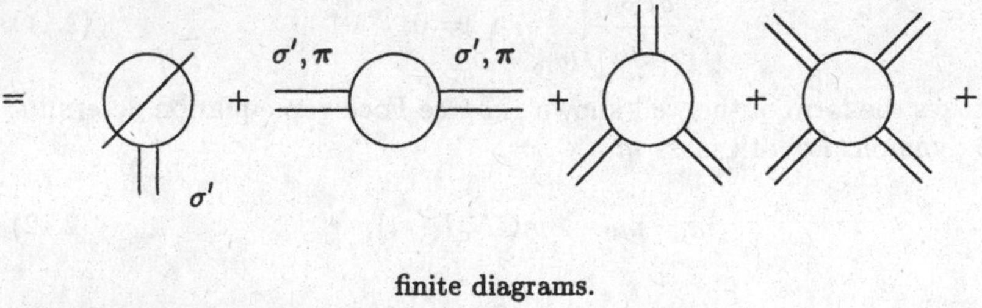

finite diagrams.

Fig. 4. Loop expansion of the fermion determinant. Solid lines in loops denote quark propagators; double lines denote the fields σ', π.

Here we have omitted the unimportant constant term $N_c \operatorname{tr} \ln \left(i\hat{\partial} - m \right)$ and used the formula $\ln (1 - x) = - \left[x + \frac{x^2}{2} + \cdots \right]$. (The tadpole diagram linear in σ' cancels by a corresponding linear term arising from the first term in (10) due to the gap equation (12).)

2. Low-momentum expansion of loop diagrams corresponding to the so-called gradient expansion of meson fields in configuration space.
3. Field renormalization,

$$(\sigma', \boldsymbol{\pi}) \to Z^{\frac{1}{2}} (\sigma', \boldsymbol{\pi}).$$

Discarding the contributions of finite n-point diagrams with $n > 4$ in the loop expansion leads to a linear σ model of composite fields (the prime in σ' is now omitted) [3]

$$\mathcal{L}_{\text{Eff.}}(\sigma, \boldsymbol{\pi}) = \frac{1}{2}\sigma \left(-\Box - m_\sigma^2 \right) \sigma + \frac{1}{2}\boldsymbol{\pi} \left(-\Box - m_\pi^2 \right) \boldsymbol{\pi} -$$

$$- g_{\sigma\pi\pi}\sigma \left(\sigma^2 + \boldsymbol{\pi}^2 \right) - g_{4\pi} \left(\sigma^2 + \boldsymbol{\pi}^2 \right)^2 . \tag{2.14}$$

Notice that the masses and coupling constants of mesons are fixed by the quark model parameters m_0, G and the (finite) momentum cut-off Λ,

$$m_\pi^2 = m_0 \frac{g_{\pi qq}^2}{mG}, \quad m_\sigma^2 = m_\pi^2 + 4m^2, \tag{2.15}$$

$$g_{\pi qq} = g_{\sigma qq} = \left(\frac{Z}{N_c} \right)^{\frac{1}{2}} = \{4N_c I_2\}^{-\frac{1}{2}} = O \left(\frac{1}{\sqrt{N_c}} \right),$$

$$g_{\sigma\pi\pi} = g_{3\sigma} = \frac{m}{(N_c I_2)^{\frac{1}{2}}}, \tag{2.16}$$

$$g_{4\pi} = g_{4\sigma} = \frac{1}{8N_c I_2},$$

where

$$I_2 = -\mathrm{i} \int^\Lambda \frac{\mathrm{d}^4 k}{(2\pi)^4} \frac{1}{(k^2 - m^2)^2}.$$

It is worth emphasizing that in the chiral limit $m_0 \to 0$, the pion becomes the massless Goldstone boson associated to SBχS. Introducing electroweak gauge bosons W_μ^i at the quark level leads to the additional diagram shown in Fig.5.

Fig.5 leads to the interaction Lagrangian

$$\Delta\mathcal{L} = \frac{g}{2}W_\mu^i \left(-\frac{m}{g_{\pi qq}} \partial^\mu \pi^i \right), \tag{2.17}$$

which yields just the PCAC meson current. Here the ratio $m/g_{\pi qq}$ has the meaning of the pion decay constant F_π. We thus obtain the Goldberger-Treiman relation

$$F_\pi = \frac{m}{g_{\pi qq}}, \tag{2.18}$$

valid at the quark level.

In the following subsection we shall show how one can derive the related nonlinear version of the σ model.

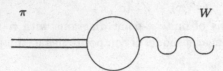

Fig. 5. Quark diagram describing the weak transition $\pi \to W$. The wavy line denotes the W-boson of weak interactions.

2.2   Nonlinear σ Model

In the low-energy region the quantum fluctuations of the σ-meson are suppressed with respect to those of light pions because of the heavier mass.
"Freezing" (i.e. neglecting) the corresponding degrees of freedom at low energies leads then to the so-called nonlinear realization of chiral symmetry by the pion field. The resulting nonlinear transformation laws for fields and covariant derivatives have in particular been used for constructing nonlinear chiral meson Lagrangians. Nonlinear chiral Lagrangians have been introduced a long time ago in the literature [20-22]. They provide a compact and extremely useful recipe to summarize low-energy theorems of nowadays QCD and are therefore widely used in hadron physics [1,23].

Let us now demonstrate how they can be derived from a NJL quark model. To this end, it is convenient to use in (6) an exponential parametrization of the meson fields

$$\sigma + \mathrm{i}\gamma_5 \boldsymbol{\tau} \cdot \boldsymbol{\pi} = (m + \tilde{\sigma})\, \mathrm{e}^{-\mathrm{i}\gamma_5 \frac{\boldsymbol{\tau} \cdot \boldsymbol{\varphi}}{F_\pi}}, \qquad (2.19)$$

which yields

$$\sigma^2 + \boldsymbol{\pi}^2 = (m + \tilde{\sigma})^2. \qquad (2.20)$$

Let us absorb the exponential in (19) by introducing chirally rotated "constituent" quarks χ,

$$q = \mathrm{e}^{\mathrm{i}\gamma_5 \frac{\boldsymbol{\tau} \cdot \boldsymbol{\xi}}{2}} \chi; \quad \boldsymbol{\xi} = \frac{\boldsymbol{\varphi}}{F_\pi} \qquad (2.21)$$

leading to the Lagrangian[1]

---

[1] Generally there arises also an additional anomalous Wess-Zumino term from the fermion measure of the path integral [4] which, however, vanishes in the case of chiral $U(2)$ symmetry.

$$\mathcal{L}_{\text{NJL}}^{\chi M} = -\frac{1}{2G}\left(m + \bar{\sigma}\right)^2 + \Delta\mathcal{L}_{\text{sb}} +$$

$$+\bar{\chi}\left[i\gamma_\mu\left(\partial^\mu + e^{-i\gamma_5\frac{\tau\cdot\xi}{2}}\partial^\mu e^{i\gamma_5\frac{\tau\cdot\xi}{2}}\right) - m - \bar{\sigma}\right]\chi, \tag{2.22}$$

with

$$\Delta\mathcal{L}_{\text{sb}} = \frac{m + \bar{\sigma}}{16G}m_0\text{tr}\left(e^{-i\gamma_5\tau\cdot\xi} + \text{h.c.}\right) \tag{2.23}$$

being a symmetry-breaking mass term.

Note the nonlinear transformation law of the meson field $\xi(x)$ under global chiral transformations $g \in \text{SU}(2)_A \times \text{SU}(2)_V$ [21,22],

$$ge^{i\gamma_5\frac{\tau\cdot\xi(x)}{2}} = e^{i\gamma_5\frac{\tau\cdot\xi'(x)}{2}}h(x), \tag{2.24}$$

where

$$h(x) = e^{i\frac{\tau}{2}\cdot\mathbf{u}(\xi(x),g)} \in \text{SU}(2)_{V,\text{loc.}}$$

is now an element of a *local* vector group. It is convenient to introduce the Cartan decomposition

$$e^{-i\gamma_5\frac{\tau\cdot\xi}{2}}\partial_\mu e^{i\gamma_5\frac{\tau\cdot\xi}{2}} = i\gamma_5\frac{\tau}{2}\cdot\boldsymbol{\mathcal{A}}_\mu\left(\xi\right) + i\frac{\tau}{2}\cdot\boldsymbol{\mathcal{V}}_\mu\left(\xi\right). \tag{2.25}$$

It is then easy to see that the fields transform under the local group $\text{SU}(2)_{V,\text{loc.}}$ as follows ($\mathcal{A}_\mu \equiv \frac{\tau}{2}\cdot\boldsymbol{\mathcal{A}}_\mu$, etc.) [22]

$$\chi \to \chi' = h(x)\chi$$

$$\mathcal{V}_\mu \to \mathcal{V}'_\mu = h(x)\mathcal{V}_\mu h^\dagger(x) - h(x)i\partial_\mu h^\dagger(x)$$

$$\mathcal{A}_\mu \to \mathcal{A}'_\mu = h(x)\mathcal{A}_\mu h^\dagger(x). \tag{2.26}$$

Thus, $\mathcal{V}_\mu$ is a gauge field with respect to $\text{SU}(2)_{V,\text{loc.}}$. This allows one to define the following chiral-covariant derivative of the quark field χ

$$D_\mu\chi = \left(\partial_\mu + i\mathcal{V}_\mu\right)\chi. \tag{2.27}$$

Using (22),(25) and (27), the inverse propagator of the rotated χ field takes the form

$$S_\chi^{-1} = i\hat{D} - m - \bar{\sigma} - \hat{\mathcal{A}}\gamma_5. \tag{2.28}$$

The nonlinear $\bar{\sigma}$ model is now obtained by "freezing" the $\bar{\sigma}$ field, performing the path integral over the χ field and then using again the loop expansion for the quark determinant $\det S_\chi^{-1}$. Choosing a gauge-invariant regularization, the loop diagram with two external $\mathcal{A}_\mu$ fields generates a mass-like term for the axial $\mathcal{A}_\mu(\xi)$ field contributing to the effective Lagrangian[2]

$$\mathcal{L}^\sigma_{\text{nlin.}} = \frac{m^2}{g^2_{\pi qq}} \text{tr}_F \mathcal{A}^2_\mu + \Delta\mathcal{L}_{\text{sb}}. \tag{2.29}$$

Here, the symmetry-breaking term is taken over from (22), and the first constant term has been omitted.

As has been shown in [22], $\mathcal{A}_\mu$ is just the chiral-covariant derivative of the ξ field admitting the expansion

$$\mathcal{A}^i_\mu \equiv D_\mu \xi^i = \partial_\mu \xi^i + O\left(\xi^3\right) = \frac{1}{F_\pi} \partial_\mu \varphi^i + O\left(\varphi^3_i\right). \tag{2.30}$$

Thus, we obtain the nonlinear σ model

$$\mathcal{L}^\sigma_{\text{nlin.}} = \frac{F^2_\pi}{2} D_\mu \xi^i D^\mu \xi^i + \Delta\mathcal{L}_{\text{sb}} = \tag{2.31}$$

$$= \frac{1}{2} \varphi \left(-\Box - m^2_\pi\right) \varphi + O\left(\varphi^3_i\right)$$

reproducing Weinberg's result [20].

3 Conclusions and Outlook

In this talk I have shown you that the path integral bosonization approach applied to QCD-motivated NJL models is a powerful tool in order to derive low-energy effective meson Lagrangians corresponding to the (nonperturbative) chiral sector of QCD.

The above considerations have further been extended to calculate higher-order derivative terms in meson fields by applying heat-kernel techniques to the evaluation of the quark determinant [4]. This allows, in particular, to estimate all the low-energy structure constants L_i introduced by Gasser and Leutwyler [23]. Moreover, it is not difficult to include vector and axial-vector currents into the NJL model and to consider the chiral group $U(3) \times U(3)$. In a series of papers [1,3,4,24] it has been shown that the low-energy dynamics of light pseudoscalar, vector and axial-vector mesons is described surprisingly well by effective chiral Lagrangians resulting from the bosonization of QCD-motivated NJL models. These Lagrangians embody the soft-pion theorems, vector dominance, Goldberger-Treiman and KSFR relations and the integrated chiral anomaly.

Finally, we have investigated the path integral bosonization of an extended NJL model including SBχS of light quarks and heavy quark symmetries of

---

[2] Note that usually generated field strength terms $-\frac{1}{4}\mathcal{A}^2_{\mu\nu}$, $-\frac{1}{4}\mathcal{V}^2_{\mu\nu}$ vanish for Cartan fields (25). A mass-like term $\sim \mathcal{V}^2_\mu$ does not appear due to gauge-invariant regularization.

heavy quarks [25] (see also [26]). This enables one to derive an effective Lagrangian of pseudoscalar, vector and axial-vector D or B mesons interacting with light π, ρ and a_1 mesons. Note that the use of the low-momentum expansion in the evaluation of the quark determinant restricts here the applicability of the resulting effective Lagrangian to such (decay) processes where the momentum of the light mesons is relatively small.

Summarizing, I thus hope to have convinced you that the path integral bosonization approach to QCD (in its bilocal or simpler local formulation) is a very interesting *analytical* nonperturbative method which, being complementary to numerical studies of lattice QCD, is worth to be developed further. In the next lecture, I will consider the related path integral "hadronization" of baryons.

References

[1] For a recent review with further references see: Ebert, D., Reinhardt, H., and Volkov, M.K.: Progr. Part. Nucl. Phys. **33** (1994),1.

[2] Nambu, Y. and Jona-Lasinio, G.: Phys. Rev. **122** (1961), 345, ibid. **124** (1961), 246.

[3] Ebert, D. and Volkov, M.K.: Yad. Fiz. **36** (1982), 1265, Z. Phys. **C16** (1983), 205.

[4] Ebert, D. and Reinhardt, H.: Nucl. Phys. **B271** (1986), 188.

[5] Bardeen, J., Cooper, L.W., and Schriffer, J.R.: Phys. Rev. **106** (1957), 162.

[6] Bogoliubov, N.N.: Zh. Eksp. Teor. Fiz. **34** (1958), 73.

[7] Hubbard, J.: Phys. Rev. Lett. **3** (1959), 77.

[8] Stratonovich, R.L.: Sov. Phys. Dokl. **2** (1957), 416.

[9] Ebert, D. and Pervushin, V.N.: Teor. Mat. Fiz. **36** (1978), 313; Ebert, D. and Kaschluhn, L.: Nucl. Phys. **B355** (1991), 123.

[10] Ebert, D., Reinhardt, H., and Pervushin, V.N.: Sov. J. Part. Nucl. **10** (1979), 444.

[11] Kleinert, H.: Phys. Lett. **B62** (1976), 77 and Erice Lectures (1978).

[12] Cahill, R.T., Praschifka, J., and Roberts, D.: Phys. Rev. **D36** (1987), 209.

[13] Efimov, G. and Nedelko, S.: Phys. Rev. **D51** (1995), 176.

[14] Ebert, D. and Reinhardt, H.: Teor. Mat. Fiz. **41** (1979), 139.

[15] Azakov, S.I.: "Two-dimensional Hubbard Model and Heisenberg Antiferromagnet" (in Lecture Notes, IASBS, Zanjan (1997)).

[16] Wolff, U.: Nucl. Phys. **B225** (1983), 391.

[17] Mandelstam, S.: Phys. Rev. **D11** (1975), 3026; see also: Coleman, S.: Phys. Rev. **D11** (1975), 2088; Luther, A. and Peschel, I.: Phys. Rev. **B9** (1974), 2911.

[18] Witten, E.: Comm. Math. Phys. **92** (1984), 455.

[19] Date, G.D., Frishman, Y., and Sonnenschein, J.: Nucl. Phys. **B283** (1987), 365; Frishman, Y. and Sonnenschein, J.: Nucl. Phys. **B294** (1987), 801.

[20] Weinberg, S.: Phys. Rev. Lett. **18** (1967), 188.

[21] Coleman, S., Wess, J., and Zumino, B.: Phys. Rev. **177** (1969), 2239; Callan, G.G. et al.: Phys. Rev. **177** (1969), 2247.

[22] Ebert, D. and Volkov, M.K.: Forschr. Phys. **29** (1981), 35.

[23] Gasser, J. and Leutwyler, H.: Nucl. Phys. **B250** (1985), 465, 517, 539.

[24] Ebert, D., Bel'kov, A.A., Lanyov, A.V., and Schaale, A.: Int. J. Mod. Phys. **A8** (1993), 1313.

[25] Ebert, D., Feldmann, T., Friedrich, R., and Reinhardt, H.: Nucl. Phys. **B434** (1995), 619; Ebert, D., Feldmann, T., and Reinhardt, H.: Phys. Lett. **B388** (1996), 154.

[26] Bardeen, W.A. and Hill, C.T.: Phys. Rev. **D49** (1993), 409; Novak, M.A., Rho, M., and Zahed, I.: Phys. Rev. **D48** (1993), 4370.

Hadronization in Particle Physics

Dietmar Ebert

Institut für Physik, Humboldt-Universität zu Berlin,
Invalidenstrasse 110, D-10115, Berlin, Germany

Abstract. The method of path integral hadronization is applied to a local quark-diquark toy model in order to derive an effective chiral meson-baryon Lagrangian. Further generalizations to models including both scalar and axial-vector diquarks as well as nonlocal interactions are discussed.

1   Introduction

In the first lecture (referred to as I in the following) I have shown that QCD-motivated effective quark models of the NJL type can be reformulated as effective theories given in terms of composite bosonic objects, mesons. An analogous treatment of baryons as relativistic bound systems leads us in the case $N_c = 3$ to the concept of diquarks which are coloured objects with spin-parity $J^P = 0^+, 1^+$. Diquarks as effective degrees of freedom have been investigated earlier both in two-dimensional QCD [1] and in four-dimensional QCD-type models [2].

In this talk I am now going to discuss baryons as colourless composite particles consisting of a quark and a diquark. By applying path integral methods to an NJL-type model with two-body $(q\bar{q})$ and (qq) forces it was, in particular, possible to derive Faddeev equations determining the spectrum of composite baryons [3-5]. It is a further challenge to derive the well known effective chiral meson-baryon Lagrangians [6, 7], which reproduce low energy characteristica of hadron physics directly by the path integral hadronization method.

Let me in this second lecture demonstrate how this can be done by choosing, for illustrations, a simple quark-diquark toy model which contains a local interaction of elementary scalar diquarks with quarks and discards axial-vector diquarks. Finally, possible generalizations to more realistic but also more complicated models containing both scalar and axial-vector diquarks (taken as elementary fields [8] or composite fields) and including nonlocal interactions mediated by quark exchange [3-5,9] are discussed.

2   Hadronization of a Quark-Diquark Toy Model

Let us consider a chiral-invariant Lagrangian containing the semi-bosonized meson-quark Lagrangian $\mathcal{L}^{\chi M}_{NJL}$ described in lecture I (see (22)) supplemented by a diquark and an interaction term

$$\mathcal{L}^{\chi \text{MD}} = \mathcal{L}^{\chi \text{M}}_{\text{NJL}} + D^\dagger \left(-\Box - M_D^2 \right) D + \tilde{G} \left(\bar{\chi} D^\dagger \right) (D\chi). \tag{2.1}$$

Here $D(x)$ is the field of an (elementary) scalar isoscalar diquark of mass M_D, and $\tilde{G}$ is a coupling constant.

Analogously to the introduction of collective meson fields into the NJL model (Cf. (4) of I) let us now introduce collective baryon (nucleon) fields B by using the identity

$$e^{i \int d^4 x \tilde{G}(\bar{\chi} D^\dagger)(D\chi)} = \mathcal{N}' \int \mathcal{D}B \mathcal{D}\bar{B} e^{i \int d^4 x (-\frac{1}{\tilde{G}} \bar{B} B - \bar{\chi} D^\dagger B - \bar{B} D\chi)}. \tag{2.2}$$

The "hadronization" of the generating functional of the Lagrangian (1) will now be performed step by step by integrating over the microscopic quark and diquark fields. We obtain

$$\mathcal{Z} = \mathcal{N}_1 \int \mathcal{D}\mu \left(\tilde{\sigma}, \varphi_i, B, D \right) e^{i \int d^4 x \left[-\text{itr} \ln S_\chi^{-1} - \frac{1}{\tilde{G}} \bar{B} B \right]} \times$$

$$e^{i \int \int d^4 x d^4 y \left[D^\dagger(x) \left(\Delta^{-1} - \bar{B} S_\chi B \right)_{(x,y)} D(y) \right]}, \tag{2.3}$$

$$\mathcal{Z} = \mathcal{N}_2 \int \mathcal{D}\mu \left(\tilde{\sigma}, \varphi_i, B \right) \exp \left\{ i \int d^4 x \left[-\text{itr} \ln S_\chi^{-1} - \frac{1}{\tilde{G}} \bar{B} B + \right. \right.$$

$$\left. \left. i \ln \left(1 - \bar{B} S_\chi \Delta B \right)_{(x,x)} \right] \right\}, \tag{2.4}$$

where the trace tr runs over Dirac, isospin, and colour indices, $\mathcal{D}\mu$ denotes the integration measure of fields, $\Delta^{-1} = -\Box - M_D^2$ is the inverse diquark propagator, and S_χ is the quark propagator defined by (Cf. (28) of I)

$$S_\chi^{-1} = i\hat{D} - m - \tilde{\sigma} - \hat{A}\gamma_5. \tag{2.5}$$

Expanding now the logarithms in power series at the one-loop level (see Fig.1) and performing a low-momentum expansion of Feynman diagrams (corresponding to a derivative expansion in configuration space), one describes both the generation of kinetic and mass terms of the composite baryon field B. This yields the expression

$$\int d^4 x d^4 y \bar{B}(x) \left[-\left(\frac{1}{\tilde{G}} + Z_1^{-1} \frac{\tau}{2} \hat{V} + g_A \frac{\tau}{2} \hat{A}\gamma_5 \right) \delta^4 (x - y) - \Sigma (x - y) \right] B(y). \tag{2.6}$$

The vertex renormalization constant Z_1^{-1} and the axial coupling g_A arise from the low-momentum (low derivative) expansion of the vertex diagrams of Fig.1 b) and c), respectively. The nucleon self-energy Σ has in momentum

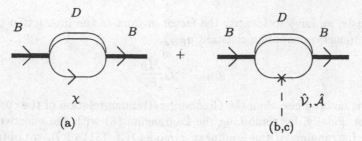

Fig. 1. Baryon self-energy diagram Σ (a) and vertex diagrams (b,c) arising from the loop expansion of the logarithms in Eq. (4). Solid lines in loops denote quark propagators; the double line denotes the diquark propagator.

space the decomposition $\Sigma(p) = \hat{p}\Sigma_\mathrm{V}\left(p^2\right) + \Sigma_\mathrm{S}\left(p^2\right)$. Its low momentum expansion generates a kinetic term $\left(\sim Z^{-1}\hat{p}\right)$ and, together with the constant term $-1/\tilde{G}$, a mass M_B given by the equation

$$\frac{1}{\tilde{G}} + M_B \Sigma_\mathrm{V}\left(M_B^2\right) + \Sigma_\mathrm{S}\left(M_B^2\right) = 0. \qquad (2.7)$$

In terms of renormalized fields defined by $B = Z^{\frac{1}{2}} B_\mathrm{r}$, where Z is the wave function renormalization constant satisfying the QED Ward identity $Z = Z_1$, we obtain from (6) the effective chiral meson-baryon Lagrangian [10]

$$\mathcal{L}_\mathrm{eff.}^\mathrm{MB} = \bar{B}_\mathrm{r}\left(\mathrm{i}\hat{D} - M_B\right) B_\mathrm{r} - g_A^\mathrm{r} \bar{B}_\mathrm{r} \gamma_\mu \gamma_5 \frac{\tau_i}{2} B_\mathrm{r} \mathcal{A}_i^\mu, \qquad (2.8)$$

with $\hat{D} = \hat{\partial} + \mathrm{i}\hat{\mathcal{V}}$ and $g_A^\mathrm{r} = Z g_A$ being the renormalized axial coupling constant. Note that expression (8) completely coincides in structure with the famous phenomenological chiral Lagrangians introduced at the end of the Sixties when considering nonlinear realizations of chiral symmetry [6,7]. However, in our case these Lagrangians are not obtained on the basis of symmetry arguments alone, but derived from an underlying microscopic quark-diquark picture which allows us to estimate masses and coupling constants of composite hadrons. Note that, due to $\mathcal{A}_\mu^i = \frac{1}{F_\pi}\partial_\mu\varphi^i + \cdots$, the second term in (8) leads to a derivative coupling of the pion field φ with the axial-vector baryon current. In order to get rid of the derivative and to reproduce the standard γ_5 coupling, it is convenient to redefine the baryon field $B_\mathrm{r} \to \tilde{B}$ by

$$B_\mathrm{r} = \mathrm{e}^{-\mathrm{i}g_A^\mathrm{r}\gamma_5\frac{\tau\cdot\varphi}{2F_\pi}} \tilde{B}. \qquad (2.9)$$

Inserting (9) into (8) and performing a power series expansion in φ leads to the expression

$$\mathcal{L}_\mathrm{eff.}^\mathrm{MB} = \bar{\tilde{B}}\left(\mathrm{i}\hat{\partial} - M_B\right)\tilde{B} + g_A^\mathrm{r}\frac{M_B}{F_\pi}\bar{\tilde{B}}\mathrm{i}\gamma_5\tau\cdot\varphi\tilde{B} + O\left(\varphi^2\right). \qquad (2.10)$$

Obviously, we have to identify the factor in front of the interaction term as the pion-nucleon coupling constant $g_{BB\varphi}$,

$$g_{BB\varphi} = g_A^{\text{r}} \frac{M_B}{F_\pi}, \tag{2.11}$$

which is nothing else than the Goldberger-Treiman relation of the composite nucleon. Finally, by combining the Lagrangian (8) with the effective chiral meson Lagrangian of the nonlinear σ model (Cf. (31) of I), we obtain the complete meson-baryon Lagrangian

$$\mathcal{L}_{\text{eff.,tot.}}^{\text{MB}} = \mathcal{L}_{\text{nlin.}}^{\sigma} + \mathcal{L}_{\text{eff.}}^{\text{MB}}. \tag{2.12}$$

It is further possible to estimate magnetic moments as well as electric and magnetic radii of composite protons and neutrons by introducing electromagnetic interactions into the toy Lagrangian (1). The obtained pattern of predicted low-energy characteristica of nucleons has been shown to describe data at best qualitatively [11]. Obviously, in order to get better agreement with data, it is necessary to consider more realistic, but also more complicated models with nonlocal quark-diquark interactions containing both scalar and axial-vector diquarks [3-5,12].

3   Further Extensions

3.1   Heavy Baryons with Scalar and Axial-Vector Diquarks

The above considerations can be easily generalized to heavy-light baryons $B \sim (Qqq)$, where $Q = c, b$ is a heavy quark and the light quarks $q = u, d, s$ form scalar (D) and axial-vector diquarks (F_μ) of the flavour group $SU(3)_\text{F}$. The quantum numbers of the light diquarks with respect to the spin, flavour, and colour follow here from the decomposition

$$\text{spin} : \frac{1}{2} \times \frac{1}{2} = 0_\text{a} + 1_\text{s}$$

$$\text{flavour} : 3_\text{F} \times 3_\text{F} = \bar{3}_{\text{F,a}} + 6_{\text{F,s}}$$

$$\text{colour} : 3_\text{c} \times 3_\text{c} = \bar{3}_{\text{c,a}} \left(+6_{\text{c,s}}\right),$$

where the indices s,a refer to symmetry, antisymmetry of the respective wave functions under interchange of quark indices. According to the Pauli principle the fields of the scalar and axial-vector diquarks must then form a flavour (anti)triplet or sextet, respectively: $D_{\bar{3}_\text{F}}, F_{6_\text{F}}^\mu$. The baryons as bound states of diquarks with a quark Q are evidently colourless, since the product representation $\bar{3}_\text{c} \times 3_\text{c}$ contains the colour singlet $\mathbb{1}$. In Ref. [8] we have studied an extended quark-diquark model given by the Lagrangian

$$\mathcal{L} = \mathcal{L}_0 + \mathcal{L}_{\text{int.}}, \tag{3.13}$$

$$\mathcal{L}_0 = \bar{q}\left(i\hat{\partial} - m_0\right)q + \bar{Q}_v iv \cdot \partial Q_v + \text{tr}\left[\partial_\mu D^\dagger \partial^\mu D - M_D^2 D^\dagger D\right]$$

$$- \frac{1}{4}\text{tr}\, F_{\mu\nu}^\dagger F^{\mu\nu} + \text{tr}\, M_F^2 F_\mu^\dagger F^\mu, \tag{3.14}$$

$$\mathcal{L}_{\text{int.}} = \tilde{G}_1 \text{tr}\left(\bar{Q}_v D^\dagger\right)(DQ_v) - \tilde{G}_2 \text{tr}\left(\bar{Q}_v F^{\dagger\mu}\right) P_{\mu\nu}^\perp \left(F^\nu Q_v\right), \tag{3.15}$$

where D^{ij}, F_μ^{ij} are antisymmetric/symmetric 3×3 flavour matrices, Q_v is a heavy quark spinor of 4-velocity v_μ (using the notation of Heavy Quark Effective Theory [13]), $P_{\mu\nu}^\perp = g_{\mu\nu} - v_\mu v_\nu$ is a transverse projector, and $\tilde{G}_1, \tilde{G}_2$ are coupling constants.

In order to bilinearize the interaction term (15) one now needs two types of baryon fields

$$T_v\left(\frac{1}{2}^+\right) \sim Q_v D, \quad S_{v\mu}\left(\frac{1}{2}^+, \frac{3}{2}^+\right) \sim Q_v F_\mu,$$

where $S_{v\mu}$ is a superfield of spin-symmetry partners $B_v\left(\frac{1}{2}^+\right), B_{v\mu}^*\left(\frac{3}{2}^+\right)$ admitting the decomposition

$$S_{v\mu} = \frac{1}{\sqrt{3}}\gamma_5\left(\gamma_\mu - v_\mu\right)B_v + B_{v\mu}^*. \tag{3.16}$$

The interaction term (15) can then be rewritten as a sum of terms bilinear in the baryon field and a Yukawa interaction term. Integrating successively over the quark fields q, Q and then over the diquark fields D, F_μ leads to determinants containing the baryon fields T and S. Finally, by employing again a loop-expansion and taking into account only lowest order derivative terms, leads to the free effective baryon Lagrangian of heavy flavour type

$$\mathcal{L}_{\text{eff.}}^0 = \text{tr}\bar{T}_v\left(iv \cdot \partial - \Delta M_T\right)T_v - \text{tr}\bar{S}_{v\mu}\left(iv \cdot \partial - \Delta M_S\right)S_v^\mu, \tag{3.17}$$

where the mass differences $\Delta M_{T,S} \equiv M_{T,S} - m_Q$ are calculable. Moreover, taking into account vertex diagrams analogously to those shown in Fig.1 (b,c) leads to the inclusion of interactions with the $SU(3)_F$-octet of light pseudoscalar mesons φ_i [8].

3.2   Composite Diquarks

For simplicity, we have considered up to now only models containing diquarks as elementary fields. Clearly, it is desirable to treat diquarks on the same footing as composite mesons as composite particles. This has been done in Ref. [9] by considering an extended NJL model for light ($q = u, d, s$) and heavy quarks ($Q = c, b$) containing 2-body interactions of diquark-type (qq) and (qQ)

$$\mathcal{L}_{\text{NJL}}^{\text{int.}} = G_1 \left(\bar{q}^c \Gamma^\alpha q \right) \left(\bar{q} \Gamma^\alpha q^c \right) + G_2 \left(\bar{q}^c \Gamma_v^\alpha Q_v \right) \left(\bar{Q}_v \Gamma_v^\alpha q \right). \qquad (3.18)$$

Here $q^c = C\bar{q}^T$ denotes a charged-conjugated quark field, and $\Gamma^\alpha, \Gamma_v^\alpha$ are flavour and Dirac (spin) matrices. The interaction term (18) can again be rewritten in terms of a Yukawa coupling of light composite scalar and axial-vector diquark fields $D\left(0^+, 1^+\right)$ or heavy diquarks $D_v\left(J^\mathrm{P} = 0^+, 1^+\right)$ to quarks. Notice that these models lead to a nonlocal quark-diquark interaction mediated by quark exchange shown in Fig.2.

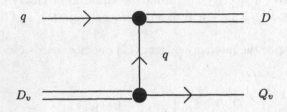

Fig. 2. Quark-exchange diagram leading to a nonlocal quark-diquark interaction with composite diquarks.

As shown in a series of papers [3-5,9], the baryon spectrum can then be found by solving Faddeev-type equations for quark-diquark bound states. Finally, the above method of path integral hadronization can also be applied to investigate 3-body interactions of the type $\mathcal{L} \sim \left(\bar{q}\bar{q}\bar{q}\right)\left(qqq\right)$ [5].

I hope that these few examples are sufficient to show you that path integral hadronization is indeed a powerful nonperturbative method in particle physics.

References

[1] Ebert, D. and Pervushin, V.N.: Teor. Mat. Fiz. **36** (1978), 759; Ebert, D. and Kaschluhn, L.: Nucl. Phys. **B355** (1991), 123.
[2] Cahill, R.T., Praschifka, J., and Burden, C.J.: Aust. J. Phys. **42** (1989), 147, 161; Kahana, D. and Vogl, U.: Phys. Lett. **B244** (1990), 10; Ebert, D., Kaschluhn, L., and Kastelewicz, G.: Phys. Lett. **B264** (1991), 420.
[3] Cahill, R.T.: Aust. J. Phys. **42** (1989), 171.
[4] Reinhardt, H.: Phys. Lett. **B244** (1990), 316.
[5] Ebert, D. and Kaschluhn, L.: Phys. Lett. **B297** (1992), 367.
[6] Coleman, S., Wess, I., and Zumino, B.: Phys. Rev. **177** (1969), 2239; ibid. 2247.

[7] Ebert, D. and Volkov, M.K.: Fortschr. Phys. **29** (1981), 35.

[8] Ebert, D., Feldmann, T., Kettner, C., and Reinhardt, H.: Z. Phys. **C71** (1996) 329.

[9] Ebert, D., Feldmann, T., Kettner, C., and Reinhardt, H.: Preprint DESY-96-010, to appear in Int. J. Mod. Phys. **A**.

[10] Ebert, D. and Jurke, T.: Preprint HUB-EP-97/74; hep-ph/9710390 (1997).

[11] Jurke, T.: Diploma Thesis, Humboldt University, Berlin, (1997).

[12] Keiner, V.: Z. Phys. **A354** (1996), 87.

[13] Isgur, N. and Wise, M.: Phys. Rev. **D41** (1990), 151; Neubert, M.: Phys. Rept. **245** (1994), 259.

The Hybrid Monte Carlo Algorithm
for Quantum Chromodynamics

Thomas Lippert

Department of Physics, University of Wuppertal, D-42097 Wuppertal, Germany

Abstract. The Hybrid Monte Carlo (HMC) algorithm currently is the favorite scheme to simulate quantum chromodynamics including dynamical fermions. In this talk—which is intended for a non-expert audience—I want to bring together methodical and practical aspects of the HMC for full QCD simulations. I will comment on its merits and shortcomings, touch recent improvements and try to forecast its efficiency and rôle in future full QCD simulations.

1 Introduction

The Hybrid Monte Carlo algorithm (Duane et al. 1987) is—for the present—a culmination in the development of practical simulation algorithms for full quantum chromodynamics (QCD) on the lattice. QCD is the theory of the strong interaction. In principle, QCD can describe the binding of quarks by gluons, forming the hadrons with their masses, as well as other hadronic properties. As QCD cannot be evaluated satisfactorily by perturbative methods, one has to recourse to *non-perturbative* stochastic simulations of the quark and gluon fields on a discrete 4-dimensional space-time lattice (Creutz 1983). In analogy to simulations in statistical mechanics, in a Markov chain, a canonical ensemble of field configurations is generated by suitable Monte Carlo algorithms. As far as full QCD lattice simulations are concerned, the HMC algorithm is the method of choice as it comprises several important advantages:

- The evolution of the gluon fields through phase space is carried out simultaneously for all *d.o.f.*, as in a *molecular dynamics* scheme, using the leap-frog algorithm or higher order symplectic integrators.
- *Dynamical fermion loops*, represented in the path-integral in form of a determinant of a huge matrix of dimension $O(10^7)$ elements, *i.e.* a highly non-local object that is not directly computable, can be included by means of a stochastic representation of the fermionic determinant. This approach amounts to the solution of a huge system of linear equations of rank $O(10^7)$ that can be solved efficiently with modern iteration algorithms, so-called Krylov-subspace methods (Frommer et al. 1994), (Fischer et al. 1996).
- As a consequence, the computational complexity of HMC is a number $O(V)$, *i.e.*, one complete sweep (update of all V *d.o.f.*) requires $O(V)$

operations, as it is the case for Monte Carlo simulation algorithms of local problems.

- HMC is *exact*, *i. e.* systematic errors arising from finite time steps in the molecular dynamics are eliminated by a *global* Monte Carlo decision.
- HMC is *ergodic* due to *Langevin*-like stochastic elements in the field update.
- HMC shows surprisingly short *autocorrelation times*, as recently demonstrated (SESAM collaboration 1997). The autocorrelation determines the statistical significance of physical results computed from the generated ensemble of configurations.
- HMC can be fully *parallelized*, a property that is essential for efficient simulations on high speed parallel systems.
- HMC is computation dominated, in contrast to memory intensive alternative methods (Luescher 1994), (Slavnov 1996). Future high performance SIMD (single addressing multiple data) systems presumably are memory bounded.

In view of these properties, it is no surprise that all large scale lattice QCD simulations including dynamical Wilson fermions as of today are based on the HMC algorithm. Nevertheless, dynamical fermion simulations are still in their infancy. The computational demands of full QCD are huge and increase extremely if one approaches the *chiral limit* of small quark mass, *i. e.* the physically relevant mass regime of the light u and d quarks. The central point is the solution of the linear system of equations by iterative methods. The iterative solver, however, becomes increasingly inefficient for small quark mass. We hope that these demands can be satisfied by parallel systems of the upcoming Teracomputer class (Schilling 1997).

The HMC algorithm is a general *global* Monte Carlo procedure that can evolve all *d.o.f.* of the system at the same instance in time. Therefore it is so useful for QCD where due to the inverse of the local fermion matrix in the stochastic representation of the fermionic determinant the gauge fields must be updated all at once to achieve $O(V)$ complexity. The trick is to stay close to the surface of constant Hamiltonian in phase space, in order to achieve a large acceptance rate in the global Monte Carlo step.

HMC can be applied in a variety of other fields. A promising novel idea is the merging of HMC with the multi-canonical algorithm (Berg & Neuhaus 1992) which is only parallelizable within global update schemes. The parallel multi-canonical procedure, can be applied at the (first-order) phase transitions of compact QED and Higgs-Yukawa model. Another example is the Fourier accelerated simulation of polymer chains as discussed in Anders Irbäck's contribution to these proceedings, where HMC well meets the non-local features of Fourier acceleration leading to a *multi-scale* update process.

The outline of this talk is as follows: In section 2, a minimal set of elements and notions from QCD, necessary for the following, is introduced. In section 3, the algorithmic ingredients and computational steps of HMC are described.

In section 4, I try to evaluate the computational complexity of HMC and suggest a scaling rule of the required CPU-time for vanishing Wilson quark mass. Using this rule, I try to give a prognosis as to the rôle of HMC in future full QCD simulations in relation to alternative update schemes.

2 Elements of Lattice QCD

I intend to give a pedagogical introduction into the HMC evaluation of QCD in analogy to Monte Carlo simulations of statistical systems. Therefore, I avoid to focus on details. I directly introduce the physical elements on the discrete lattice that are of importance for the HMC simulation. For the following, we do not need to discuss their parentage and relation to continuum physics in detail.

QCD is a constituent element of the standard model of elementary particle physics. Six quarks, the flavors up, down, strange, charm, bottom, and top interact via gluons. In 4-dimensional space-time, the fields associated with the quarks, $\psi_a^\alpha(\mathbf{x})$ have four Dirac components, $\alpha = 1, \ldots, 4$, and three color components, $a = 1, \ldots, 3$. The 'color' degree of freedom is the characteristic property reflecting the non-abelian structure of QCD as a gauge theory. This structure is based on local SU(3) gauge group transformations acting on the color index.

The gluon field $A_\mu^a(\mathbf{x})$ consists of four Lorentz-vector components, $\mu = 1, \ldots, 4$. Each component carries an index a running from 1 to 8. It refers to the components of the eight gluon field in the basis of the eight generators λ_a of the group SU(3). The eight 3×3 matrices $\lambda_a/2$ are traceless and hermitean defining the algebra of SU(3) by $[\frac{\lambda_i}{2}, \frac{\lambda_j}{2}] = i f_{ijk} \frac{\lambda_k}{2}$.[1]

On the lattice, the quark fields $\psi_\mathbf{n}$ are considered as approximations to the continuum fields $\psi(\mathbf{x})$, with $\mathbf{x} = a\mathbf{n}$, $\mathbf{n} \in \mathbf{N}^4$ (All lattice quantities are taken dimensionless in the following.). As shown in Fig. 1, they 'live' on the sites. Their fermionic nature is expressed by anti-commutators,

$$[\psi_\mathbf{n}^\alpha, \psi_\mathbf{m}^\beta]_+ = [\psi_\mathbf{n}^{\dagger\alpha}, \psi_\mathbf{m}^\beta]_+ = [\psi_\mathbf{n}^{\dagger\alpha}, \psi_\mathbf{m}^{\dagger\beta}]_+ = 0, \qquad (2.1)$$

characterizing the quark fields as Grassmann variables. The gluon fields in the 4-dimensional discretized world are represented as bi-local objects, the so-called links $U_\mu(\mathbf{n})$. They are the bonds between site $\mathbf{n}$ and site $\mathbf{n} + \mathbf{e}_\mu$, with $\mathbf{e}_\mu$ being the unit vector in direction μ. Unlike the continuum gluon field, the gluon in discrete space is $\in$ SU(3). $U_\mu(\mathbf{n})$ is a discrete approximation to the parallel transporter known from continuum QCD, $U(x,y) = \exp\left(ig_s \int_\mathbf{x}^\mathbf{y} dx'^\mu A_\mu^a(\mathbf{x}')\lambda_a/2\right)$, with g_s being the strong coupling constant.

QCD is defined via the action $S = S_g + S_f$ that consists of the pure gluonic part and the fermionic action. The latter accounts for the quark

---

[1] For the explicit structure constants f_{ijk} and the generators $\lambda_i/2$ see Cheng & Li, 1989.

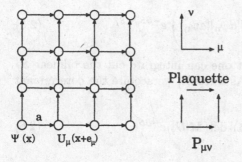

Fig. 1. 2-dimensional projection of the 4-dimensional euclidean space-time lattice.

gluon interaction and the fermion mass term. Taking the link elements from above one can construct a simple quantity, the plaquette $P_{\mu\nu}$, see Fig. 1:

$$P_{\mu\nu}(\mathbf{n}) = U_\mu(\mathbf{n})U_\nu(\mathbf{n} + \mathbf{e}_\mu)U_\mu^\dagger(\mathbf{n} + \mathbf{e}_\nu)U_\nu^\dagger(\mathbf{n}). \tag{2.2}$$

The Wilson gauge action is defined by means of the plaquette:

$$\beta S_g = \frac{6}{g_s^2} \sum_{\mathbf{n},\mu,\nu} \left[1 - \frac{1}{2}\mathrm{Tr}(P_{\mu\nu}(\mathbf{n}) + P_{\mu\nu}^\dagger(\mathbf{n})) \right]. \tag{2.3}$$

In the limit of vanishing lattice spacing, one can recover the continuum version of the gauge action, $-\int d^4\mathbf{x}\frac{1}{4}F_{\mu\nu}(\mathbf{x})F^{\mu\nu}(\mathbf{x})$. The deviation from the continuum action due to the finite lattice spacing a is of $O(a^2)$.

The discrete version of the fermionic action cannot be constructed by a simple differencing scheme, as it would correspond to 16 fermions instead of 1 fermion in the continuum limit. One method to get rid of the doublers is the addition of a second order derivative term, $(\psi_{\mathbf{n}+\mathbf{e}_\mu} - 2\psi_\mathbf{n} - \psi_{\mathbf{x}-\mathbf{e}_\mu})/2$, to the standard first order derivative $\gamma^\mu\partial_\mu\psi(\mathbf{x}) \to \gamma^\mu(\psi_{\mathbf{n}+\mathbf{e}_\mu} - \psi_{\mathbf{n}-\mathbf{e}_\mu})/2$. This scheme is called Wilson fermion discretization. The fermionic action can be written as a bilinear form, $S_f = \bar\psi_\mathbf{n} M_{\mathbf{n}\mathbf{y}}\psi_\mathbf{m}$, with the Wilson matrix M,

$$M_{\mathbf{n}\mathbf{m}} = \delta_{\mathbf{n}\mathbf{m}} - \kappa \sum_{\mu=1}^{4} \left[(1 - \gamma_\mu)U_\mu(\mathbf{n})\,\delta_{\mathbf{n},\mathbf{m}-\mathbf{e}_\mu} + (1 + \gamma_\mu)U_\mu^\dagger(\mathbf{n} - \mathbf{e}_\mu)\delta_{\mathbf{n},\mathbf{m}+\mathbf{e}_\mu} \right]. \tag{2.4}$$

The stochastic simulation of QCD starts from the analogy of the path-integral—the quantization prescription—to a partition sum as known from statistical mechanics. As it is oscillating, it would be useless for stochastic evaluation. The appropriate framework for stochastic simulation of QCD is that of Euclidean field theory. Therefore, one performs a rotation of the time direction $t \to i\tau$. The ensuing effect is a transformation of the Minkowski metrics into a Euclidean metrics, while a positive definite Boltzmann weight $\exp(-\beta S_g)$ is achieved. This form of the path-integral, *i.e.* the partition function, is well known from statistical mechanics:

$$Z = \int \left(\prod_{\mathbf{n},\mu} [dU_\mu(\mathbf{x})][d\bar{\psi}_\mathbf{n}][d\psi_\mathbf{n}] \right) e^{-\beta S_g - S_f}. \tag{2.5}$$

It is important for the following that one can integrate out the bilinear S_f over the Grassmann fermion fields. As a result, we acquire the determinant of the fermionic matrix:

$$Z = \int \prod_{\mathbf{n},\mu} [dU_\mu(\mathbf{n})] \det(M[U]) e^{-\beta S_g}. \tag{2.6}$$

3   Hybrid Monte Carlo

The Euclidean path-integral, Eq. (2.6), can in principle be evaluated by Monte Carlo techniques. We see that the fermionic fields do not appear in Z after the integration[2]. Hence, it suffices to generate a representative ensemble of fields $\{U_i\}$, $i = 1, \ldots, N$, and subsequently, to compute any observable along with the statistical error according to

$$\langle O \rangle = \frac{1}{N} \sum_{i=1}^N O_i[U_i] \quad \text{and} \quad \sigma_O^2 = \frac{2\tau_{\text{int}}}{N} \left(\frac{1}{N} \sum_{i=1}^N |O_i[U_i]|^2 - \langle O \rangle^2 \right). \tag{3.7}$$

The integrated autocorrelation time τ_{int} reflects the fact that the members of the ensemble are generated by importance sampling in a Markov chain. Therefore, a given configuration is correlated with its predecessors, and the actual statistical error of a result is increased compared to the naive standard deviation. The length of the autocorrelation time is a crucial quantity for the efficiency of a simulation algorithm.

3.1 $O(V)$ Algorithms for Full QCD

If we want to generate a series of field configurations $U_1, U_2, U_3 \ldots$ in a Markov process, besides the requirement for ergodicity, it is sufficient to fulfill the condition of detailed balance to yield configurations according to a canonical probability distribution:

$$e^{-S} P(U \to U') = e^{-S'} P(U' \to U). \tag{3.8}$$

$P(U \to U')$ is the probability to arrive at configuration U' starting out from U. Let us for the moment forget about $\det(M[U])$, i.e., we set $\det(M)$ equal to 1 in Eq. (2.6). In that case, the action is purely gluonic (pure gauge theory), and local. Therefore, using the rules of Metropolis et al. we can update each link independently one by one by some (reversible!) stochastic modification $U_\mu(\mathbf{n}) \to U'_\mu(\mathbf{n})$, while only local changes in the action are induced.

---

[2] Similarly, one can perform the computation of any correlation function of $\bar{\psi}$ and ψ, leading to products of the quark propagator, i.e. the inverse of M^{-1}.

One 'sweep' is performed if all links are updated once. By application of the Metropolis rule, $P(U \rightarrow U') = \min[1, \exp(-\Delta S_g)]$, detailed balance is fulfilled, and we are guaranteed to reach the canonical distribution. Starting from a random configuration, after some thermalization steps, we can assume hat the generated configurations belong to an equilibrium distribution. Without dynamical fermions—*i. e.* in the quenched approximation—standard Metropolis shows a complexity $O(V)$, with V being the number of *d.o.f.*

However, if we try to use Metropolis for full QCD, the decision $P(U \rightarrow U') = \min\left[1, \exp(-\Delta S_g) \frac{\det(M[U'])}{\det(M[U])}\right]$ would imply the evaluation of the fermionic determinant for each $U_\mu(\mathbf{n})$ separately. A direct computation of the determinant requires $O(V^3)$ operations and therefore, the total computational complexity would be a number $O(V^4)$.

These implications for the simulation of full QCD with dynamical fermions have been recognized very early. In a series of successful steps, the computational complexity could be brought into the range of quenched simulations[3]. The following table gives an (incomplete) picture of this struggle towards exact, ergodic, practicable and parallelizable $O(V)$ algorithms for full QCD. A key step was the introduction of the fermionic determinant by a Gaussian

Table 1. Towards exact and ergodic $O(V)$ algorithms.

Method	order	exact	ergodic	year
Metropolis	V^4	yes	yes	Metropolis et al. 1953
Pseudo Fermions	V^2	no	yes	Fucito et al. 1981
Gauss Representation	V^2	yes	yes	Petcher, Weingarten 1981
Langevin	V	no	yes	Parisi, Wu 1981
Microcanonical	V	no	no	Polonyi at al. 1982
Hybrid Molecular Dynamics	V	no	yes	Duane 1985
HMC	V	yes	yes	Duane et al. 1987
Local Bosonic Algorithm	V	no	yes	Lüscher 1994
Exact LBA	V	yes	yes	DeForcrand et al. 1995
5-D Bosonic Algorithm	V	no	yes	Slavnov 1996

integral. As a synthesis of several ingredients, HMC is a mix of Langevin simulation, micro-canonical molecular dynamics, stochastic Gauss representation of the fermionic determinant, and Metropolis.

3.2 Hybrid Monte Carlo: Quenched Case

For simplicity, I first discuss the quenched approximation, *i. e.* $\det(M) =$ const. Each sweep of the HMC is composed of two steps:

---

[3] Take this *cum grano salis*. Two $O(V)$ algorithms can extremely differ in the coefficient of V.

1. The gauge field is evolved through phase space by means of (micro-canonical) molecular dynamics. To this end, an artificial guidance Hamiltonian $\mathcal{H}$ is introduced adding the quadratic action of momenta to S_g, "conjugate" to the gauge links. The micro-canonical evolution proceeds in the artificial time direction as induced by the Hamiltonian. Choosing random momenta at the begin of the trajectory, ergodicity is guaranteed, as it is by the stochastic force in the Langevin algorithm. In contrast to Langevin, HMC carries out many integration steps between the refreshment of the momenta.

2. The equations of motion are chosen to conserve $\mathcal{H}$. In practice, a numerical integration can conserve $\mathcal{H}$ only approximately. However, the change $\Delta\mathcal{H} = \mathcal{H}_f - \mathcal{H}_i$ is small enough to lead to high acceptances of the Metropolis decision—rendering HMC exact, the essential improvement of HMC compared to the preceding hybrid-molecular dynamics algorihm.

With

$$\mathcal{H} = S_g[U] + \frac{1}{2} \sum_{n,\mu,color} \operatorname{Tr} H_\mu^2(n) \quad \text{and} \quad Z = \int [dH][dU]e^{-\mathcal{H}}, \qquad (3.9)$$

expectation values of observables are not altered with respect to Eq. (2.6), if the momenta are chosen from a Gaussian distribution. A suitable H is found using the fact that $U \in SU(3)$ under the evolution. Taylor expansion of $U(\tau + \Delta\tau)$ leads to $U(\tau)\dot{U}^\dagger(\tau) + \dot{U}(\tau)U^\dagger(\tau) = 0$. This differential equation is fulfilled choosing the first equation of motion as

$$\dot{U} = iHU, \qquad (3.10)$$

with H represented by the generators of SU(3) and thus being hermitean and traceless, $H_\mu(n) = \sum_{a=1}^{8} \lambda_a h_\mu^a(n)$. Each component h is a Gaussian distributed random number. As $\mathcal{H}$ should be a constant of motion, $\dot{\mathcal{H}} = 0$, we get

$$\dot{\mathcal{H}} = \sum_{n,\mu} \operatorname{Tr}\left\{ H_\mu(n)\dot{H}_\mu(n) - \frac{\beta}{6}[\dot{U}_\mu(n)V_\mu(n) + h.c.] \right\} = 0$$

$$\dot{\mathcal{H}} = \sum_{n,\mu} \operatorname{Tr}\left\{ H_\mu(n)\left[\dot{H}_\mu(n) - i\frac{\beta}{6}(U_\mu(n)V_\mu(n) - h.c.) \right] \right\} = 0. \qquad (3.11)$$

We note that $[]\propto 1$ since $\{\}$ must be traceless. Since $\dot{H}$ must stay explicitly traceless under the evolution it follows that $[] = 0$. The second equation of motion reads:

$$i\dot{H}(n) = -\frac{\beta}{6}\left\{ U_\mu(n)V_\mu(n) - h.c. \right\}. \qquad (3.12)$$

The quantities $V_\mu(n)$ corresponding to a gluonic force term are the staples, i.e. the incomplete plaquettes that arise in the differentiation,

$$V_\mu(\mathbf{n}) = \sum_{\nu \neq \mu} \left\{ \begin{array}{c} \overset{\mathbf{x}+\nu}{\boxed{}} \\ \mathbf{x} \quad \mathbf{x}+\mu \end{array} \oplus \begin{array}{c} \overset{\mathbf{x} \quad \mathbf{x}+\mu}{\boxed{}} \\ \mathbf{x}-\nu \end{array} \right\}. \tag{3.13}$$

For exact integration, the Hamiltonian $\mathcal{H}$ would be conserved. However, numerical integration only can stay close to $\mathcal{H} = const.$ Therefore, one adds a global Metropolis step,

$$P_{\mathrm{acc}} = \min(1, e^{-\Delta \mathcal{H}}), \tag{3.14}$$

to reach a canonical distribution for $\{U\}$. As a necessary condition for detailed balance the integration scheme must lead to a *time reversible* trajectory and fulfill *Liouville's theorem*, *i. e.* preserve the phase-space volume. *Symplectic integration* is the method of choice. It is stable as far as energy drifts are concerned.

3.3   Including Dynamical (Wilson) Fermions

Dynamical fermions are included in form of a stochastic Gaussian representation of the fermionic determinant in Eq. (2.6). In order to ensure convergence of the Gauss integral, the interaction matrix must be hermitean. Since the Wilson fermion matrix M is a complex matrix, it cannot be represented directly. A popular remedy is to consider the two light quarks u and d as mass degenerate. With the identity $\det^2(M) = \det(M^\dagger M)$ the representation reads

$$\det(M^\dagger M) = \int \left(\prod_{\mathbf{n}} [d\bar{\phi}_{\mathbf{n}}][d\phi_{\mathbf{n}}] \right) e^{-\phi_{\mathbf{x}}^* (M^\dagger M)_{\mathbf{n},\mathbf{m}}^{-1} \phi_{\mathbf{m}}}. \tag{3.15}$$

The bosonic field ϕ can be related to a vector R of Gaussian random numbers. In a heat-bath scheme, it is generated using the standard Muller-Box procedure, and with $\phi = M^\dagger R$, we arrive at $R^\dagger R$, the desired starting distribution, equivalent to $\phi^*(M^\dagger M)^{-1}\phi = \phi^* X$. Adding the fermionic action to $\mathcal{H}$, its time derivative reads:

$$\frac{dS_f}{d\tau} = \kappa \sum_{\mathbf{n},\mu} \mathrm{Tr}[\dot{U}_\mu(\mathbf{n}) F_\mu(\mathbf{n}) + h.c.],$$

$$F_\mu(\mathbf{n}) = [MX]_{\mathbf{n}+\mathbf{e}_\mu} X_{\mathbf{n}}^\dagger (1 + \gamma_\mu) + X_{\mathbf{n}+\mathbf{e}_\mu} [MX]_{\mathbf{n}}^\dagger (1 - \gamma_\mu). \tag{3.16}$$

F is the fermionic force that modifies the second equation of motion to

$$i\dot{H}(\mathbf{n}) = -\frac{\beta}{6} \left\{ U_\mu(\mathbf{n}) V_\mu(\mathbf{n}) + \kappa \mathrm{Tr}\, F_\mu(\mathbf{n}) - h.c. \right\}. \tag{3.17}$$

3.4   Numerical Integration and Improvements

The finite time-step integration of the equation of motion must be reversible and has to conserve the phase-space volume, while it should deviate little from the surface $\mathcal{H} = const$. The leap-frog scheme can fulfill these requirements. It consists of a sequence of triades of the following form:

$$H_\mu(\mathbf{n}, \tau + \frac{\Delta\tau}{2}) = H_\mu(\mathbf{n}, \tau) + \frac{\Delta\tau}{2}\dot{H}_\mu(\mathbf{n}, \tau)$$

$$U_\mu(\mathbf{n}, \tau + \Delta\tau) = e^{i\Delta\tau H_\mu(\mathbf{n}, \tau + \frac{\Delta\tau}{2})}U_\mu(\mathbf{n}, \tau)$$

$$H_\mu(\mathbf{n}, \tau + \Delta\tau) = H_\mu(\mathbf{n}, \tau + \frac{\Delta\tau}{2}) + \frac{\Delta\tau}{2}\dot{H}_\mu(\mathbf{n}, \tau + \Delta\tau). \qquad (3.18)$$

It can be shown that the leap frog scheme approximates $\mathcal{H}$ correctly up to $O(\Delta t^2)$ for each triade. As a rule of thumb, the time step and the number of integration steps, N_{md}, should be chosen such that the length of a trajectory in fictitious time is $N_{md} \times \Delta\tau \simeq O(1)$ at an acceptance rate $> 70\%$. It is easy to see from the discrete equations of motion (EOM) that the phase space volume $[dH][dU]$ is conserved: loosely speaking, dU is conserved as the first EOM amounts to a rotation in group space, and from the second EOM follows that $dH' = dH$. In order to improve the accuracy of the numerical integration, one can employ higher order symplectic integrators[4]. As the integration part of HMC is not specific for QCD, higher order integrators could be very useful for other applications as the Fourier accelerated HMC introduced by A. Irbäck.

Despite of the reduction of the computational complexity to $O(V)$, the repeated determination of the large "vector" X, $X = (M^\dagger M)^{-1}\phi$, renders the simulation of QCD with dynamical fermions still computationally extremely intensive. The size of the vector X is about 1 - 20 $\times 10^6$ words. The code stays more than 95 % of execution time in this phase. Since typical simulations run several months in dedicated mode on fast parallel machines, any percent of improvement is welcome. Traditionally, the system was solved by use of Krylov subspace methods such as conjugate gradient, minimal residuum or Gauss-Seidel. In the last three years, improvements could be achieved by introduction of the BiCGstab solver (Frommer et al. 1994) and by use of novel parallel preconditioning techniques (Fischer et al. 1996) called local-lexicographic SSOR (symmetric successive over-relaxation). Further improvements have been achieved through refined educated guessing, where the solution X of previous steps in molecular dynamics time is fed in to accelerate the current iteration (Brower et al. 1997). Altogether, a factor of about 4 up to 8 could be gained by algorithmic research.

---

[4] This strategy has been used so far only for fine-resolved integration of the gauge fields, and coarse resolved integration of the fermions (sparing inversions). For small quark masses, this approach can fail, however.

4   Efficiency and Scaling

Apart from purely algorithmic issues, the efficiency of a Monte Carlo simulation is largely determined by the autocorrelation of the Markov chain. A significant determination of autocorrelation times of HMC in realistic full QCD with Wilson fermions could not be carried out until recently (SESAM collaboration 1997). The length of the trajectory samples in these simulations was around 5000 (Here, with 'trajectory' we denote a new field configuration at the end of a Monte Carlo decision.). The lattice sizes were $16^3 \times 32$ and $24^3 \times 40$.

The finite time-series approximation to the true autocorrelation function for an observable $\mathcal{O}_t$, $t = 1, \ldots, t_{MC}$, is defined as

$$C^{\mathcal{O}}(t) = \frac{\sum\limits_{s=1}^{t_{MC}-t} \mathcal{O}_s \mathcal{O}_{s+t} - \frac{1}{t_{MC}-t}\left(\sum\limits_{s=1}^{t_{MC}} \mathcal{O}_s\right)^2}{t_{MC}-t}. \tag{4.19}$$

The definition of correlation in an artificial time is made in analogy to connected correlation functions in real time. The *integrated* autocorrelation time is defined as $\tau_{int}^{\mathcal{O}} = \frac{1}{2} + \sum_{t'=1}^{t_{MC} \to \infty} \frac{C^{\mathcal{O}}(t')}{C^{\mathcal{O}}(0)}$. In equilibrium, $\tau_{int}^{\mathcal{O}}$ characterizes the statistical error of the observable $\mathcal{O}$.

The integrated autocorrelation times have been determined from several observables, such as the plaquette and the smallest eigenvalue of M. They are smaller than anticipated previously, and their length is between 10 and 40 trajectories. Therefore, one can consider configurations as decorrelated that are separated by $\tau_{int}^{\mathcal{O}}$ trajectories.

The quality of the data allowed to address the issue of critical slowing down for HMC, approaching the chiral limit of vanishing u and d quark mass, where the pion correlation length $\xi_\pi = 1/m_\pi a$ is growing. The autocorrelation time is expected to scale with a power of ξ_π, $\tau = \epsilon \xi_\pi^z$, z is called *dynamical critical exponent*. As a result the dynamical critical exponent of HMC is located between $z = 1.3$ and 1.8 for local and extended observables, respectively.

Finally let me try to give a conservative guess of the computational effort required with HMC for de-correlation. The pion correlation length ξ_π must be limited to $V^{\frac{1}{4}}/\xi_\pi \approx 4$ to avoid finite size effects as the pion begins to feel the periodic boundary of the lattice. With ξ_π fixed, the volume factor goes as ξ_π^4. Furthermore the compute effort for BiCGstab increases $\propto \xi^{-2.7}$ (SESAM collaboration 1997). In order to keep the acceptance rate constant, the time step has been reduced (from 0.01 to 0.004) with increasing lattice size ($16^3 \times 32$ to $24^3 \times 40$), while the number of time steps was increased from 100 to 125. Surprisingly, the autocorrelation time of the 'worst case' observable, the minimal eigenvalue of M, goes down by 30 % compensating the increase in acceptance rate cost! In a conservative estimate, the total time scales as m_π^{-8} to $m_\pi^{-8.5}$. As a result, for Wilson fermions, the magic limit of $\frac{m_\pi}{m_\rho} < 0.5$, will be in reach on $32^3 \times L_t$ lattices—on a Teracomputer.

Alternative schemes like the local bosonic algorithm or the 5-dimensional bosonic scheme are by far more memory consuming than HMC. Here, a promising new idea might be the Polynomial HMC (Frezzotti & Jansen). The autocorrelation times of these alternative schemes in realistic simulations are not yet known accurately, however. In view of the advantages of HMC mentioned in the introduction, and the improvements achieved, together with the our new findings as to its critical dynamics, I reckon HMC to be the method of choice for future full QCD simulations on tera-computers.

Acknowledgments. I thank the members of the SESAM and the TχL collaborations and A. Frommer for many useful discussions.

References

Duane, S., Kennedy, A., Pendleton, B., Roweth, D. (1987): Phys. Lett. **B195**, 216
Creutz, M. (1983): *Lattice Gauge Theory* (Cambridge University Press, Cambridge)
Frommer, A. et al. (1994): Int. J. Mod. Phys. **C5**, 1073
Fischer, S. et al. (1996): Comp. Phys. Comm. **98**, 20-34
Sesam collaboration, to appear
R. C. Brower, et. al. (1997): Nucl. Phys. **B484**, 353-374
Luescher, M. (1994): Nucl. Phys. **B418**, 637-648
Slavnov, A. (1996): Preprint SMI-20-96, 6pp
Schilling, K. (1997): TERAcomputing in Europa: Quo Vamus? Phys. Bl. **10**, 976-978
Berg, B. A. and Neuhaus, T. (1992): Phys. Rev. Lett. **68**, 9-12
Cheng, T.-P. Li, L.-F. (1989): *Gauge Theory of Elementary Particle Physics* (Clarendon Press, Oxford)
Frezzotti, R., Jansen, K. (1997) Phys. Lett. **B402**, 328-334

The Hybrid Monte Carlo Method for Polymer Chains

Anders Irbäck

Complex Systems Group, Department of Theoretical Physics, University of Lund, Sölvegatan 14A, S-223 62 Lund, Sweden

Abstract. The use of the Hybrid Monte Carlo method in simulating off-lattice polymer chains is discussed. I focus on the problem of finding efficient algorithms for long flexible chains. To speed up the simulation of such chains the Fourier acceleration technique is used. Numerical results are presented for four models with different repulsive interactions between the monomers.

1 Introduction

The Hybrid Monte Carlo (HMC) method (Duane et al. 1987) was originally developed for Quantum Chromodynamics (QCD), and was specifically designed for simulations including the effects of dynamical quarks. It is based on a hybrid Langevin/molecular dynamics evolution in a fictitious time. In this way the whole system can be updated in parallel, which is important in QCD simulations since the effective action for QCD with dynamical quarks is highly nonlocal. By including a global accept/reject step, an "exact" Monte Carlo algorithm is obtained, free from finite-step-size errors.

The HMC method is very general and has later on been applied to various systems other than QCD. In this paper I discuss the use of HMC in simulating polymer chains (for a discussion of HMC for QCD, see e.g. Thomas Lippert's contribution to these proceedings). I focus on single linear chains with repulsive interactions between the monomers, such as screened or unscreened Coulomb interactions.

A HMC update of such a chain with N monomers requires a computer time of order N^2. This is fast for an update of the whole system, but plain HMC is still a bad choice of algorithm for these systems, compared to, for example, the pivot algorithm (Lal 1969; for a recent review on simulation methods for polymer systems, see Binder 1995). The reason is that short-wavelength modes tend to evolve much faster than long-wavelength modes. As a result, to have stability at short wavelengths, one must choose the step size so small that a very large number of steps are needed in order to generate significant changes in global quantities such as the end-to-end distance.

However, the HMC scheme leaves room for improvement. In particular, one has considerable freedom in defining the auxiliary Hamiltonian that governs the Langevin/molecular dynamics evolution. One method for speeding up the evolution is the Fourier acceleration method, in which the different

Fourier modes are assigned different step sizes (or masses). This technique was introduced for Langevin simulations of field theories (Batrouni et al. 1985), and can be applied to HMC simulations too.

A Fourier accelerated algorithm is designed so that it speeds up the simulation of a given quadratic potential; the eigenmodes of this potential are updated with step sizes that are inversely proportional to their frequencies.

In this paper I present numerical results for two Fourier acceleration schemes, corresponding to different quadratic potentials. Most of the calculations were performed using a scheme corresponding to the potential with nearest-neighbor interactions only and uniform strength along the chain (Irbäck 1994). This algorithm is equivalent to performing plain HMC updates of the bond vectors connecting adjacent monomers along the chain. It turns out that this somewhat arbitrary choice of scheme gives a dramatic improvement of the efficiency for all the chains studied compared to plain HMC. An attractive feature of the algorithm is that short- and long-wavelength structure evolve at a similar speed. The efficiency is found to be somewhat higher for the unscreened Coulomb potential than for the other potentials studied, which have shorter range.

Some of the systems were also simulated using a scheme that was obtained by an optimization procedure. Here the quadratic potential defining the scheme was determined variationally, for each system studied. Although the number of variational parameters is large, it turns out that this can be carried out relatively fast by using the method of Jönsson et al. (1995). This optimization procedure was tested for unscreened and screened Coulomb chains. The results were very similar to those obtained earlier, which shows that the first, simpler scheme might be close to optimal for these chains.

The plan of this paper is as follows. In Sect. 2 I define the models studied. The HMC method and the Fourier acceleration technique are described in Sects. 3 and 4, respectively. Numerical results are presented in Sect. 5.

2   The Models

Throughout this paper I consider a linear chain of N sites or monomers, with positions in three-dimensional space given by $\mathbf{x}_i$, $i = 1, \ldots, N$. Alternatively, the system can be described by the $N - 1$ bond vectors $\mathbf{b}_i = \mathbf{x}_{i+1} - \mathbf{x}_i$, $i = 1, \ldots, N - 1$, connecting adjacent monomers (see Fig. 1).

The energy function of the system has the form

$$E = \sum_{1 \leq i < j \leq N} \left(\frac{1}{2} \delta_{i,j-1} r_{ij}^2 + v(r_{ij}) \right) , \qquad (2.1)$$

where $r_{ij} = |\mathbf{x}_i - \mathbf{x}_j|$ is the distance between monomers i and j. The first term in (2.1) is just a harmonic attraction that enforces the chain structure. The behavior of the system at temperature T is defined by the partition function

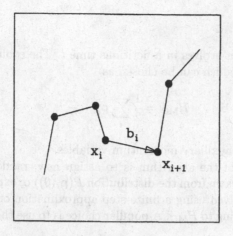

Fig. 1. Description of the chain.

$$Z = \int d^N \mathbf{x}\, \delta(\mathbf{x}_{cm}) \exp(-E/T) \; , \qquad (2.2)$$

where the overall translational degree of freedom is eliminated by holding the center of mass, $\mathbf{x}_{cm}$, fixed at the origin.

The system will be studied for four different choices of the potential $v(r)$ in (2.1). First I consider a simple model of a polyelectrolyte in a solution at finite salt concentration c_s. Here $v(r)$ is a screened Coulomb potential with inverse Debye screening radius $\kappa \propto \sqrt{c_s}$,

$$v(r) = \frac{\exp(-\kappa r)}{r} \; . \qquad (2.3)$$

This model is studied both with and without screening. Following Jönsson et al. (1995), I take $(T, \kappa) = (0.838, 0)$ (no screening) and $(T, \kappa) = (0.838, 1.992)$.

In addition to the unscreened and screened Coulomb potentials, I consider two potentials of intermediate range, given by

$$v(r) = \frac{1}{r^\lambda} \qquad (2.4)$$

with $\lambda = 2$ and 2.5, respectively.

These choices of $v(r)$ lead to different values of the swelling exponent ν, which describes the scaling of the end-to-end distance r_{ee} with N; $r_{ee} \sim N^\nu$, $N \to \infty$. The Flory result $\nu = 0.6$ is approximately correct for the short-range screened Coulomb potential, while $\nu = 1$ for the unscreened Coulomb potential. For the potential (2.4) with $2 \leq \lambda < 3$, Bouchaud et al. (1991) have predicted $\nu = 2/\lambda$ using a variational approach. For $\lambda = 2$ they also predicted a logarithmic correction to the power law behavior: $r_{ee} \sim N(\ln N)^{-\alpha}$ with $\alpha = 1/2$.

3   HMC

In HMC the system evolves in a fictitious time t. The evolution is governed by a Hamiltonian which can be chosen as

$$H_{\mathrm{MC}} = \frac{1}{2} \sum_i \mathbf{p}_i^2 + \frac{E}{T} \ , \tag{3.5}$$

where the $\mathbf{p}_i$'s are auxiliary momentum variables.

The first step of the algorithm is to assign new, random values to the momenta, $\mathbf{p}_i(0)$, drawn from the distribution $P(\mathbf{p}_i(0)) \propto \exp(-\mathbf{p}_i^2(0)/2)$. The system is then evolved using a finite-step approximation of the equations of motion corresponding to H_{MC}. A popular choice is to use the leapfrog scheme

$$\mathbf{x}_i(t + \frac{\epsilon}{2}) = \mathbf{x}_i(t) + \frac{\epsilon}{2}\mathbf{p}_i(t) \tag{3.6}$$

$$\mathbf{p}_i(t + \epsilon) = \mathbf{p}_i(t) - \frac{\epsilon}{T}\nabla_i E(t + \frac{\epsilon}{2}) \tag{3.7}$$

$$\mathbf{x}_i(t + \epsilon) = \mathbf{x}_i(t + \frac{\epsilon}{2}) + \frac{\epsilon}{2}\mathbf{p}_i(t + \epsilon) \tag{3.8}$$

After n such leapfrog iterations, often referred to as one trajectory, the new configuration, $\{\mathbf{x}_i(n\epsilon)\}$, is subjected to a Metropolis accept/reject step, where the probability of acceptance is given by $\min(1, \exp(-\Delta H_{\mathrm{MC}}))$, $\Delta H_{\mathrm{MC}} = H_{\mathrm{MC}}(n\epsilon) - H_{\mathrm{MC}}(0)$.

Two important properties of the leapfrog scheme are that it is time-reversible and preserves phase space area. Given that the discretized equations of motion have these two properties, it can be shown that the scheme described above samples the desired distribution without finite-step-size corrections.

The algorithm has two tunable parameters, the step size ϵ and the number of leapfrog steps in each trajectory, n. The trajectory length is $n\epsilon$. The choice $n = 1$ corresponds to a corrected Langevin algorithm.

4   Fourier Acceleration

In this section I give a brief description of HMC for quadratic potentials. The performance of the algorithm can in this case be studied analytically (see e.g. Kennedy and Pendleton 1991). This analysis provides the background for the Fourier acceleration technique.

Suppose that the algorithm described in the previous section is applied to the chain with potential given by (2.1) with $v(r) = 0$,

$$\frac{E}{T} = \frac{1}{2} \sum_i r_{i\,i+1}^2 \tag{4.9}$$

(for a quadratic potential the T dependence is trivial so I set $T = 1$). The calculation of acceptance rate and autocorrelation times is most easily performed by transforming to the Fourier variables

$$\tilde{\mathbf{x}}_k = \sqrt{\frac{2}{N}} \sum_i \mathbf{x}_i \cos \frac{\pi k(i - 1/2)}{N} \ . \tag{4.10}$$

This orthogonal transformation diagonalizes the Hamiltonian,

$$H_{\text{MC}} = \frac{1}{2} \sum_i \mathbf{p}_i^2 + \frac{E}{T} = \frac{1}{2} \sum_k (\tilde{\mathbf{p}}_k^2 + \omega_k^2 \tilde{\mathbf{x}}_k^2) \ , \tag{4.11}$$

where

$$\omega_k^2 = 4 \sin^2 \frac{\pi k}{2N} \ . \tag{4.12}$$

The k-dependent autocorrelation function $C_k(t) = \langle \tilde{\mathbf{x}}_k(t) \cdot \tilde{\mathbf{x}}_k(0) \rangle$ is easy to calculate in the limit $\epsilon \to 0$, where the $\tilde{\mathbf{x}}_k$'s evolve as a set of uncoupled oscillators and the accept/reject step can be ignored since H_{MC} is conserved. For trajectory length $t_0 = n\epsilon$, one finds that $|C_k(t)| \propto \exp(-t/\tau_k)$ with autocorrelation time

$$\tau_k = \frac{t_0}{- \ln |\cos(\omega_k t_0)|} \ . \tag{4.13}$$

In particular, this implies that the autocorrelation times for long-wavelength modes grow as N^2 for large N.

The average acceptance rate is found to be $P_{\text{acc}} = \text{erfc}(\frac{1}{2} \langle \Delta H_{\text{MC}} \rangle^{1/2})$ (Gupta et al. 1990), where erfc is the complementary error function and

$$\langle \Delta H_{\text{MC}} \rangle \sim \frac{\epsilon^4}{32} \sum_k \omega_k^4 \sin^2(\omega_k t_0) \qquad (\epsilon \to 0) \tag{4.14}$$

is the average energy change in one trajectory. This shows that it is primarily the short-wavelength modes that determine how small the step size must be taken in order to have a reasonable acceptance rate.

Since the step size must be small to maintain stability at short wavelengths, the evolution of long-wavelength modes becomes very slow; a large number of leapfrog steps is needed to produce significant changes in global observables. This problem can be overcome by introducing a k-dependent step size $\epsilon_k = \tilde{\epsilon}/\omega_k$. This technique is called Fourier acceleration. After modifying the algorithm in this way, one finds that $\tau_k = -\tilde{t}_0 / \ln |\cos \tilde{t}_0|$, where $\tilde{t}_0 = n\tilde{\epsilon}$, independent of k. Here the autocorrelation times are expressed in units of simulation time rather than leapfrog steps. In estimating computational effort, one has to take into account that $\tilde{\epsilon}$ must be decreased with increasing N in order to keep a constant acceptance. However, this N dependence is weak, $\tilde{\epsilon} \sim N^{-1/4}$.

This Fourier accelerated algorithm can be formulated in a simple and useful alternative way by using the bond vectors $\mathbf{b}_i$. In fact, this algorithm is equivalent to a plain HMC in the b_i's with uniform step size. This is for linear chains. An advantage of the bond vector formulation is that it can be applied to branched structures too. The Fourier variable formulation, on the other hand, is, of course, well suited for cyclic chains.

The algorithm discussed so far speeds up simulations for the potential (4.9). For a general quadratic potential, one can proceed in the same way: diagonalize the potential and update the eigenmodes using step sizes that are inversely proportional to their frequencies.

In the next section two different Fourier acceleration schemes are tested for the models defined in Sect. 2. The first of these is that described above, corresponding to the potential (4.9). The other was obtained by an optimization procedure, starting from the general ansatz

$$\frac{E}{T} = \frac{1}{2} \sum_{ij} G_{ij}^{-1} b_i \cdot b_j \; , \tag{4.15}$$

where the parameters G_{ij}^{-1} are elements of the inverse correlation matrix. For a given model and system size, these parameters were determined variationally by using the method of Jönsson et al. (1995). In implementing this algorithm, it is convenient to express the symmetric and positive definite matrix G^{-1} as the product of a matrix and its transpose, $G^{-1} = WW^T$.

5 Numerical Tests

In this section I discuss numerical tests of the Fourier acceleration method. I begin with results obtained using the scheme corresponding to the potential (4.9), which was tested for all the four models defined in Sect. 2. I then discuss some results obtained using the "optimized" scheme corresponding to (4.15) with variationally determined parameters G_{ij}^{-1}.

Figure 2 shows the evolution of the end-to-end distance r_{ee} in two simulations of an unscreened Coulomb chain, one with and one without Fourier acceleration. Both runs were carried out using $N = 16$, $n = 1$, and a step size such that $P_{acc} \approx 0.80$. From Fig. 2 it is evident that the evolution of r_{ee} is much faster in the simulation with Fourier acceleration. Due to the coordinate transformations (4.10), which were performed by using fast Fourier transform, the cost of each iteration is slightly higher for the Fourier accelerated algorithm, but this difference is negligible.

To study how the efficiency of the Fourier accelerated algorithm depends on N, simulations were performed for N up to 512. The integrated autocorrelation time, which controls the statistical error, was calculated for a number of different observables. For details of these simulations, see Irbäck (1994).

An important issue is to what extent the efficiency depends on the length scale considered. To investigate this, the integrated autocorrelation time $\tau_{int,k}$ for $\tilde{x}_k^2$ was measured for all possible k. As an example, Fig. 3 shows the results

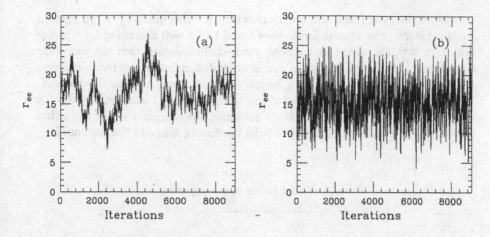

Fig. 2. The Monte Carlo evolution of the end-to-end distance in two HMC simulations of an unscreened Coulomb chain for $N = 16$: (a) without and (b) with Fourier acceleration.

obtained for the chain with potential $v(r) = 1/r^2$ and $N = 512$. As can be seen from this figure, the k dependence of $\tau_{int,k}$ is weak. Similar results were obtained for other chains.

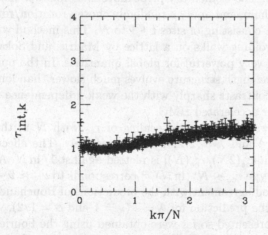

Fig. 3. The autocorrelation time $\tau_{int,k}$ in units of trajectories ($n = 104$) against k for a $N = 512$ chain with potential $v(r) = 1/r^2$.

Having seen that the k dependence is weak, let us now focus on one observable, the end-to-end distance. To get a measure of computational effort,

I consider the integrated autocorrelation time for this quantity, E, in units of leapfrog steps. The results for E were found to be well described by a power law, $E \propto N^{z'}$, for all the four models studied. This means that the computer time required to generate a given number of independent measurements grows as $N^{2+z'}$, since the cost of each leapfrog step scales quadratically with N. The fitted values of the exponent z' are given in Table 1 and lie between 0.6 and 0.9. Although the differences are not large, the efficiency of the algorithm shows a clear tendency to improve with increasing range of the potential.

Table 1. Fitted values of the exponent z'.

$v(r)$	z'
$1/r$	0.66(2)
$\exp(-\kappa r)/r$	0.82(2)
$1/r^2$	0.76(3)
$1/r^{2.5}$	0.79(3)

A popular method for simulations of long flexible chains is the pivot algorithm (Lal 1969). The elementary pivot move is as follows: choose a random site i along the chain, and apply a randomly chosen rotation/reflection to the part of the chain consisting of sites $i + 1$ to N. This method was thoroughly tested for self-avoiding walks on a lattice by Madras and Sokal (1988), and was found to be very powerful for global quantities. In the pure pivot algorithm, short-wavelength structure evolves much slower than long-wavelength structure. This contrasts sharply with the weak k dependence that was seen above for Fourier accelerated HMC.

Figure 4 shows results for the scaling of r_{ee} with N in the two models with $v(r) = 1/r^\lambda$ and $\lambda = 2$ and 2.5, respectively. The effective exponent $\nu_N = (2 \ln 2)^{-1} \ln[r_{ee}^2(2N)/r_{ee}^2(N)]$ is plotted against $1/\ln N$. An asymptotic behavior of the type $r_{ee} \sim N^\nu (\ln N)^{-\alpha}$ corresponds to $\nu_N \approx \nu - \alpha/\ln N$. The results are in good agreement with the predictions of Bouchaud et al. (1991). The line shows the prediction for $\lambda = 2$ ($\nu = 1$ and $\alpha = 1/2$).

The results presented so far were obtained using the Fourier acceleration scheme corresponding to (4.9). As mentioned above, this scheme is equivalent to a plain HMC in the bond vectors $\mathbf{b}_i$, which corresponds to the choice $G_{ij}^{-1} = \delta_{ij}$ in (4.15). I now turn to the optimization procedure, where the parameters G_{ij}^{-1} are determined variationally. The algorithm obtained in this way was tested for unscreened and screened Coulomb chains. The efficiency was found to be very similar to that of the previous algorithm. At least partly, this can be explained by the fact the variations in frequency among

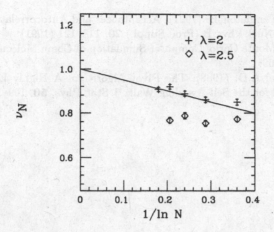

Fig. 4. The effective exponent ν_N against $1/\ln N$ for the two models with $v(r) = 1/r^\lambda$ and $\lambda = 2$ and 2.5, respectively.

the eigenmodes of the optimized $\mathbf{G}^{-1}$ turn out to be fairly small. For example, for the screened Coulomb chain the inverse frequencies vary between 0.96 and 1.21 for $N = 16$, and between 0.96 and 1.49 for $N = 64$.

Acknowledgements

I would like to thank Bo Jönsson, Carsten Peterson and Bo Söderberg for useful discussions.

References

Batrouni, G.G., Katz, G.R., Kronfeld, A.S., Lepage, G.P., Svetitsky, B., Wilson, K.G. (1985): Langevin Simulation of Lattice Field Theories. Phys. Rev. D **32**, 2736–2747

Bouchaud, J.-P., Mézard, M., Parisi, G., Yedidia, J.S. (1991): Polymers with Long-Ranged Self-Repulsion: A Variational Approach. J. Phys. A **24**, L1025–L1030

Binder, K., ed. (1995): *Monte Carlo and Molecular Dynamics Simulations in Polymer Science* (Oxford University Press, New York)

Duane, S., Kennedy, A.D., Pendleton, B.J., Roweth, D. (1987): Hybrid Monte Carlo. Phys. Lett. B **195**, 216–222

Gupta, S., Irbäck, A., Karsch, F., Petersson, B. (1990): The Acceptance Probability in the Hybrid Monte Carlo Method. Phys. Lett. B **242**, 437–443

Irbäck, A. (1994): Hybrid Monte Carlo Simulation of Polymer Chains. J. Chem. Phys. **101**, 1661–1667

Jönsson, B., Peterson, C., Söderberg, B. (1995): A Variational Approach to the Structure and Thermodynamics of Linear Polyelectrolytes with Coulomb and Screened Coulomb Interactions. J. Phys. Chem. **99**, 1251–1266

142   Anders Irbäck

Kennedy, A.D., Pendleton, B. (1991): Acceptances and Autocorrelations in Hybrid Monte Carlo. Nucl. Phys. B (Proc. Suppl.) **20**, 118–121 (1991)
Lal, M. (1969): 'Monte Carlo' Computer Simulation of Chain Molecules. Mol. Phys. **17**, 57–64 (1969)
Madras, N., Sokal, A.D. (1988): The Pivot Algorithm: A Highly Efficient Monte Carlo Method for the Self-Avoiding Walk. J. Stat. Phys. **50**, 109–186

Simulations of Toy Proteins

Anders Irbäck

Complex Systems Group, Department of Theoretical Physics, University of Lund, Sölvegatan 14A, S-223 62 Lund, Sweden

Abstract. Folding properties of two simple off-lattice protein models in two and three dimensions, respectively, are analyzed numerically by using the simulated-tempering method. Both models have two types of "amino acids", hydrophobic and hydrophilic. In the two-dimensional model, a total of 300 randomly selected sequences with 20 monomers are studied. About 10% of these meet criteria for good folders. A statistical analysis of the distribution of hydrophobic monomers along the chains is performed, both for the good folders in this model and for functional proteins. This analysis convincingly shows that the hydrophobicity distribution is nonrandom for functional proteins. Furthermore, qualitatively similar deviations from randomness are found for good folders in the model. The study of the three-dimensional model demonstrates the importance of local interactions.

1 Introduction

A protein is a linear chain of amino acids, often consisting of 100 to 500 amino acids. A major class of proteins, globular proteins, fold into compact native structures. Each globular protein has a unique native structure, determined by its amino-acid sequence (Anfinsen 1973, Creighton 1993). Understanding the mechanism by which this structure is selected among the huge number of possible conformations is one of the great challenges in molecular biology. The task of predicting the three-dimensional native structure from knowledge of the linear amino-acid sequence is often referred to as the protein folding problem.

There are two major theoretical approaches to protein folding. One is molecular dynamics simulations of fairly detailed models, in which each interaction center represents one or a few atoms (for a review, see e.g. Karplus and Šali 1995). Such simulations can be used for exploring details of the energy landscape near the native structure, and for studying unfolding from the native state. However, the evolution of the system is much too slow for a proper exploration of the full conformational space.

In this situation, simplified "coarse-grained" models have become an increasingly popular complementary tool. These models are often lattice-based and the basic entities are amino acids rather than atoms. Their motivation is roughly analogous to that behind the Ising model for magnetism. They are used to study general characteristics of possible folding mechanisms. Needless to say, they are not intended to reproduce the detailed behavior of specific amino-acid sequences.

Simplified models for protein folding have been studied both analytically and numerically. Useful insights into the phase diagram of such models have been obtained by using ideas from the theory of spin glasses (for a recent review, see Garel et al. 1997). However, in this approach one calculates averages over the form of the disorder, which in this case is the sequence information. This is a limitation since the behavior of sequences that fold well may be different from the average. By using numerical simulations averages over sequence can be avoided.

Most of the simplified models that have been studied numerically are lattice-based with nearest-neighbor contact interactions only. Such simple models have been found to exhibit interesting and nontrivial properties, which is encouraging (for reviews, see e.g. Karplus and Šali 1995 and Dill et al. 1995). However, this does not imply that the approximations involved are understood. Although these models have advantages in terms of computational efficiency, it is therefore important to pursue the study of alternative models.

In this paper I discuss numerical results obtained in two simple off-lattice models in two (2D) and three (3D) dimensions, respectively. The simulations were performed by using the simulated-tempering method (Marinari and Parisi 1992). Both models have only two types of "amino acids", hydrophobic and hydrophilic, instead of the 20 that occur naturally. The monomers interact through sequence-dependent Lennard-Jones potentials that favor the formation of a hydrophobic core. In addition, the models contain sequence-independent local interactions.

This paper is organized as follows. In Sect. 2 I define the two models and give a brief description of the simulation-tempering method. Section 3 deals with a study of the folding properties of 300 randomly selected sequences in the 2D model. In Sect. 4 the statistical distribution of hydrophobic monomers along chains is examined, both for functional proteins and for good folding sequences in the 2D model. The 3D model is discussed in Sect. 5, focusing on the effects of the local interactions. A brief summary is given in Sect. 6.

2   The Models and the Algorithm

The models studied contain two kinds of monomers, A (hydrophobic) and B (hydrophilic). These are linked by rigid bonds of unit length to form linear chains. A sequence is specified by a choice of monomer types at each position along the chain, $\{\sigma_i\}$, where σ_i takes the values A and B and i is a monomer index. The structure of a chain with N monomers is described by $N-1$ unit bond vectors, $\{b_i\}$, where b_i connects monomers i and $i+1$.

The energy function consists of sequence-independent local interactions and sequence-dependent Lennard-Jones potentials. The latter are responsible for the compactification of the chain, and are chosen so as to favor the formation of a core of A monomers.

In the 2D model, introduced by Stillinger et al. (1993), the energy function is given by

$$E = \frac{1}{4} \sum_{i=1}^{N-2} (1 - \mathbf{b}_i \cdot \mathbf{b}_{i+1}) + 4 \sum_{i=1}^{N-2} \sum_{j=i+2}^{N} \left(\frac{1}{r_{ij}^{12}} - \frac{C(\sigma_i, \sigma_j)}{r_{ij}^6} \right) , \qquad (2.1)$$

where r_{ij} is the distance between sites i and j. The first term in (2.1) consists of bend potentials that favor alignment of adjacent bond vectors. The species-dependent coefficient $C(\sigma_i, \sigma_j)$ in the Lennard-Jones potential is taken to be 1 for an AA pair, 1/2 for a BB pair, and -1/2 for an AB pair.

The energy function of the 3D model (Irbäck et al. 1997b) is given by

$$E = -\kappa_1 \sum_{i=1}^{N-2} \mathbf{b}_i \cdot \mathbf{b}_{i+1} - \kappa_2 \sum_{i=1}^{N-3} \mathbf{b}_i \cdot \mathbf{b}_{i+2} + 4 \sum_{i=1}^{N-2} \sum_{j=i+2}^{N} \epsilon(\sigma_i, \sigma_j) \left(\frac{1}{r_{ij}^{12}} - \frac{1}{r_{ij}^6} \right) ,$$
$$\qquad (2.2)$$

where $\epsilon(\sigma_i, \sigma_j)$ is 1 for an AA pair, and 1/2 for BB and AB pairs. The parameters κ_1 and κ_2 determine the strength of the local interactions. The model will be studied for the three choices $(\kappa_1, \kappa_2) = (0, 0)$, $(-1, 0)$ and $(-1, 0.5)$.

Thermodynamic simulations of these models have been performed for different sequences and temperatures, by using the simulated-tempering method (Marinari and Parisi 1992). At low temperatures conventional Monte Carlo methods tend to become extremely time-consuming, due to the presence of high free-energy barriers. In simulated tempering one tries to overcome this problem by allowing the system to visit higher temperatures where the barriers are lower. More precisely, rather than simulating the Boltzmann distribution at a fixed temperature, one simulates the joint probability distribution

$$P(\mathbf{b}, k) \propto \exp(-g_k - E/T_k) , \qquad (2.3)$$

where T_k, $k = 1, \ldots, K$, are the allowed temperatures. The g_k's are tunable parameters that govern the probabilities of visiting the different temperatures. Hence, they must be chosen carefully, which is done by means of trial runs. The joint distribution (2.3) is then simulated by using separate, ordinary updates of $\{\mathbf{b}_i\}$ and k. Such a simulation directly generates Boltzmann-weighted configurations for each T_k.

Simulated tempering turns out be of great help in simulating these models. For the 2D model speedup factors of 10^3–10^4 were observed, compared with conventional methods (Irbäck and Potthast 1995).

In simulated tempering the temperature becomes a dynamical parameter. It is, of course, possible to apply this idea also to other parameters of the model. An algorithm where the sequence is treated as a dynamical parameter was tested for the 2D model with some success (Irbäck and Potthast 1995). Other dynamical-parameter algorithms have been successfully applied to, for example, the Potts model (Kerler and Weber 1993) and U(1) gauge theory (Kerler et al. 1995).

A method closely related to simulated tempering is the method of multiple Markov chains (Geyer and Thompson 1994, Tesi et al. 1996), also called parallel tempering. Here a set of K allowed temperatures are simulated using K copies of the system, which evolve in parallel. This method has the advantage that there are no g_k parameters to be tuned. The efficiency of this method was tested for the 3D model, with results very similar to those for simulated tempering.

3   Good Folding Sequences

In this section I discuss a study of 300 randomly selected sequences with 14 A and 6 B monomers in the 2D model defined in Sect. 2 (Irbäck et al. 1997a). Both thermodynamic and kinetic simulations were performed for each sequence. The thermodynamic properties are found to be strongly sequence dependent in contrast to the kinetic ones. Hence, criteria for good folding sequences are formulated entirely in terms of thermodynamic properties.

The thermodynamic simulations were performed by using simulated tempering with 13 allowed temperatures, ranging from 0.15 to 0.60. As the temperature is decreased from 0.60 to 0.15, the size of the chains, as measured by e.g. the radius of gyration, decreases substantially. This compactification is found to take place gradually. The variations in size among the different sequences studied, at a fixed temperature, are found to be small, which is expected since they have the same composition (14 A and 6 B).

The low-temperature behavior shows, nevertheless, a strong sequence dependence. This can be clearly seen from the probability distribution of the mean-square distance between different configurations, δ^2. For two configurations with monomer positions $\{x_i^{(a)}\}$ and $\{x_i^{(b)}\}$, respectively, this is defined by

$$\delta^2 = \delta_{ab}^2 = \min \frac{1}{N} \sum_{i=1}^{N} |x_i^{(a)} - x_i^{(b)}|^2 \; , \tag{3.4}$$

where the minimum is taken over translations, rotations and reflections. Figure 1 shows three examples of δ^2 distributions at $T = 0.15$ and 0.60. The corresponding three sequences are given in Table 1. At $T = 0.60$ the three distributions are similar and very broad, showing that the fluctuations in shape are large. At $T = 0.15$ the δ^2 distribution is, by contrast, strongly sequence dependent. One of the three chains (sequence 81) has a well-defined shape at this temperature.

In particular, this implies that these sequences have different folding temperatures T_f. The folding temperature is the temperature where the dominance of a single state sets in. Sequence 81 has highest T_f (> 0.15) among the three sequences in Fig. 1.

The energy level spectrum was studied by using a quenching procedure; for a large number of different initial configurations, obtained from the simulated-

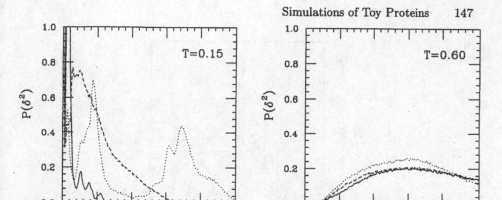

Fig. 1. δ^2 distributions in the 2D model at $T = 0.15$ and 0.60 for the sequences in Table 1; 81 (solid line), 10 (dashed line) and 50 (dotted line). At $T = 0.15$ the distribution for sequence 81 has two narrow peaks at small δ^2 which extend outside the figure with a maximum value of around 20.

Table 1. Three of the sequences studied in the 2D model.

81 AAABAAABAAABBAABBAAA
10 ABAAABBAABAAAAAAAABB
50 BABAAAAAAABAAAABAABB

tempering run, the system was brought to a nearby local energy minimum by means of a conjugate gradient method.

In the subsequent kinetic simulations the mean-square distance δ_0^2 to the minimum energy configuration was monitored. As a criterion of successful folding the condition $\delta_0^2 < 0.3$ was used. The probability of successful folding within a given number of Monte Carlo steps, called the foldicity, was estimated using 25 random initial configurations for each sequence and temperature. Figure 2 shows the foldicity against temperature for the three sequences in Table 1. As expected, the foldicity is found to be low both at high temperatures, where the search is random, and at low temperatures, where the search is hindered by the ruggedness of the free-energy landscape. The freezing temperature where the slow low-temperature behavior sets in, is found to have a weak sequence dependence compared to the folding temperature T_f.

A good folder is a sequence that, at some temperature, exists in a unique and kinetically accessible state with well-defined shape. Having found that the foldicity shows a relatively weak sequence dependence, simplified criteria

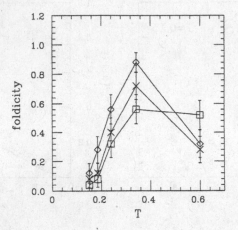

Fig. 2. Foldicity against T in the 2D model for the sequences 81 ($\times$), 10 ($\diamond$) and 50 ($\square$); see Table 1.

for good folders were formulated, based entirely on the δ^2 distribution. With these criteria, 37 of the 300 sequences were classified as good folders.

4   Nonrandom Hydrophobicity Patterns

Hydrophobicity plays a central role in the formation of protein structures. To understand the statistical distribution of hydrophobic amino acids along proteins is therefore of great interest. In this section I discuss a study of this distribution (Irbäck et al. 1996) based on the SWISS-PROT protein sequence data bank (Bairoch and Boeckmann 1994) and binary hydrophobicity assignments (± 1). The same analysis is carried out for good folding sequences in the 2D model (see Sect. 3).

The hydrophobicity distribution is analyzed by two different methods, based on block and Fourier variables, respectively. By using these variables rather than the raw sequence variables σ_i, the analysis becomes more sensitive to long-range correlations along the sequence.

Analyzing the behavior of block variables is a widely used and fruitful technique in statistical mechanics, and this application turns out to be no exception. As a first step the sequence is divided into blocks of size s; the block variable $\sigma_k^{(s)}$ is then defined as the sum of the s σ_i's in the block labeled k. The behavior of the block variables will be studied using the normalized mean-square fluctuation $\psi^{(s)}$, defined as

$$\psi^{(s)} = \frac{1}{K}\frac{s}{N}\sum_{k=1}^{N/s}(\sigma_k^{(s)} - s\bar{\sigma})^2 \ , \tag{4.5}$$

where N is the length of the sequence, and $\bar{\sigma}$ denotes the average of σ_i over the whole sequence. The normalization factor K is chosen such that the average of $\psi^{(s)}$ over all possible sequences with N_+ positive σ_i's and length N, is given by

$$\langle \psi^{(s)} \rangle_{N,N_+} = s \; , \tag{4.6}$$

independent of N and N_+.

The Fourier analysis is based on a random walk representation of the sequences,

$$\rho_n = \sum_{i=1}^{n} \sigma_i - n \frac{2N_+ - N}{N} \; , \tag{4.7}$$

which is defined so that $\rho_0 = \rho_N = 0$. With these boundary conditions, the sine transform of ρ_n, to be called f_k, is considered. The average of f_k^2 over all sequences with fixed N and N_+ is found to be

$$\langle f_k^2 \rangle_{N,N_+} = \frac{2N_+(N - N_+)}{N-1} \left(2 \sin \frac{\pi k}{2N} \right)^{-2} . \tag{4.8}$$

Using this, the normalized quantity $\tilde{f}_k^2 = f_k^2 / \langle f_k^2 \rangle_{N,N_+}$ is formed.

Before applying these methods to protein sequences, there are two important observations to be made. First, the data originating from the ends of the sequences display a different behavior than the data from the rest of the sequences. Therefore, the analysis is performed using the central 70% of the sequences only. Second, sequences with different fractions of hydrophobic amino acids tend to behave in different ways. Because of this, the analysis is performed for a fixed range of the variable

$$X = \frac{N_+ - Np}{\sqrt{Np(1-p)}} \; , \tag{4.9}$$

where p is the average of N_+/N over all sequences.

Figure 3a shows the results of the blocking analysis for three different regions in X: $|X| < 0.5$, $|X| > 3$ and all X. The straight line represents random sequences, (4.6). The results for large $|X|$ lie above this line, while those for small $|X|$ show the opposite behavior. The stability of the results was tested by performing the same analysis both for different intervals in N and for a selected set of nonredundant sequences. Figure 3b shows the results for the good folders in the 2D model. These data show deviations from randomness that are qualitatively similar to those for functional proteins with small $|X|$.

Figure 4a shows the results of the Fourier analysis for $|X| < 0.5$. As one might have expected, there is a peak around the wavelength corresponding to α-helix structure, $2N/k = 3.6$. In addition, there are also clear deviations from randomness at long wavelengths. Such components are suppressed. The results for the good folders in the 2D model are shown in Fig. 4b. Although

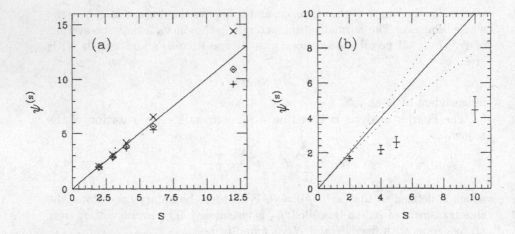

Fig. 3. $\psi^{(s)}$ against block size s. (a) Functional proteins for $|X| < 0.5$ (+; 10154 qualifying proteins), $|X| > 3$ (×; 4928), and all X (◇; 36765). The straight line is the result for random sequences. (b) Good folders in the 2D model. Also shown are the mean s (full line) and the $s \pm \sigma$ band (bounded by dotted lines) for random sequences.

the statistical errors are somewhat large, there is a clear suppression of the long-wavelength components in this case also.

Here the results have been compared to those for an uncorrelated system. As an illustration of the effects of correlations on the observables $\psi^{(s)}$ and $\tilde{f}_k^2$ one may consider the one-dimensional Ising model. For antiferromagnetic coupling one finds, as for functional proteins with $|X| < 0.5$, that $\psi^{(s)} < s$ and that $\tilde{f}_k^2$ is suppressed for small k. For ferromagnetic coupling the behavior is the opposite, which is what one observes for $|X| > 3$.

The analysis of good folding sequences in the 2D model shows that there exist substantial correlations between certain sequence variables and the folding properties. This suggests that, to some extent, it should be possible to predict folding properties given the sequence. To test this, a feedforward artificial neural network was trained to predict the mean of the δ^2 distribution, at $T = 0.15$, with sequence information only as input (Irbäck 1997a). The results were promising and are likely to improve with a larger data set.

5   Local Interactions

I now turn to the 3D model defined in Sect. 2, focusing on the effects of the local interactions. Both the local structure of the chains and the δ^2 distributions are discussed, for the three parameter choices $(\kappa_1, \kappa_2) = (0,0)$, $(-1, 0)$ and $(-1, 0.5)$. The numerical results were obtained using six different

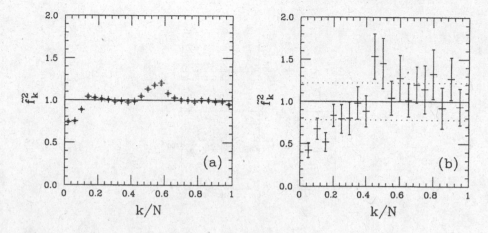

Fig. 4. $\tilde{f}_k^2$ against k/N. (a) Functional proteins for $|X| < 0.5$. (b) Good folders in the 2D model. The full line and dots are as in Fig. 3b.

sequences of length $N = 20$ (Irbäck et al. 1997b), which were deliberately chosen to represent different types of behavior.

In a 3D off-lattice model, it is possible to check the local properties against those for functional proteins in a direct way. To this end, I consider the structure defined by the protein backbone of C^α atoms. The shape of a chain with N monomers can be specified by $N - 2$ (virtual) bond angles τ_i and $N - 3$ torsional angles α_i. In Fig. 5 the distributions of the angles τ_i and α_i are shown, using data from the Protein Data Bank (Bernstein et al. 1977). From these distributions it is evident that there are strong regularities in the local structure. There are two favored regions in the (τ_i, α_i) plane. One of these corresponds to right-handed α-helix, $\tau_i \in [85°, 100°]$ and $\alpha_i \in [35°, 70°]$, and the other to β-sheet, $\tau_i \in [105°, 145°]$ and $\alpha_i \in [170°, 250°]$. Needless to say, the model is not intended to reproduce the precise form of these distributions. It turns out, however, that there are substantial differences in the behaviors for the different (κ_1, κ_2). In particular, the α_i distribution is much flatter, and less protein-like, for $(\kappa_1, \kappa_2) = (0,0)$ than for $(\kappa_1, \kappa_2) = (-1,0)$ and $(-1, 0.5)$. Also, two-bond correlations were found to decay very fast with separation along the chain for $(\kappa_1, \kappa_2) = (0,0)$. This shows that the pure Lennard-Jones potential is not sufficient to produce the strong regularities in the local structure observed for proteins.

Next I turn to the question of how the choice of (κ_1, κ_2) affects the overall stability of the chains. To study this the δ^2 distributions were computed (see Sect. 3). In contrast to the local quantities discussed above, the δ^2 distribution is strongly sequence dependent. Nevertheless, some general trends were observed. First, the pure Lennard-Jones potential yields very broad δ^2 distributions. Second, the stability tends to be somewhat higher

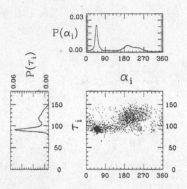

Fig. 5. Virtual bond (τ_i) and torsional (α_i) angle distributions for functional proteins.

for $(\kappa_1, \kappa_2) = (-1, 0.5)$ than for $(\kappa_1, \kappa_2) = (-1, 0)$. Formation of a compact and well-defined structure was observed for two of the six sequences for $(\kappa_1, \kappa_2) = (-1, 0.5)$.

These results show that local interactions are necessary for structural stability in the 3D model. In the 2D model, where the movements are hampered by compressing one dimension, the local interactions were found to be less crucial. Simulations without the local interaction term were performed for 15 of the 300 sequences, and relatively small changes in the δ^2 distributions were observed.

It should be stressed that the sequences studied contain only two types of monomers. In the work by Yue et al. (1995) a number of two-letter sequences were studied in a 3D lattice model with contact interactions only. The two-letter code was found to be insufficient in the sense that these sequences did not have unique native structures. In models with a larger alphabet more of the possible sequences have unique native structures (Shakhnovich 1994). In the model we have discussed here, with sequence-independent local interactions, it is possible to find two-letter sequences that have unique native structures (see also Irbäck and Sandelin 1997).

6   Summary

Two simple off-lattice models for protein folding have been studied. In the 2D model the termodynamic δ^2 distribution was found to have a much stronger sequence dependence than the kinetic foldicity. Hence criteria for good folding sequences were devised solely in terms of the δ^2 distribution. Approximately 10% of the 300 sequences studied meet these criteria.

Using blocking and Fourier analysis, it was shown that the hydrophobicity distribution of good folders in the 2D model is nonrandom. Qualitatively similar deviations from randomness were observed for low-$|X|$ functional proteins.

In the extension to 3D of the 2D model, the sequence-independent local interactions were found to play a crucial role. It was shown that, in the presence of these, it is possible to find two-letter sequences that have compact, unique native structures. This is in contrast to the results of an earlier study of a 3D lattice model with two-letter code and contact interactions only, where no such sequences were found.

The simulations of these systems were carried out by using simulated tempering, which gives a dramatic improvement of the efficiency compared to conventional methods. Still, as it stands, the results for each $N = 20$ chain in the 3D model require 70 CPU hours on a DEC Alpha 200, so there is room for further improvement.

References

Anfinsen, C.B. (1973): Principles that Govern the Folding of Protein Chains. Science **181**, 223–230

Bairoch, A., Boeckmann, B. (1994): The SWISS-PROT Protein Sequence Data Bank: Current Status. Nucleic Acids Res. **22**, 3578–3580

Bernstein, F.C., Koetzle, T.F., Williams, G.J.B., Meyer, E.F., Brice, M.D., Rodgers, J.R., Kennard, O., Shimanouchi, T., Tasumi, M. (1977): The Protein Data Bank: A Computer-Based Archival File for Macromolecular Structure. J. Mol. Biol. **112**, 535–542

Creighton, T.E. (1993): *Proteins: Structures and Molecular Properties* (W.H. Freeman, New York)

Dill, K.A., Bromberg, S., Yue, K., Fiebig, K.M., Yee, D.P. Thomas, P.D., Chan, H.S. (1995): Principles of Protein Folding — A Perspective from Simple Exact Models. Protein Science **4**, 561–602

Garel, T., Orland, H., Pitard, E. (1997): Saclay preprint T97-003, to appear in *Spin Glasses and Random Fields*, A.P. Young (ed.) (World Scientific, Singapore)

Geyer, C.J., Thompson, E.A. (1994): Annealing Markov Chain Monte Carlo with Applications to Pedigree Analysis. Preprint, University of Minnesota

Irbäck, A., Potthast, F. (1995): Studies of an Off-Lattice Model for Protein Folding: Sequence Dependence and Improved Sampling at Finite Temperature. J. Chem. Phys. **103**, 10298–10305

Irbäck, A., Peterson, C., Potthast, F. (1996): Evidence for Nonrandom Hydrophobicity Structures in Protein Chains. Proc. Natl. Acad. Sci. USA **93**, 9533–9538

Irbäck, A., Peterson, C., Potthast, F. (1997a): Identification of Amino Acid Sequences with Good Folding Properties in an Off-Lattice Model. Phys. Rev. E **55**, 860–867

Irbäck, A., Peterson, C., Potthast, F., Sommelius, O. (1997b): Local Interactions and Protein Folding: A 3D Off-Lattice Approach. J. Chem. Phys. **107**, 273–282

Irbäck, A., Sandelin, E. (1997): Local Interactions and Protein Folding: A Model Study on the Square and Triangular Lattices. LU TP 97-17, cond-mat/9708049

Karplus, M., Šali, A. (1995): Theoretical Studies of Protein Folding and Unfolding. Curr. Opin. Struct. Biol. **5**, 58–73

Kerler, W., Weber, A. (1993): Cluster Dynamics for First-Order Phase Transitions in the Potts Model. Phys. Rev. B **47**, 11563–11566

Kerler, W., Rebbi, C., Weber, A. (1995): Phase Transition and Dynamical-Parameter Method in U(1) Gauge Theory. Nucl. Phys. B **450**, 452–460

Marinari, E., Parisi, G. (1992): Simulated Tempering: A New Monte Carlo Scheme. Europhys. Lett. **19**, 451–458

Shakhnovich, E.I. (1994): Proteins with Selected Sequences Fold into Unique Native Conformations. Phys. Rev. Lett. **72**, 3907–3910

Tesi, M.C., Janse van Rensburg, E.J., Orlandini, E., Whittington, S.G. (1996): Monte Carlo Study of the Interacting Self-Avoiding Walk Model in Three Dimensions. J. Stat. Phys. **82**, 155–181

Stillinger, F.H., Head-Gordon, T., Hirschfeld, C.L. (1993): Toy Model for Protein Folding. Phys. Rev. E **48**, 1469–1477

Yue, K., Fiebig, K.M., Thomas, P.D., Chan, H.S., Shakhnovich, E.I., Dill, K.A. (1995): A Test of Lattice Protein Folding Algorithms. Proc. Natl. Acad. Sci. USA **92**, 325–329

Two Lectures on Phase Mixing: Nucleation and Symmetry Restoration

Marcelo Gleiser[1,2]

[1] Department of Physics and Astronomy, Dartmouth College,
Hanover, NH 03755, USA
[2] Osservatorio Astronomico di Roma, Viale del Parco Mellini, 84
Roma I-00136, Italy

Abstract. The dynamics of phase transitions plays a crucial rôle in the so-called interface between high energy particle physics and cosmology. Many of the interesting results generated during the last fifteen years or so rely on simplified assumptions concerning the complex mechanisms typical of nonequilibrium field theories. After reviewing well-known results concerning the dynamics of first and second order phase transitions, I argue that much is yet to be understood, in particular in situations where homogeneous nucleation theory does not apply. I present a method to deal with departures from homogeneous nucleation, and compare its efficacy with numerical simulations. Finally, I discuss the interesting problem of matching numerical simulations of stochastic field theories with continuum models.

LECTURE I

1 Homogeneous Nucleation

The fact that the gauge symmetries describing particle interactions can be restored at high enough temperatures has led, during the past 15 years or so, to an active research program on the possible implications that this symmetry restoration might have had to the physics of the very early Universe. One of the most interesting and popular possibilities is that during its expansion the Universe underwent a series of phase transitions, as some higher symmetry group was successively broken into products of smaller groups, up to the present standard model described by the product $SU(3)_C \otimes SU(2)_L \otimes U(1)_Y$. Most models of inflation and the formation of topological (and nontopological) defects are well-known consequences of taking the existence of cosmological phase transitions seriously [1].

One, but certainly not the only, motivation of the works addressed in this talk comes from the possibility that the baryon asymmetry of the Universe could have been dynamically generated during a first order electroweak phase transition [2]. As is by now clear, a realistic calculation of the net baryon number produced during the transition is a formidable challenge. We probably must invoke physics beyond the standard model (an exciting prospect for most people), push perturbation theory to its limits (and beyond, due to the nonperturbative nature of magnetic plasma masses that regulate the perturbative expansion in the symmetric phase), and we must deal with nonequilibrium aspects of the phase transition. Here I will focus on the latter problem, as it seems to me to be the least discussed of the pillars on which most baryon number calculations are built upon. To be more specific, it is possible to separate the nonequilibrium aspects of the phase transition into two main subdivisions. If the transition proceeds by bubble nucleation, we can study the propagation of bubbles in the hot plasma and the transport properties of particles through the bubble wall. A considerable amount of work has been devoted to this issue, and the reader can consult the works of Ref. [3] for details. These works assume that homogeneous nucleation theory is adequate to investigate the evolution of the phase transition, at least for the range of parameters of interest in the particular model being used to generate the baryon asymmetry. This brings us to the second important aspect of the nonequilibrium dynamics of first order phase transitions, namely the validity of homogeneous nucleation theory to describe the approach to equilibrium. This is the issue addressed in this talk.

Nucleation theory is a well-studied, but far from exhausted, subject. Since the pioneering work of Becker and Döring on the nucleation of droplets in supercooled vapor [4], the study of first order phase transitions has been of interest to investigators in several fields, from meteorology and materials science to quantum field theory and cosmology. Phenomenological field theories were developed by Cahn and Hilliard and by Langer [5, 6] in the context of coarse-grained time-dependent Ginzburg-Landau models, in which an expression for the decay rate per unit volume was obtained by assuming a steady-state probability current flowing through the saddle-point of the free-energy functional [6, 7]. The application of metastable decay to quantum field theory was initiated by Voloshin, Kobzarev, and Okun [8], and soon after put onto firmer theoretical ground by Coleman and Callan [9]. The generalization of these results for finite temperature field theory was first studied by Linde [10], and has been the focus of much recent attention [11].

The crucial ingredient in the evaluation of the decay rate is the computation of the imaginary part of the free energy. As shown by Langer [6], the decay rate $\mathcal{R}$ is proportional to the imaginary part of the free energy $\mathcal{F}$,

$$\mathcal{R} = \frac{|E_-|}{\pi T} \mathrm{Im} \mathcal{F} , \qquad (1.1)$$

where E_- is the negative eigenvalue related to metastability, which depends on nonequilibrium aspects of the dynamics, such as the growth rate of the critical bubble. Since $\mathcal{F} = -T \ln Z$, where Z is the partition function, the computation for the rate boils down to the evaluation of the partition function for the system comprised of critical bubbles of the lower energy phase inside the metastable phase.

If we imagine the space of all possible field configurations for a given model, there will be different paths to go from the metastable to the ground state. We can think of the two states as being separated by a hill of a given "height". The energy barrier for the decay is then related to the height of this hill. At the top of the hill, only one direction leads down to the ground state, the unstable direction. Fluctuations about this direction will grow, with rate given by the negative eigenvalue which appears in the above formula. All other directions are positively curved, and fluctuations about them give rise to positive eigenvalues which do not contribute to the decay rate. The path which will cost less energy is the one which will dominate the partition function, the so-called critical bubble or bounce. It is simply the field configuration that interpolates between the two stable points in the *energy landscape*, the metastable and ground state. The energy barrier for the decay is the energy of this particular field configuration.

For a dilute gas of bubbles only, the partition function for several bubbles is given by [12, 6],

$$
Z \simeq Z(\varphi_f) + Z(\varphi_f) \left[\frac{Z(\varphi_b)}{Z(\varphi_f)} \right] + Z(\varphi_f) \frac{1}{2!} \left[\frac{Z(\varphi_b)}{Z(\varphi_f)} \right]^2 + \dots
$$
$$
\simeq Z(\varphi_f) \exp \left[\frac{Z(\varphi_b)}{Z(\varphi_f)} \right] , \tag{1.2}
$$

where φ_f is the metastable vacuum field configuration and φ_b is the bubble configuration, the bounce solution to the $O(3)$-symmetric Euclidean equation of motion. We must evaluate the partition functions above. This is done by the saddle-point method, expanding the scalar field $\phi(\mathbf{x}, \tau)$, such that $\phi(\mathbf{x}, \tau) \rightarrow \varphi_f + \zeta(\mathbf{x}, \tau)$ for $Z(\varphi_f)$, and $\phi(\mathbf{x}, \tau) \rightarrow \varphi_b(\mathbf{x}) + \eta(\mathbf{x}, \tau)$ for $Z(\varphi_b)$, where $\zeta(\mathbf{x}, \tau)$ and $\eta(\mathbf{x}, \tau)$ are small fluctuations about equilibrium.

It is crucial to note that the saddle-point, or Gaussian, method only gives good results if indeed the fluctuations about equilibrium are sufficiently small that nonlinear terms in the fields can be neglected. Even though the method sums over all amplitude fluctuations, it does so by assuming that the functional integral is well approximated by truncating the expansion of the action to second order. The efficiency of the method relies on the fact that higher amplitudes will be suppressed fast enough that their contribution to the partition function will be negligible. One can visualize this by comparing a sharp parabolic curve with a flatter one with minimum at x_0, and investigating when $\int dx e^{-f(x)}$ will be well approximated by writing $f(x) \simeq f(x_0) + \frac{1}{2}(x - x_0)^2 f''(x_0)$. For a sharp curve, larger amplitude fluctuations will be strongly suppressed and thus give a negligible contribution to the integral over all amplitudes. Clearly, this will not be the case for flatter curves.

Skipping details [11], using the saddle-point method one obtains for the ratio of partition functions, $\frac{Z(\varphi_b)}{Z(\varphi_f)}$,

$$\frac{Z(\varphi_b)}{Z(\varphi_f)} \overset{saddle-point}{\simeq} \left[\frac{\det(-\Box_E + V''(\varphi_b))_\beta}{\det(-\Box_E + V''(\varphi_f))_\beta}\right]^{-\frac{1}{2}} e^{-\Delta S}, \qquad (1.3)$$

where $[\det(M)_\beta]^{-\frac{1}{2}} \equiv \int D\eta \exp\left\{-\int_0^\beta d\tau \int d^3x \frac{1}{2}\eta[M]\eta\right\}$ and $\Delta S = S_E(\varphi_b) - S_E(\varphi_f)$ is the difference between the Euclidean actions for the field configurations φ_b and φ_f. [Note that $S_E(\varphi)$, and hence ΔS, does not include any temperature corrections. It would if one had summed over other fields coupled to φ.] Thus, the free energy of the system is,

$$\mathcal{F} = -T \left[\frac{\det(-\Box_E + V''(\varphi_b))_\beta}{\det(-\Box_E + V''(\varphi_f))_\beta}\right]^{-\frac{1}{2}} e^{-\Delta S}. \qquad (1.4)$$

Let me stress again the assumptions that go into computing the free energy. First, that the partition function is given by Eq. 1.2 within the dilute gas approximation, and second, that the partition function is evaluated approximately by assuming *small* fluctuations about the homogeneous metastable state φ_f. It is clear that for situations in which there are large amplitude fluctuations about the metastable equilibrium state the above formula must break down. Thus the breakdown of the expression for the rate is intimately connected with the question of how well-localized the system is about the metastable state as the temperature drops below the critical temperature T_c. Homogeneous nucleation, as its name already states, is only accurate when the metastable state is sufficiently homogeneous. In the presence of inhomogeneities, there is no reason to expect that the decay rate formula will apply. The question then is to quantify *when* does it break down and how can we incorporate nonperturbative corrections to the decay rate in a consistent way.

2   Nonperturbative Corrections to Decay Rates

In order to investigate the importance of large-amplitude fluctuations in the description of first order phase transitions, I have developed numerical simulations in two [13] and, with J. Borrill, three [14] spatial dimensions, which measured the fraction of the volume of the system in the initial phase as a function of the barrier height. Since these have been documented elsewhere, here I quickly describe the main idea and results.

Imagine a scalar field with a degenerate double-well potential. The field is coupled to a thermal bath at temperature T through a Langevin-like equation which assumes that the bath is Markovian, *i.e.*, the noise is white and additive. The system is artificially divided into two regions, left and right of the maximum of the potential, call it the negative and positive regions, respectively. The system is prepared initially in one of the regions, say, the negative region with $\phi(\mathbf{x}, t = 0) = -1$. The coupling to the bath will then drive fluctuations of the field around this minimum and we measure the fraction of the total volume in each of the two regions as a function of the parameters controlling the height of the potential barrier, usually the temperature and/or a coupling constant.

We observed that while for large enough potential barriers the system remained well-localized around its initial state, a sharp change of behavior occurred for a critical value of the specific control parameter being varied. In the case examined in Ref. [14], the control parameter was the quartic coupling of the scalar field, λ. We showed that for $\lambda > \lambda_c$ the system became completely mixed, in that the volume was equally shared by the positive and negative regions. In other words, for $\lambda > \lambda_c$, the effective potential describing the system is not a degenerate double-well, but a parabolic curve centered at $\langle \phi \rangle = 0$; Thermal fluctuations have "restored" the symmetry of the system.

The challenge was thus to model the large amplitude fluctuations which were responsible for this phase mixing. In what follows I review the so-called subcritical bubbles method which can model *quantitatively* the dynamics of large, nonperturbative, thermal fluctuations in scalar field theories.

LECTURE II

2.1 Modeling Nonperturbative Fluctuations: Symmetry Restoration and Phase Mixing

As was stressed before, the computation of decay rates based on homogeneous nucleation theory assumes a smooth metastable background over which critical bubbles of the lower free energy phase will appear, grow and coalesce, as the phase transition evolves. However, as the results from the numerical simulations indicate, the assumption of smoothness is not always valid. To the skeptical reader, I point out that several condensed matter experiments indicate that homogeneous nucleation fails to describe the transition when the nucleation barrier $(\Delta S/T)$ becomes too small. Furthermore, the agreement between theory and experiment has a long and problematic history [15]. Homogeneous nucleation has to be used with care, in a case by case basis.

The basic idea is that in a hot system, not only small but also large amplitude fluctuations away from equilibrium will, in principle, be present. Small amplitude fluctuations are perturbatively incorporated in the evaluation of the finite temperature effective potential, following well-known procedures. Large amplitude fluctuations probing the nonlinearities of the theory are not. Whenever they are important, the perturbative effective potential becomes unreliable. In an ideal world, we should be able to sum over all amplitude fluctuations to obtain the exact partition function of the model, and thus compute the thermodynamic quantities of interest. However, we can only to this perturbatively, and will always miss information coming from the fluctuations not included in its evaluation. If large amplitude fluctuations are strongly suppressed, they will not contribute to the partition function, and we are in good shape. But what if they are important, as argued above? We can try to approach this question avoiding complicated issues related to the evaluation of path integrals beyond the Gaussian approximation by obtaining a kinetic equation which describes the fraction of volume populated by these large amplitude fluctuations. In order to keep the treatment simple, and thus easy to apply, several assumptions are made along the way, which I believe are quite sensible. In any case, the strength of the method is demonstrated when the results are compared with the numerical experiments described before.

Large amplitude fluctuations away from equilibrium are modelled by Gaussian-profile spherically-symmetric field configurations of a given size and amplitude. They can be thought of as being coreless bubbles. Keeping with the notation of the numerical experiment, fluctuations away from the 0-phase [called the "negative phase" above], and into the 0-phase are written respectively as,

$$\phi_c(r) = \phi_c e^{-r^2/R^2} , \quad \phi_0(r) = \phi_c \left(1 - e^{-r^2/R^2}\right) , \qquad (2.5)$$

where R is the radial size of the configuration, and ϕ_c is the value of the amplitude at the bubble's core, away from the 0-phase. In previous treatments (cf. Refs. [16] and [17]), it was assumed that $\phi_c = \phi_+$, that is, that the configuration interpolated between the two minima of the effective potential, and that $R = \xi(T)$, where $\xi(T) = m(T)^{-1}$ is the mean-field correlation length. But in general, one should sum over all radii and amplitudes above given values which depend on the particular model under study. This will become clear as we go along.

Define $dn(R, \phi, t)$ as the number density of bubbles of radius between R and $R + dR$ at time t, with amplitudes $\phi \geq \phi_c$ between ϕ and $\phi + d\phi$. By choosing to sum over bubbles of amplitudes ϕ_c and larger, we are effectively describing the system as a "two-phase" system. For example, in the numerical simulation above it was assumed that the negative-phase was for amplitudes $\phi \leq \phi_{\max}$, and the positive-phase was for amplitudes $\phi > \phi_{\max}$. Clearly, for a continuous system this division is artificial. However, since the models we are interested in have two local minima of the free energy, this division becomes better justified. Fluctuations with small enough amplitude about the minima are already summed over in the computation of the effective potential. It is the large amplitude ones which are of relevance here. To simplify the notation, from now on I will denote by "+ phase" all fluctuations with amplitudes $\phi > \phi_c$ and larger. The choice of ϕ_c is model-dependent, as will be clear when we apply this formalism to specific examples.

The fact that the bubbles shrink will be incorporated in the time dependence for the radius R^1. Here, I will only describe a somewhat simplified approach to the dynamics. More details are provided in the work by Gleiser, Heckler, and Kolb [19]. The results, however, are essentially identical.

The net rate at which bubbles of a given radius and amplitude are created and destroyed is given by the kinetic equation,

$$\frac{\partial n(R,t)}{\partial t} = -\frac{\partial n(R,t)}{\partial R}\left(\frac{dR}{dt}\right) + \left(\frac{V_0}{V_T}\right)\Gamma_{0\to+}(R)$$
$$- \left(\frac{V_+}{V_T}\right)\Gamma_{+\to 0}(R) \qquad (2.6)$$

Here, $\Gamma_{0\to+}(R)$ $(\Gamma_{+\to 0}(R))$ is the rate per unit volume for the thermal nucleation of a bubble of radius R of positive-phase within the 0-phase (0-phase within the positive-phase). $V_{0(+)}$ is the volume available for nucleating bubbles of the $+(0)$ phase. Thus we can write, for the total volume of the system, $V_T = V_0 + V_+$, expressing the fact that the system has been "divided" into two available phases, related to the local minima of the free energy density. It is convenient to define the fraction of volume in the $+$ phase, γ, as

$$\frac{V_+}{V_T} \equiv \gamma = 1 - \frac{V_0}{V_T}. \qquad (2.7)$$

In order to compute γ we must sum over *all* bubbles of different sizes, shapes, and amplitudes within the $+$ phase, *i.e.*, starting with $\phi_{\min} \geq \phi_c$. Clearly, we cannot compute γ exactly. But it turns out that a very good approximation is obtained by assuming that the bubbles are spherically symmetric, and with radii above a given minimum radius, $R_{\min}$. The reason we claim that the approximation is good comes from comparing the results of this analytical approach with numerical simulations. The approximation starts to break down as the background becomes more and more mixed, and the morphology of the "bubbles" becomes increasingly more important, as well as other terms in the kinetic equation which were ignored. For example, there should be a term which accounts for bubble coalescence, which increases the value of γ. This term becomes important when the density of bubbles is high enough for the probability of two or more of them coalescing to be non-negligible. As we will see, by this point the mixing is already so pronounced that we are justified in neglecting this additional complication to the kinetic equation. As a bonus, we will be able to solve it analytically. The expression for γ is,

---

[1] Of course, the amplitude ϕ will also be time-dependent. However, its time-dependence is coupled to that of the radius, as recent studies have shown [18]. In order to describe the effect of shrinking on the population of bubbles it is sufficient to include only the time dependence of the radius.

$$\gamma \simeq \int_{\phi_{\min}}^{\infty} \int_{R_{\min}}^{\infty} \left(\frac{4\pi R^3}{3}\right) \frac{\partial^2 n}{\partial \phi \partial R} d\phi dR \ . \tag{2.8}$$

The attentive reader must have by now noticed that we have a coupled system of equations; γ, which appears in the rate equation for the number density n, depends on n itself. And, to make things even worse, they both depend on time. Approximations are in order, if we want to proceed any further along an analytical approach. The first thing to do is to look for the equilibrium solutions, obtained by setting $\partial n/\partial t = 0$ in the kinetic equation. In equilibrium, γ will also be constant[2]. If wished, after finding the equilibrium solutions one can find the time-dependent solutions, as was done in Ref. [17]. Here, we are only interested in the final equilibrium distribution of subcritical bubbles, as opposed to the approach to equilibrium.

The first approximation is to take the shrinking velocity of the bubbles to be constant, $dR/dt = -v$. This is in general not the case (cf. Ref. [18]), but it does encompass the fact that subcritical bubbles shrink into oblivion. The strength of the thermodynamic approach is that details of how the bubbles disappear are unimportant, only the time-scale playing a rôle. The second approximation is to assume that the rates for creation and destruction of subcritical fluctuations are Boltzmann suppressed, so that we can write them as $\Gamma = AT^4 e^{-F_{sc}/T}$, where A is an arbitrary constant of order unity, and $F_{sc}(R, \phi_c)$ is the cost in free energy to produce a configuration of given radius R and core amplitude ϕ_c. For the Gaussian ansatz we are using, F_{sc} assumes the general form, $F_{sc} = \alpha R + \beta R^3$, where $\alpha = b\phi_c^2$ (b is a combination of π's and other numerical factors) and β depends on the particular potential used. In practice, the cubic term can usually be neglected, as the free energy of small $(R \sim \xi)$ subcritical bubbles is dominated by the gradient (linear) term. We chose to look at the system at the critical temperature T_c. For this temperature, the creation and destruction rates, $\Gamma_{0 \to +}$ and $\Gamma_{+ \to 0}$ are identical. Also, for T_c, the approximation of neglecting the cubic term is very good (in fact it is better and better the larger the bubble is) even for large bubbles, since for degenerate vacua there is no gain (or loss) of volume energy for large bubbles. Finally, we use that $V_+/V_T = \gamma$ in the $\Gamma_{+ \to 0}$ term. A more sophisticated approach is presented in Ref. [19].

We can then write the equilibrium rate equation as,

$$\frac{\partial n}{\partial R} = -cf(R) \ , \tag{2.9}$$

where,

$$c \equiv (1 - 2\gamma)AT^4/v, \quad f(R) \equiv e^{-F_{sc}/T} \ . \tag{2.10}$$

---

[2] This doesn't mean that thermal activity in or between the two phases is frozen; equilibrium is a statement of the average distribution of thermodynamical quantities. Locally, bubbles will be created and destroyed, but always in such a way that the average value of n and γ are constant.

Integrating from $R_{\min}$ and imposing that $n(R \to \infty) = 0$, the solution is easily found to be,

$$n(R) = \frac{c}{\alpha(\phi_c)/T} e^{-\alpha(\phi_c)R/T} . \tag{2.11}$$

Not surprisingly, the equilibrium number density of bubbles is Boltzmann suppressed. But we now must go back to γ, which is buried in the definition of c. We can solve for γ perturbatively, by plugging the solution for n back into Eq. 2.8. After a couple of fairly nasty integrals, we obtain,

$$\gamma = \frac{g\left(\alpha(\phi_{\min}), R_{\min}\right)}{1 + 2g\left(\alpha(\phi_{\min}), R_{\min}\right)} , \tag{2.12}$$

where,

$$g\left(\alpha(\phi_{\min}), R_{\min}\right) = \frac{4\pi}{3}\left(\frac{AT^4}{v}\right)\left(\frac{T}{\alpha}\right)^3 \frac{e^{-\alpha R_{\min}/T}}{\alpha/T} \times$$

$$\times \left[6 + \left(\frac{\alpha R_{\min}}{T}\right)^3 + 3\frac{\alpha R_{\min}}{T}\left(2 + \frac{\alpha R_{\min}}{T}\right)\right] \tag{2.13}$$

We can now apply this formalism to any model we wish. The first obvious application is to compare γ obtained from the numerical experiment with the value obtained from the kinetic approach. From the definition of the equilibrium fractional population difference, $\Delta F_{\mathrm{EQ}}(\theta_c) \equiv f_0^{\mathrm{eq}} - f_+^{\mathrm{eq}}$,

$$\Delta F_{\mathrm{EQ}}(\theta_c) = 1 - 2\gamma . \tag{2.14}$$

Thus, it is straightforward to extract the value of γ from the numerical simulations as a function of λ. Also, as we neglected the volume contribution to the free energy of subcritical bubbles, we have,

$$F_{sc} = \alpha(\phi_c)R_{\min} = \frac{3\sqrt{2}}{8}\pi^{3/2}X_-^2(\theta_c)R_{\min} , \tag{2.15}$$

where, as you recall, X_- is the position of the maximum of the mean-field potential used in the simulations. So, we must sum over all amplitudes with $X \geq X_-$, and all radii with $R \geq 1$ (in dimensionless units), as we took the lattice spacing to be $\ell = 1$. That is, we sum over all possible sizes, down to the minimum cut-off size of the lattice used in the simulations. In practice, we simply substitute $\phi_c = X_-$ and $R_{\min} = 1$ in the expression for γ. In Fig. 1, we compare the numerical results for γ (dots) with the results from the analytical integration of the kinetic equation. The plots are for different values of the parameter A/v. Up to the critical value for $\lambda \simeq 0.025$, the agreement is very convincing. As we increase λ into the mixed phase region of the diagram, the kinetic approach underestimates the amount of volume in the $+$-phase. This is not surprising, since for these values of λ the density of subcritical bubbles is high enough that terms not included in the equation become important, as I mentioned before. However, the lack of agreement for higher values of λ is irrelevant, if we are interested in having a measure of the smoothness of the background; clearly, the rise in γ is sharp enough that homogeneous nucleation should not be trusted for $\lambda > 0.024$ or so, as the fraction of volume occupied by the +-phase is already around 30% of the total volume. Subcritical bubbles give a simple and quantitatively accurate picture of the degree of inhomogeneity of the background, offering a guideline as to when homogeneous nucleation theory can be applied with confidence, or, alternatively, when the effective potential needs higher order corrections.

2.2   Modeling Nonperturbative Fluctuations: "Inhomogeneous" Nucleation

Now we apply the subcritical bubbles method to the decay of metastable states in the case that the homogeneous nucleation formalism (section I) does not apply. Details of this work can be found in Ref. [21].

As mentioned before, if there is significant phase mixing in the background metastable state, its free-energy density is no longer $V(\phi = 0)$, where I assume the potential has a metastable state at $\phi = 0$. One must also account for the free-energy density of the nonperturbative, large-amplitude fluctuations. Since there is no formal way of deriving this contribution outside improved perturbative schemes, we propose to estimate the corrections to the background free-energy density by following another route. We start by writing the free energy density of the metastable state as $V(\phi = 0) + \mathcal{F}_{sc}$, where $\mathcal{F}_{sc}$ is the nonperturbative contribution to the free-energy density due to the large amplitude fluctuations, which we assume can be modelled by subcritical bubbles. We will calculate $\mathcal{F}_{sc}$ further below.

We thus define the effective free-energy difference between the two minima, ΔV_{cg}, which includes corrections due to phase mixing, as

$$\Delta V_{cg} = \Delta V_0 + \mathcal{F}_{sc} \tag{2.16}$$

which is the sum of the free-energy difference calculated in the standard way, and the "extra" free-energy density due to the presence of subcritical bubbles. Henceforth, the subscript 'cg' will stand for "coarse-grained".

Since for degenerate potentials (temperature-dependent or not) no critical bubbles should be nucleated, taking into account subcritical bubbles must lead to a change in the coarse-grained free-energy density (or potential) describing the transition. Thus, it should be possible to translate the "extra" free energy available in the system due to the presence of subcritical bubbles in the background into a corrected potential for the scalar order parameter. We will write this corrected potential as $V_{cg}(\phi)$.

The standard coarse-grained free energy is calculated by integrating out the short-wavelength modes (usually up to the correlation length) from the

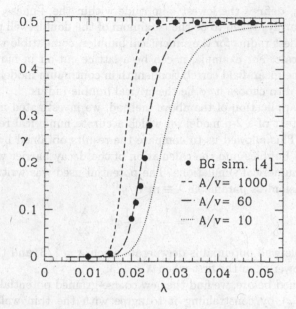

Fig. 1. The fraction of the volume in the + phase. The dots are from numerical simulations of Ref. 14, while the lines are the solutions of the Boltzmann eq. for different values of the parameter $A/|v|$.

partition function of the system, and is approximated by the familiar form [20]

$$F_{cg}(\phi) = \int d^3r \left(\frac{1}{2}(\nabla\phi)^2 + V_{cg}(\phi) \right) . \tag{2.17}$$

How do we estimate V_{cg}? One way is to simply constrain it to be consistent with the thin wall limit. That is, as $V_{cg}(\phi)$ approaches degeneracy (*i.e.* $\Delta V_{cg}(\phi) \to 0$), it must obey the thin wall limit of eq. (2.16). Note that with a simple rescaling, a general polynomial potential (to fourth order) can be written in terms of one free parameter. Thus, the thin wall constraint can be used to express the corrected value of this parameter in terms of $\mathcal{F}_{sc}$ in appropriate units. The free energy of the critical bubble is then obtained by finding the bounce solution to the equation of motion $\nabla^2\phi - dV_{cg}(\phi)/d\phi = 0$ by the usual shooting method, and substituting this solution into eq. (2.17).

In order to determine V_{cg}, we must first calculate the free-energy density $\mathcal{F}_{sc}$ of the subcritical bubbles. From the formalism presented in the previous subsection,

$$\mathcal{F}_{sc} \approx \int_{\phi_{min}}^{\infty} \int_{R_{min}}^{R_{max}} F_{sb} \frac{\partial^2 n(R,t)}{\partial R \partial \phi_A} dR d\phi_A, \tag{2.18}$$

where ϕ_{min} defines the lowest amplitude within the +phase, typically (but not necessarilly) taken to be the maximum of the double-well potential. R_{min} is the smallest radius for the subcritical bubbles, compatible with the coarse-graining scale. For example, it can be a lattice cut-off in numerical simulations, or the mean-field correlation length in continuum models. As for R_{max}, it is natural to choose it to be the critical bubble radius.

As an application of the above method, we investigated nucleation rates in the context of a 2-d model for which accurate numerical results are available [22]. This allowed us to compare the results obtained by incorporating subcritical bubbles into the calculation of the decay barrier with the results from the numerical simulations. The potential used was written in terms of one dimensionless parameter $\lambda \equiv m^2 h/g^2$,

$$V(\phi) = \frac{1}{2}\phi^2 - \frac{1}{6}\phi^3 + \frac{\lambda}{24}\phi^4. \tag{2.19}$$

This double-well potential is degenerate when $\lambda = 1/3$, and the second minimum is lower than the first when $\lambda < 1/3$.

As argued before, we find the new coarse-grained potential V_{cg} (or, equivalently, λ_{cg}) by constraining it to agree with the thin wall limit. Simple algebra from eqs. (2.16) and (2.19) yields, to first order in the deviation from degeneracy,

$$\lambda_{cg} = \lambda - \frac{\tilde{\mathcal{F}}_{sc}}{54} \tag{2.20}$$

where $\tilde{\mathcal{F}}_{sc} = \frac{g^2}{m^6}\mathcal{F}_{sc}$ is the dimensionless free-energy density in subcritical bubbles. The new potential V_{cg} is then used to find the bounce solution and the free energy of the critical bubble.

In Fig. 2 we show that the calculation of the nucleation barrier including the effects of subcritical bubbles is consistent with data from lattice simulations, whereas the standard calculation overestimates the barrier by a large margin. In fact, the inclusion of subcritical bubbles provides a reasonable explanation for the anomalously high nucleation rates observed in the simulations close to degeneracy.

3   Matching Numerical Simulations to Continuum Field Theories

As a final topic to be discussed in this lecture, I would like to change gears and briefly turn to the issue of how to match numerical simulations of field theories with their continuum counterparts. In particular, I am interested in situations where a degree of stochasticity is present in the simulations, as for example happens when we model the coupling of fields to a thermal or quantum bath via a Langevin-like equation, or even in the form of noisy initial conditions.

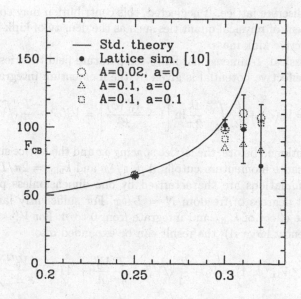

Fig. 2. Comparison between numerical data and theoretical predictions for the decay barrier with and without the inclusion of subcritical bubbles. The parameter a is related to an extra term in the Boltzmann eq. which can be safely neglected.

Although field theories are continuous and usually formulated in an infinite volume, lattice simulations are discrete and finite, imposing both a maximum ("size of the box" L) and a minimum (lattice spacing δx) wavelength that can be probed by the simulation. When the system is coupled to an external thermal (or quantum) bath, fluctuations will be constrained within the allowed window of wavelengths, leading to discrepancies between the continuum formulation of the theory and its lattice simulations; the results will be dependent on the choice of lattice spacing.

Parisi suggested that if proper counterterms were used, this depedence on lattice spacing could be attenuated [23]. Recently, Borrill and Gleiser (BG) have examined this question within the context of 2-d critical phenomena [24]. They have computed the counterterms needed to render the simulations indepedent of lattice spacing and have obtained a match between the simulations and the continuum field theory, valid within the one-loop approximation used in their approach. Here, I want to focus mostly on the application of these techniques to 1-d field theories, in particular to the problem of thermal nucleation of kink-antikink pairs. [This is based on work of Ref. [25].]

Even though 1-d field theories are free of ultra-violet divergences, the ultra-violet cutoff imposed by the lattice spacing will generate a *finite* contribution to the effective potential which must be taken into account if we are to obtain a proper match between the theory and its numerical simulation on a discrete lattice. If neglected, this contribution may compromise the measurement of physical quantities such as the density of kink-antikink pairs or the effective kink mass.

For classical, 1-dimensional finite-temperature field theories, the one-loop corrected effective potential is given by the momentum integral [23]

$$V_{1L}(\phi) = V_0(\phi) + \frac{T}{2} \int_0^\infty \frac{dk}{2\pi} \ln \left[1 + \frac{V_0''(\phi)}{k^2} \right] = V_0(\phi) + \frac{T}{4} \sqrt{V_0''(\phi)} . \quad (3.21)$$

As mentioned before, the lattice spacing δx and the lattice size L introduce long and short momentum cutoffs $\Lambda = \pi/\delta x$ and $k_{\min} = 2\pi/L$, respectively. Lattice simulations are characterized by one dimensionless parameter, the number of degrees of freedom $N = L/\delta x$. For sufficiently large L one can neglect the effect of $k_{\min}$ and integrate from 0 to Λ. For $V_0'' \ll \Lambda^2$ (satisfied for sufficiently large Λ), the result can be expanded into

$$V_{1L}(\phi, \Lambda) = V_0 + \frac{T}{4} \sqrt{V_0''} - \frac{T}{4\pi} \frac{V_0''}{\Lambda} + \Lambda T \, \mathcal{O} \left(\frac{V_0''^2}{\Lambda^4} \right) . \quad (3.22)$$

As is to be expected for a 1-dimensional system, the limit $\Lambda \to \infty$ exists and is well-behaved; there is no need for renormalization of ultra-violet divergences. However, the effective one-loop potential is lattice-spacing dependent through the explicit appearance of Λ, and so are the corresponding numerical simulations. In order to remove this dependence on δx, we follow the renormalization procedure given by BG [24]; it is irrelevant if the Λ-dependent terms are ultra-violet finite ($d = 1$) or infinite ($d \geq 2$). In the lattice formulation of the theory, we add a (finite) counterterm to the tree-level potential V_0 to remove the lattice-spacing dependence of the results,

$$V_{ct}(\phi) = \frac{T}{4\pi} \frac{V_0''(\phi)}{\Lambda} . \quad (3.23)$$

There is an additional, Λ-independent, counterterm which was set to zero by an appropriate choice of renormalization scale. The lattice simulation then uses the corrected potential

$$V_{latt}(\phi) = V_0(\phi) + \frac{T\delta x}{4\pi^2} V_0''(\phi) . \quad (3.24)$$

Note that the above treatment yields two novel results. First, that the use of V_{latt} instead of V_0 gets rid of the dependence of simulations on lattice spacing. [Of course, as $\delta x \to 0$, $V_{latt} \to V_0$. However, this limit is often not computationally efficient.] Second, that the effective interactions that are simulated must be compared to the one-loop corrected potential $V_{1L}(\phi)$ of Eq. (3.21); once the lattice formulation is made independent of lattice spacing by the addition of the proper counterterm(s), it simulates, within its domain of validity, the thermally corrected one-loop effective potential.

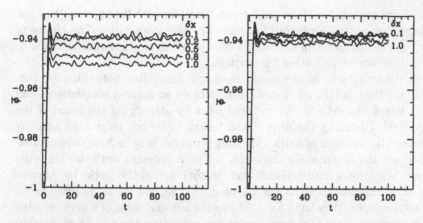

Fig. 3. Average field value $\bar{\phi}(t)$ for $T = 0.1$ using the tree-level potential, left, and the corrected potential, right. The filter cutoff is $\Delta L = 3$.

Applying this method to the formation of kink-antikink pairs, we get a corrected potential,

$$V_{\text{latt}}(\phi) = V_0(\phi) + \frac{3}{4\pi^2}\lambda T \delta x \phi^2 \; ; \tag{3.25}$$

simulations using V_{latt} will, in principle, match the continuum theory

$$V_{1L}(\phi) = V_0(\phi) + \frac{T\sqrt{\lambda}}{4}\sqrt{3\phi^2 - \phi_0^2} \; , \tag{3.26}$$

which has (shifted) minima at $\pm\phi_{\min}(T)$, with $\phi_{\min}(T) < \phi_0$.

As a first test of our procedure, we investigate the mean field value $\bar{\phi}(t) = (1/L)\int \phi(x,t)dx$ *before* the nucleation of a kink-antikink pair, *i.e.*, while the field is still well localized in the bottom of the well. In Fig. 3 we show the ensemble average of $\bar{\phi}$ (after 100 experiments) for different values of δx, ranging from 1 down to 0.1, at $T = 0.1$. The simulations leading to the left graphs use the "bare" potential V_0, whereas the right graphs are produced employing V_{latt} (Eq. 3.25).

Perhaps the most difficult task when counting the number of kink-antikink pairs that emerge during a simulation is the identification of what precisely is a kink-antikink pair at different temperatures. Typically, we can identify three "types" of fluctuations: i) small amplitude, perturbative fluctuations about one of the two minima of the potential; ii) full-blown kink-antikink pairs interpolating between the two minima of the potential; iii) nonperturbative fluctuations which have large amplitude but not quite large enough to satisfy the boundary conditions required for a kink-antikink pair. These latter fluctuations are usually dealt with by a smearing of the field over a certain length scale. Basically, one chooses a given smearing length ΔL which will be large enough to "iron out" these "undesirable" fluctuations but not too large that actual kink-antikink pairs are also ironed-out. The choice of ΔL is, in a sense, more an art than a science, given our ignorance of how to handle these nonperturbative fluctuations.

The smearing was implemented as a low pass filter with filtering cutoff ΔL; the field is Fourier transformed, filtered at a given wavelength, and Fourier transformed back. We counted pairs by identifying the zeros of the filtered field. Choosing the filter cutoff length to be too large may actually undercount the number of pairs. Choosing it too low may include nonpertubative fluctuations as pairs. We chose $\Delta L = 3$ in the present work, as this is the smallest "size" for a kink-antikink pair. In contrast, in the works by Alexander et al. a different method was adopted, that looked for zero-crossings for eight lattice units (they used $\delta x = 0.5$) to the left and right of a zero crossing [26]. We have checked that our simulations reproduce the results of Alexander et al. if we: i) use the bare potential in the lattice simulations and ii) use a large filter cutoff length ΔL. Specifically, the number of pairs found with the bare potential for $T = 0.2$, $\delta x = 0.5$ are: $n_p = 36$, 30, and 27, for $\Delta L = 3$, 5, and 7 respectively. Alexander et al. found (for our lattice length) $n_p = 25$. Comparing results for different ΔL, it is clear that the differences between our results and those of Alexander et al. come from using a different potential in the simulations, *viz.* a dressed vs. a bare potential. For small δx these differences disappear.

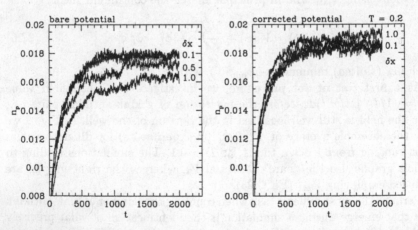

Fig. 4. Density of kink-antikinks (half of density of zeros), for $T = 0.2$ and $\delta x = 1$, 0.5, 0.2, and 0.1. The filter cutoff is $\Delta L = 3$.

Fig. 4 compares measurements of the kink-antikink pair density (half the number of zeros of the smeared field), ensemble-averaged over 100 experiments, for different lattice spacings. Again it is clear from the graphs on the left that using the tree-level potential V_0 in the simulations causes the results to be dependent on δx, whereas the addition of the finite counterterm removes this problem quite efficiently.

Another step is to establish what is the continuum theory being simulated. Due to space limitations, I refer the reader to the work of Ref. [25] for more details.

This work was written in part while the author was visiting the Osservatorio Astronomico di Roma. I thank Franco Occhionero and Luca Amendola for their kind hospitality. The author was partially supported by the National Science Foundation through a Presidential Faculty Fellows Award no. PHY-9453431 and by the National Aeronautics and Space Administration through grant no. NAGW-4270.

References

[1] E. W. Kolb and M. S. Turner (1990): *The Early Universe* (Addison-Wesley, New York, 1990).

[2] A. G. Cohen, D. B. Kaplan, and A. E. Nelson, *Annu. Rev. Nucl. Part. Sci.* **43**, 27 (1993); A. Dolgov, *Phys. Rep.* **222**, 311 (1992).

[3] B. Liu, L. McLerran, and N. Turok, *Phys. Rev.* **D46**, 2668 (1992); A. G. Cohen, D. B. Kaplan, and A. E. Nelson, *Phys. Lett.* **B336**, 41 (1994); M. B. Gavela, P. Hernández, J. Orloff, and O. Pène, *Mod. Phys. Lett.* **A9**, 795 (1994); P. Huet and E. Sather, *Phys. Rev.* **D51**, 379 (1994).

[4] R. Becker and W. Döring, *Ann. Phys.* **24**, 719 (1935).

[5] J. W. Cahn and J. E. Hilliard, *J. Chem. Phys.* **31**, 688 (1959).

[6] J. S. Langer, *Ann. Phys. (NY)* **41**, 108 (1967); *ibid.* **54**, 258 (1969).

[7] J. D. Gunton, M. San Miguel and P. S. Sahni, in *Phase Transitions and Critical Phenomena*, **Vol. 8**, Ed. C. Domb and J. L. Lebowitz (Academic Press, London, 1983).

[8] M. B. Voloshin, I. Yu. Kobzarev, and L. B. Okun', *Yad. Fiz.* **20**, 1229 (1974) [Sov. J. Nucl. Phys. **20**, 644 (1975)].

[9] S. Coleman, *Phys. Rev.* **D15**, 2929 (1977); C. Callan and S. Coleman, *Phys. Rev.* **D16**, 1762 (1977).

[10] A. D. Linde, *Phys. Lett.* **70B**, 306 (1977); *Nucl. Phys.* **B216**, 421 (1983); [Erratum: **B223**, 544 (1983)].

[11] M. Gleiser, G. Marques, and R. Ramos, *Phys. Rev.* **D48**, 1571 (1993); D. E. Brahm and C. Lee, *Phys. Rev.* **D49**, 4094 (1994); D. Boyanovsky, D. E. Brahm, R. Holman, and D.-S. Lee, *Nucl. Phys.* **B441**, 609 (1995).

[12] P. Arnold and L. McLerran, *Phys. Rev.* **D36**, 581 (1987); *ibid.* **D37**, 1020 (1988).

[13] M. Gleiser, *Phys. Rev. Lett.* **73**, 3495 (1994).

[14] J. Borrill and M. Gleiser, *Phys. Rev.* **D51**, 4111 (1995).

[15] E.D. Siebert and C.M. Knobler, *Phys. Rev. Lett.*, **52**, 1133 (1984); J.S. Langer and A.J. Schwartz, *Phys. Rev.* **A21**, 948 (1980); A. Leggett in *Helium Three*, ed. by W.P. Halperin and L.P. Pitaevskii, (North-Holland,

New York, 1990); for an (outdated) review of the situation in the early eighties see Ref. [7].

[16] M. Gleiser, E. W. Kolb, and R. Watkins, *Nucl. Phys.* **B364**, 411 (1991); G. Gelmini and M. Gleiser, *Nucl. Phys.* **B419**, 129 (1994); M. Gleiser and E. W. Kolb, Phys. Rev. Lett. **69**, 1304 (1992); N. Tetradis, *Z. Phys.* **C57**, 331 (1993).

[17] G. Gelmini and M. Gleiser in Ref. [16].

[18] M. Gleiser, Phys. Rev. **D49**, 2978 (1994); E.J. Copeland, M. Gleiser, and H.-R. Müller, Phys. Rev. **D52**, 1920 (1995).

[19] M. Gleiser, A. Heckler, and E.W. Kolb, Phys. Lett. B405 (1997) 121.

[20] J.S. Langer, Physica **73**, 61 (1974).

[21] A. Heckler and M. Gleiser, Phys. Rev. Lett. **76**, 180–183 (1996).

[22] M. Alford and M. Gleiser, Phys. Rev. **D48**, 2838–2844 (1993).

[23] G. Parisi, *Statistical Field Theory* (Addison-Wesley, New York, 1988).

[24] J. Borrill and M. Gleiser, Nucl. Phys. **B483**, 416–428 (1997).

[25] M. Gleiser and H.-R. Müller, Dartmouth preprint no. DART-HEP-97/04, hep-lat/9704005.

[26] F. J. Alexander and S. Habib, Phys. Rev. Lett. **71**, 955 (1993); F. J. Alexander, S. Habib, and A. Kovner, Phys. Rev. E **48**, 4282 (1993); S. Habib, cond-mat/9411058.

Neural Networks
and Confidence Limit Estimates

Bernd A. Berg[1,2,5] and Ion-Olimpiu Stamatescu[3,4,5]

[1] Department of Physics, Florida State University, Tallahassee, FL 32306, USA
[2] Supercomputer Computations Research Institute,
Florida State University, Tallahassee, FL 32306, USA
[3] FEST, Schmeilweg 5, D-69118 Heidelberg, Germany
[4] Institut für Theoretische Physik, Universität Heidelberg,
D-69120 Heidelberg, Germany
[5] ZIF, Wellenberg 1, D-33615 Bielefeld, Germany

Abstract. We give a brief review of neural networks. This includes views and features about the general concept, the modelling problem for Neural Networks and their treatment with statistical physics methods. Their use as Expert Systems is illustrated for the application of Multilayer Perceptrons to High Energy Physics data analysis. Finally, we focus on recent statistical insight, which under certain conditions allows to obtain rigorous confidence limit estimates from a Neural Network analysis.

1 Views and Features of Neural Networks

There are various points of view concerning neural network models (in the following we shall generally use the shortening NN, sometimes ANN when we want to stress the "artificial" aspect):
- of cognition science, as models for cognitive processes;
- of artificial intelligence, as models for information processing devices;
- of physics, as statistical mechanics models;
- of neurophysiology, as models for brain processes.

While until the eighties the dominant paradigm in Artificial Intelligence (AI) and Cognition was the so called "physical-symbol-systems", the neural networks have themselves a long history, starting with the *perceptron* concept (Rosenblatt 1962). It was, however, only in the last decades, that developments in neurophysiology, on one hand, and in the predictability and characterization of artificial networks, on the other hand, have permitted to set up a systematic modeling frame. There is very much literature, of course, in connection with neural networks and related problems (to quote only a few books: McClelland and Rumelhart 1986, Kohonen 1989, Churchland 1989, Amit 1989, Boden 1990, Bechtel and Abrahamsen 1991; Hertz, Krogh and Palmer 1991, Wasserman 1993, Domany et al. 1994, ...).

From the point of view of cognition neural networks correspond to a new paradigm, called "connectionism". In contradistinction with the "physical-symbol-systems" paradigm (corresponding to the "standard AI"), which is a top-down approach, "connectionism" represent a bottom-up approach in which "intelligent information processing" need not be *designed* but *results* from the activity of big assemblies of multiply interconnected, simple units. A cognitive science view is to consider NN as modeling "the micro-structure of cognitions" (McClelland et al. 1986). This makes NN adequate for modeling associative recognition of visual patterns (with the neurons coding the pixels), as well as for modeling semiotic processes, e.g., learning languages (with neurons coding words). In this connection the interesting features of neural networks appear to be:

- the "self – organization" aspect,
- the inherent associative character of their information flows,
- the natural way of accounting for multiple simultaneous constraints,
- fuzziness,
- stability under small disturbances,
- easiness in adaptation to new requirements, etc.

Of course, there are also disadvantages, some of them ensuing from just these advantages, e.g., the fuzziness. For applied AI neural networks represent devices which adapt themselves to the given problem. It is this aspect, together with their flexibility and their special capacity of treating problems with many constraints that have promoted ANN to important information processing devices. However, just because of this it is much more difficult to prove existence of solutions, to asses the time of convergence and to estimate the possible errors and their consequences. Some results have been obtained hereto, many of them due to application of methods from statistical mechanics.

The statistical physics interest for ANN is related to the phase structure of their state space. Some of these phases are characterized by a strongly degenerated ground state structure, typical for frustrated systems. It is just this aspect that is related to the pattern classification and associative recognition capabilities of these systems used as information processing devices. This approach has introduced a new perspective in AI, namely the statistical one. Before entering these matters, we want to sketch the "neurophysiology connection", since this is not only important for itself: in fact a good part of the interest in ANN is based on the assumed relation between their functioning and that of the brain.

Detailed information about the brain and its activity is provided by the modern neurophysiology (see, e.g., Kandel et al. 1995). The brain is a highly interconnected system (for the about 10^{11} neurons there are between a few and as many as about 10^5 synapses per neuron - one quotes an average synapse number of 10^4). We find:

- mostly threshold neurons (with essentially two states: active/inactive; the threshold behavior is however only approximative); but there are also linear response neurons;

- local as well as distant connections, which can be bundled, divergent or convergent;
- non-modifiable and modifiable synapses, and the modifiable synapses can show non-associative and associative modulation, with short term and long term effects;
- excitatory (mostly axo-dendritic), inhibitory (mostly axo-somatic) and modulatory synapses (axo-axonic: local modulation of the synapse from a second to a third neuron);
- slow and fast synapses (in fact, a wide distribution of synaptic delays);
- electric and chemical synapses, three important types of ions channels among very many others involved in the synaptic transmission, with a large variety of functions;
- biased topological mapping intermixed with functionally determined arrangements (e.g., visual field mapping and orientation columns in the visual cortex);
- various hierarchies of neurons and group structures with alternate interconnectivity and projection from one group into another, sometimes one-directional, sometimes with feedback.

Many of these characteristic features are believed to be represented in the construction of ANN. However, we would like to stress the observation – which is apparent from the above – that together with some general, more or less invariant features (like the predominance of threshold behaviour, high interconnectivity, synapse plasticity) there is a strong diversity of features, such that one could say that *for practically each problem there seem to be at least two – but typically many more – mechanisms or implementations.* To the extent that we want to use the "brain paradigm" for ANN modeling it is very important to try to asses correctly which part of the variability hinted at above is functional, which part represents stabilizing fluctuations (and therefore in a different sense is again functional) and which is uncontrollable, mostly harmless "noise". This is to say, that both the use of NN for modeling brain activity *and* the use of neuro-physiological "hints" for developing ANN is not straightforward.

In what follows we shall first briefly describe the modeling problem for ANN and the main results hereto, then consider an application to automatic analysis of noisy experimental data and see how can we – in this particular example – control the inherent fallibility in the functioning of an ANN by turning it into limits for confidence levels for the results.

2   The Modeling Problem for Neural Networks

Typical problems for ANN modeling are:
- classify, memorize and retrieve patterns;
- reproduce associative structures between complex stimuli;
- develop optimal behavior under complex constraints.

These problems are related and in fact one can characterize the capacities of ANN by restricting oneself to one of these problems, namely that of storing and associative retrieving of patterns. This is at the same time one of their primary uses as information processing devices (notice that pattern classi- fication and generalizations questions can be put in the same context – of associative recognition).

In establishing a neural network model we must define:

(a) the architecture, *i.e.* the type of units and their connections,

(b) the functioning, *i.e.* the rules controlling the activity of the network, and

(c) the learning (or training) algorithms, by which the desired response or behavior is ensured.

2.1 Architecture

The main type of neuron is the non-linear response unit. It is characterized by some kind of "sigmoid" function g, which describes the output V (one may think in terms of voltage) produced by submitting the neuron to a certain input h:

$$V = g(h) \tag{2.1}$$

e.g.

$$g(h) = c \tanh(a(h - b)) + d \tag{2.2}$$

In the limit of a very steep sigmoid we recover the two-state unit, usually termed Ising neuron if the $\sigma = \{\pm 1\}$ representation is used to describe the states *active/inactive* or McCulloch-Pitts neuron (McCulloch and Pitts 1943) if the $s = \{1, 0\}$ representation is used instead. There is, of course, a simple transformation rule between the various representations: $\sigma = 2s - 1$. How- ever, different functioning or training laws prefer different representations in the sense that they take a more simple and intuitive form in the preferred representation.

These units[1] introduce a local non-linearity by transforming a more or less linear excitation potential into a threshold controlled response. They make synapses on other units and the connectivity characteristics (fully or partially connected nets, hierarchical or sheet-type organization of the connections, symmetric or asymmetric connections, etc.) defines the architecture of the network.

---

[1] There are, of course, also other kind of units, as linear response, multi-state units, etc. We shall not discuss them here, since they either cannot be used in networks to produce complex functions – like the former – or do not bring in genuine new kind of behavior – like the latter.

2.2 Functioning

The functioning of the network is defined in the following way: the synapses made by the neurons $\{j\}$ on a neuron i produce signals $C_{ij}V_j$ depending on the outputs of j, V_j and on the "synaptic weights" C_{ij}. All these signals then add to an input for the neuron i

$$h_i = \sum_{j=1}^{N} C_{ij}V_j \tag{2.3}$$

which then responds according to (2.1). Here N is the number of neurons making synapses on i.

The biological similitude is most easily seen in the McCulloch-Pitts representation: firing of neuron j propagates and generates a presynaptic potential C_{ij} at the neuron i (C_{ij} can be positive or negative, according to whether the synapse ij is excitatory or inhibitory). The signals from all synapses at i add to form a post-synaptic potential:

$$h_i = \sum_{j=1}^{N} C_{ij}s_j, \quad s_j = 0,1 \tag{2.4}$$

This potential then may trigger a firing of neuron i according to the rule:

$$s_i = \theta\left(h_i - \theta_i\right) \tag{2.5}$$

with θ_i a threshold characterizing neuron i. In the Ising representation this reads:

$$h_i = \sum_{j=1}^{N} C_{ij}\sigma_j, \quad \sigma_j = \pm 1 \tag{2.6}$$

and correspondingly:

$$\sigma_i = \epsilon\left(h_i - \theta_i\right) \tag{2.7}$$

(of course the synapses and thresholds have to be re-adapted when changing the representation).

This deterministic procedure can be replaced by a stochastic process, e.g. by introducing a noise component in the thresholds θ_i, e.g.:

$$\theta_i = \theta_i^0 + \beta\omega_i, \quad \overline{\omega_i} = 0, \quad \overline{\omega_i\omega_j} = \delta_{ij}$$

(a construction akin to the Langevin equation of Brownian motion and therefore called sometimes Langevin dynamics) or by replacing the rule (2.7) with a probabilistic one, e.g.:

$$\sigma_i = 1 \text{ with probability } p = \frac{1}{1 + e^{-2\beta(h_i - \theta_i)}} \tag{2.8}$$

(the so called Glauber dynamics) or

$$\sigma_i = 1 \text{ with probability } p = \min\left[1, e^{-\beta(h_i - \theta_i)}\right] \tag{2.9}$$

(the so called Metropolis dynamics) – in both cases $\sigma_i = -1$ with probability $1 - p$. In (2.8) and (2.9) $1/\beta$ represents a kind of "temperature". The zero – temperature limit is the deterministic rule (2.5), (2.7).

The process can go synchronously (at each step all neurons fire simultaneously) or asynchronously, in particular, serially[2]. Nontrivial time dependence may be introduced by assigning synapse delays d_{ij} and letting the post-synaptic signal decay in time with some monotonically decreasing function $e(t)$ such that:

$$h_i(t) = \sum_j C_{ij}\sigma_j e(t - t_j - d_{ij}), \quad e(t) = 0 \text{ for } t < 0 \tag{2.10}$$

The synapses (the weights C_{ij}) may be taken symmetrical or asymmetrical. The Hopfield model is defined by using the Ising representation and assuming symmetric weights

$$C_{ij} = C_{ji}, \quad C_{ii} = 0$$

for the potential (2.6) in a fully interconnected network with N neurons. Figure 1 illustrates the connectivity of a Feed - Forward Net or multi-layer perceptron versus the connectivity of the Hopfield model. In a fully interconnected, symmetric network (*Hopfield network*) the activity patterns change each step but typically get locked after few steps in some "attractor": a fixed activity pattern which from now on will reproduce itself at each step. This happens for a whole set of starting activity patterns which do not differ too much from the "attractor". In a *feed - forward* network the neurons in the input sheet activate through the synapses relating them to the next sheet an activity pattern in the latter (this can be the output sheet or some intermediary sheet – in the latter case the procedure is repeated until it reaches the output sheet). Again, whole classes of input patterns would lead each time to equivalent output patterns. We see therefore directly the pattern recognition or pattern classification properties. Optimization, control and other problems can be implemented using the same basic functioning.

2.3   Learning

The final step in setting up an ANN model is the definition of the procedure to tune the synapses (weights) C_{ij} to solve the given problem – e.g., to recognize classes of patterns. Of course, we can fix *by hand* the weights: for instance, if we know the patterns which we want to be produced in a Hopfield net we can just use a simple transformation ("the pseudo-inverse") to fix the weights such that exactly these patterns will be attractors in the

---

[2] For a continuous-response function $g(h)$ one can also consider a continuous dynamics using time-derivatives.

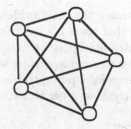 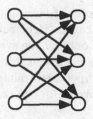

Fig. 1. Fully connected Hopfield Net and Feed - Forward Net. Neurons are depicted as circles.

functioning of the network as described above. However, in problems like that of classification, of self-organization or of optimization *we do not know the attractors*[3]. Interesting about the neural networks is that not only we do not need to know how to implement a given structure but we do not even need to know what structure to implement: we only need to know the problem, that is, *what we want the network to do*. Then we simply let it work and use an algorithm which will tune the weights such that after some time the desired behavior is reached: the terms "learning" or "training" have therefore very intuitive meanings in connection with these algorithms.

The perceptron learning rule, which is the basic supervised training algorithm, is given here for an array of input Ising neurons $\{\sigma_j\}$ projecting onto a single output Ising neuron σ. It uses for each given input the difference between the desired output and the actual one to continuously adjust the weights. If we load the input array with the patterns $\{\sigma_j^\mu\}$, $\mu = 1, 2, \ldots$ the learning rule is

$$C_j \rightarrow C_j + \epsilon \sum_\mu (\eta^\mu - \sigma^\mu)\, \sigma_j^\mu \qquad (2.11)$$

with η^μ the *desired* output to the input σ_j^μ which has lead to the *actual* output σ^μ of the output neuron (ϵ is a parameter used for further tuning of the algorithm). In problems of classification and generalization a one-layer perceptron will achieve good results for linearly separable "pattern" spaces. For nonlinear classification problems multi-layer networks are used. For feed-forward multi-layer networks the formal algorithm of "back-propagation" (see, e.g., Rumelhart, Hinton and Williams in McClelland and Rumelhart 1986) can be shown to have good convergence properties. All these are "supervised training" algorithms, where the weights ("synapses") change proceeds in accordance to the correctness of the behavior. We shall consider continuous response neurons and multilayer perceptrons in section 4, after quoting a number of results from the statistical mechanics of the Hopfield networks in the next section.

---

[3] Moreover, there is no evident biological mechanism implementing such a procedure, hence the heuristics based on the brain analogy are not effective.

From the point of view of the biological learning the most evident algorithm is the so called "Hebb rule" which asserts that there is a feedback from a firing neuron onto all the synapses which contributed positive presynaptic potentials to its firing and this feedback increases the corresponding weights. This is an "unsupervised training" algorithm which for a perceptron amounts simply to remove σ^μ in eqn.(2.11). For a fully connected net this rule takes the form

$$C_{ij} \rightarrow C_{ij} + \epsilon \sum_\mu \sigma_i^\mu \sigma_j^\mu \qquad (2.12)$$

which can be interpreted as a self-organization rule, the input being iterated. The neurophysiological mechanism of "potentiation" could be thought of staying at the basis of the Hebb rule, but on one hand the neurophysiology of the synapse plasticity is very complicated and on the other hand the simple rule (2.12) must usually be further refined to provide good learning properties. Generally, ANN will always find themselves between the alternatives of optimization (as far as formal procedures can be obtained by just mathematical reasoning) and of biological motivation (which, under circumstances, may be vital, since not all problems allow easy formalization). More refined networks and learning rules include "sparse coding", "resilient propagation" or even "evolutionary algorithms" governing the development of the architecture itself (see, e.g., the articles of Horner and Kühn, and of Menzel in Ratsch et al. 1997). Finally, one can also consider the problem of "unspecific reinforcement", where the evaluation gives only the average error for a whole set of training steps (see Stamatescu 1996).

If one sees the rules for functioning (information processing) as "program", then one can say that ANN are essentially self-programming devices: they develop their own programs according to the requirements to which they are subjected during training.

3   The Physics of Neural Networks

As already noted, the present development of neural networks is to a large extent due to their opening to concepts and methods of statistical physics. The basis of this opening, besides the fact that ANN represent typical systems with many degrees of freedom, simple units and high interconnectivity, is that the questions one wants to have answered are of statistical character. If we think of biological networks (brains!), each is involved in another set of similar experiences and confronted with a different set of similar problems. We are not interested to find out which *certain* set of patterns can be stored by which *certain* network, but, say, which average number of more or less randomly occurring patterns can be stored as function of some general characteristics of a class of networks. It is also these kind of question which statistical physics method can help answering in connection with the AI applications of ANN. In this frame computer simulations, which have already a good tradition in statistical physics, provide a kind of "intermediate phenomenology" useful both for the heuristic interpretation and for testing.

The attractors of the Hopfield model, which has been mostly studied, can be shown to represent minima of the energy function:

$$E = \frac{1}{2} \sum_{i,j} \sigma_i C_{ij} \sigma_j + \sum_i \eta_i \sigma_i \qquad (3.13)$$

The external field η is related to the thresholds θ introduced in section 2. Then the rule (2.7) is an algorithm for minimizing the energy E and the stochastic procedures (2.8), (2.9) are algorithms for minimizing the free energy F obtained from the partition function

$$Z = e^{-F} = \sum_{\sigma_1 = \pm 1} \sum_{\sigma_2 = \pm 1} \cdots \sum_{\sigma_n = \pm 1} e^{-\beta E}$$

This gives us at hand the whole arsenal which was or is developed in connection with spin glasses and other multi-connected statistical mechanics systems. The questions approached in this way concern:
- attractors, basins, stability, fluctuations and noise;
- collective phenomena, symmetries and order parameters, phase transitions, multi-scale phenomena;
- convergence questions for learning and for combined dynamics (activity and learning); ...

Summing up these problems under the general question of the structure of the state space, which bears a direct relation to that of pattern storing and retrieving, the general result (see, e.g., Amit 1989) is illustrated in figure 2, where α is the number of patterns divided by the specific number of synapses – the average number of synapses per neuron). Storing many patterns is equivalent to asking for multiple minima of the energy function E. This is typically achieved by a mixed structure of excitatory and inhibitory synapses, which provides many degenerated ground states serving as attractors for the network activity. Given a network and a number of random patterns which we want to store we use an algorithm of type (2.12) by which the patterns $\{\sigma_i^\nu\}$ are learned one after the other:

$$C_{ij}^\nu = C_{ij}^{\nu-1} + \epsilon \left(1 - \delta_{ij}\right) \sigma_i^\nu \sigma_j^\nu \qquad (3.14)$$

(in fact a refined, optimized form of such an algorithm is used). We then find:
1. a paramagnetic phase above some temperature T_g in which the activity has no fixed points;
2. a spin glass phase between temperatures T_g and T_M characterized by many meta-stable states: the activity of the network visits these states for shorter or longer times but does not find the patterns learned under (3.14);
3. a phase of retrieval states below T_M: the stored patterns can be retrieved as fixed points, they are even

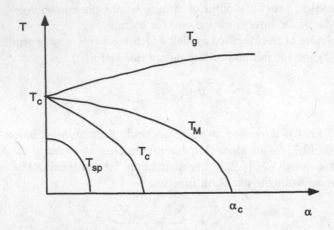

Fig. 2. The phases of the Hopfield model

4. true global minima of the free energy below T_c;
5. finally, below T_{sp} spurious fixed points appear as mixtures of the learned patterns.

As can be seen from these considerations there exists a maximal value α_c above which retrieval is no longer possible ($\alpha_c \simeq 0.14$ for $T = 0$).

New developments put forward the question of non-trivial time structures. On one hand one wants to implement the synapse delay structure observed in biological systems and obtain in this way memorization and retrieval of time series. These are realized as limiting cycles of the activity and typically involve asymmetric synapses (again, biologically justified). On the other hand a more fundamental question is whether only the average firing rates carry information or there exist well timed events in the brain. That is, the timing of spikes may be relevant. While there is little secured biological information about the latter question, strong motivation for studying such aspects come from the cognitive problems known as "binding": how do we recognize separate features in a complex structure ("pattern segmentation") and how do we realize that a number of separately recognized parts build one complex pattern ("feature linking")? Problems which are of course of interest also from the point of view of developing powerful information processing devices. The study of such non-trivial time evolutions needs an enlargement of the frame first set by statistical physics – which essentially deals with equilibrium phenomena. These developments depart in fact from the Hopfield model as introduced above. This model still remains paradigmatic for the behavior of neural networks, however, and the amount of results obtained for it represents a good knowledge basis for the study of general ANN.

Finally, an important question is that of characterizing the performance of ANN beyond average behaviour: typical fluctuation, worst case, etc, which is relevant for ANN applications. A first step is to quantify the effect of occasional "wrong behavior" (incorrect recognition, misinterpretation, false association) from the point of view of the reliability of the result. A particular example will be studied in the next sections.

4   Multilayer Perceptrons and Data Analysis

Feed-forward ANN have become an increasingly popular tool for solving feature recognition and function mapping problems in a wide range of applications. In this section we illustrate this for High Energy Physics (HEP) data analysis. To date, the most commonly used architecture is the Multilayer Perceptron (MLP), which generalizes the perceptron concept of Rosenblatt 1962. The introduction of "hidden layers" (sheets of neurons which do not have direct connection to the exterior) permits tackling the problem of non-linear classification, that is, where the classification cannot be represented by a hyperplane in the pattern space. In fact it can be shown, that for most non-linear problems one hidden layer of non-linear neurons is sufficient[4]. Notice that linear response neurons cannot be used for this job, since then the hidden layers can be transformed away by a linear transformation. For a perceptron with L input neurons, 1 hidden layer with l neurons and 1 output neuron, with continuous, non-linear response functions (sometimes also called analogue neurons) the network function is defined by

$$Y_k = g \left[\sum_{i=1}^{l} c_i \ g \left(\sum_{j=1}^{L} C_{ij} d_j^k + \Theta_j \right) + \theta \right] \qquad (4.15)$$

where $g(x)$ is a non-linear neuron activation function, e.g. (2.2) or

$$g(x) = 1/ \left[1 + \exp(-2x) \right]. \qquad (4.16)$$

The forward propagation from one layer with 3 nodes to the next layer with 3 nodes is depicted in the right part of figure 1. Also included in the definition of the network function are the activation thresholds θ, Θ_i. The subscript $k = 1, \ldots, N$ labels the input data, $\{d_j^k\}$. The forward-weights c_i of the hidden layer contract to a single node, and the input $\{d_j^k\}$ leads to the output Y_k. Usually the MLP is used in the analogue mode (continuous response function g). In this way for each data point k a single output number $0 < Y_k < 1$ is obtained, which we shall interpret after discussing the involved issues in more detail.

In experimental HEP a single data point d_i^k, $i = 1, \ldots, L$ is described by the kinematic variables recorded by the detector and already processed by some track re-construction software or hardware. For instance, according to the Standard Model each top quark of a produced $t\bar{t}$ pair preferably decays

---

[4] See, e.g. Hertz, Krogh and Palmer 1991.

into a b quark and a W boson. The all-jets channel consists of events in which both W bosons decay into light quarks observed as jets. Such jet events may be described by kinematic parameters like d_1^k centrality, d_2^k aplanarity, d_3^k average number of jets, d_4^k sum of non-leading transverse jet energies, and possibly others. Fed with such parameters (from the input layer), the task of the MLP would be to decide whether an event k is a $t\bar{t}$ signal or background. In the particular case of searching for $t\bar{t}$ in the all-jets channel, the signals are obscured by an enormous background consisting of events from other QCD and semi-weak processes. However, the method discussed does by no means depend on this example or any particular use of kinematic variables.

Quite commonly one faces a situation where the probability densities ρ_s for signals and ρ_b for background overlap. An example with two kinetic variables is depicted in figure 3, for which the optimal dividing line (cut) between signals and background is explicitly known. Plotted are 160 signals and 160 background data, generated by Monte Carlo (MC). From these data best cuts at constant $x = 0.798$ or $y = 0.619$, respectively are calculated and also depicted in the figure. Obviously it is non-trivial to find a reasonable approximation to the theoretical line from the data, even if the number of data can be significantly enlarged. The situation gets worse in realistic HEP applications, where normally a data point is described by $L > 2$ variables and a $L - 1$-dimensional hypersurface ought to be determined. Although the parametric form of the optimal hypersurface between signal and background is typically unknown, theory may still allow to generate a large number of artificial signals and background events through a MC event generator like for instance HERWIG (Marchesini et al. 1992). In the following such data are called *training sample*. The AI aspect of a MLP manifests itself in the fact that it can convert the information of the training sample into a parametric form for an estimated optimal hypersurface, dividing signals and background. This is achieved by training the MLP, what amounts to minimizing the function

$$E = \sum_{k=1}^{N} (Y_k - \delta_{bk})^2 \quad \text{where} \quad \delta_{bk} = \begin{cases} 0 \text{ for } k \text{ signal,} \\ 1 \text{ for } k \text{ background,} \end{cases} \qquad (4.17)$$

with respect to the MLP weight parameters $\{C, c\}$ and threshold parameters $\{\Theta, \theta\}$ defined in (4.15). The choice $\delta_{bk} = 1$ for background and 0 for signal is, of course, definition and could be interchanged without altering the essence of the problem. Important is that the number of data N exceeds by far the number of free parameters $\{C, c\}$ and $\{\Theta, \theta\}$. Otherwise over-fitting may result, what means that the trained MLP reflects then primarily properties of the particular training sample and not generic properties of their probability densities.

Due to the non-linearity of the MLP it is non-trivial to find the global minimum of E (4.17). However, for practical purposes this in not necessary

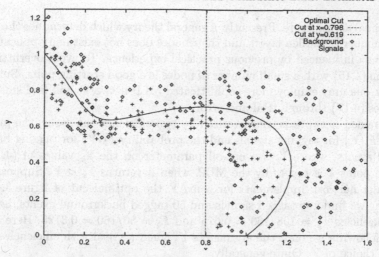

Fig. 3. *Data with overlapping signal $\rho_s(x,y)$ and background $\rho_b(x,y)$ probability densities.*

either, instead any solution will be acceptable that allows later on to disentangle signals from background. Of course, better minima will lead to an increased efficiency. For practical purposes improved variants of backpropagation algorithm (see, e.g. Hertz, Krogh and Palmer 1991, Wasserman 1993, Menzel in Ratsch et al. 1997) work well; some of them are distributed in the JETNET package (Peterson et al. 1994). Backpropagation is a gradient descent minimization procedure, which in its simplest form reads:

$$\Delta c_i = -\eta \frac{\partial E}{\partial c_i} = \eta \sum_{k=1}^{N} \delta^k V_i^k, \quad \delta^k = g'(h^k)(\delta_{bk} - Y_k) \qquad (4.18)$$

$$\Delta C_{ij} = -\eta \frac{\partial E}{\partial C_{ij}} = \eta \sum_{k=1}^{N} \Delta_i^k d_j^k, \quad \Delta_i^k = g'(H_i^k)c_i\delta^k \qquad (4.19)$$

η being an optimization parameter. Here $V_i^k = g(H_i^k)$ are the outputs of the hidden neurons when the input d_j^k is set in the input layer and

$$H_i^k = \sum_{j=1}^{L} C_{ij}d_j^k, \quad h^k = \sum_{i=1}^{l} c_iV_i^k \qquad (4.20)$$

are the corresponding inputs to the hidden, respectively output neurons. Since for large networks simple gradient descent may be rather time consuming one uses various improvements over this rule, e.g. conjugate gradient or other methods. Although in principle 1 hidden layer is sufficient for reproducing any smooth hypersurface in a classification problems, this tells nothing about the optimal architecture. Presently, a general theory which determines the optimal number of hidden layers and their nodes does not exist and application are mainly influenced by previous practical experience. To avoid overfitting, equation (4.15) with a small number of nodes is a good starting point. Subsequently, one may employ more sophisticated MLP networks and find smaller E-values (4.17) through trial and error.

A trained network is characterized by two efficiency functions: the probability $F_s(Y)$ for tagging signals and the probability $F_b(Y)$ for tagging background events, where Y is a cut-off parameter on the Y_k values of (4.15). A Data point k is *tagged* by the MLP, when it returns $Y_k < Y$. Suppose, a particular network implements for some Y the optimal cut of figure 3. By counting we find 108 tagged signals and 50 tagged background events, hence the efficiencies $F_s = 108/160 = 0.675$ and $F_b = 50/160 = 0.3125$. Here and in the following we omit the arguments Y, when we deal with efficiencies for a fixed choice of Y. Quite generally

$$0 < F_b < F_s < 1$$

is the (reasonable) range for the efficiencies and in the following we assume that their error is negligible. In practice, there are of course errors due to the finite size of the training sample and, possibly more severe, due to uncertainties in the method of generating the training sample. Their relevance is best explored by repeating the entire approach with appropriate changes. See the next section.

Once the network is trained, it may be applied to N real data and it will return

$$0 \leq N^Y \leq N \tag{4.21}$$

tagged data. From this the mean value for the number of signal in the N events is estimated to be

$$N_s = \frac{N^Y - F_b(Y)N}{F_s(Y) - F_b(Y)}. \tag{4.22}$$

Notable with this equation is that the r.h.s. depends on Y, whereas the l.h.s. does not. In practice this amounts to a stability criterium, namely the cut-off Y should be picked in a region where the r.h.s. of (4.22) defines a flat function of Y. In the next section we investigate the confidence limits of the thus estimated signal likelihood $p = N_s/N$.

5   From MLP Output to Confidence Limits

We summarize some recent work (Berg and Riedler 1997), which rigorously solves the problem of determining signal confidence limits from the output (4.21) of a trained MLP with known efficiencies. The approach works for all methods (including conventional cuts) which tag signals versus background with different, known efficiencies.

5.1   Generalized Clopper–Pearson Approach

For the situation where signals are observed directly, Clopper and Pearson (1934) (for textbooks see Brandt 1983 or van der Waerden 1971) derived a method, which relies on the binomial distribution and allows to calculate rigorous confidence bounds on the signal probability $p = N_s/N$. The probability to observe N_s signals within N measurements is given by the binomial probability density

$$b(N_s|N,p) = \binom{N}{N_s} p^{N_s} q^{N-N_s} \, , \quad q = 1 - p \, . \tag{5.23}$$

We are faced with the inverse problem: if N_s signals are observed, what is the confidence to rule out certain p? Assume that probabilities p_-^c and p_+^c are given (the superscript c stands for "confidence") and we like to find the following bounds (no superscripts) on the (unknown) exact signal likelihood p: The *largest* p_- and the *smallest* p_+ such that for every (hypothetical) p, $p < p_-$ with probability p_-^c (or less) and $p > p_+$ with probability p_+^c (or less). For instance, with $p_-^c = p_+^c = 0.159$ this guarantees $p_- \leq p \leq p_+$ with the standard one error bar confidence probability of 68.2%. In fact, for almost all p the actual confidence turns out to be better, but (due to their definition) the $p_\pm$ bounds cannot be improved without leading, for some (hypothetical) p-values, to violations of the imposed probabilities $p_\pm^c$. Clopper and Pearson proved that the thus defined lower p_- and upper p_+ bounds are given as solutions of the equations

$$p_-^c = \sum_{k=N_s}^{N} b(k|N,p_-) \quad \text{and} \quad p_+^c = \sum_{k=0}^{N_s} b(k|N,p_+) \tag{5.24}$$

with the additional convention $p_- = 0$ for $N_s = 0$ and $p_+ = 1$ for $N_s = N$ (for a graphical illustration see Berg and Riedler 1997).

We are interested in the more involved situation of the previous section where a NN (or similar device) yields statistical information by tagging signals and background data with different, known, efficiencies: F_s and F_b. Applying the network to all N data results in N^Y tagged data, composed of $N^Y = N_s^Y + N_b^Y$, where N_s^Y are the tagged signals and N_b^Y are the tagged background data. Of course, the values for N_s^Y and N_b^Y are not known. Our task is to determine confidence levels for the signal likelihood p from the sole knowledge of N^Y. We proceed by writing down the probability density of N^Y for given p and, subsequently, generalizing the Clopper-Pearson method.

First, assume fixed N_s. The probability densities of N_s^Y and N_b^Y are binomial and thus the probability density for N^Y is given by the convolution

$$P(N^Y|N_s) = \sum_{N_s^Y + N_b^Y = N^Y} b(N_s^Y|N_s, F_s)\, b(N_b^Y|N_b, F_b), \quad N_b = N - N_s \, .$$

$$\tag{5.25}$$

Summing over N_s removes the constraint and the N^Y-probability density, with N, p fixed, is

$$P(N^Y|N,p) = \sum_{N_s=0}^{N} b(N_s|N,p)\, P(N^Y|N_s)\,. \qquad (5.26)$$

For given p_-^c, p_+^c and N^Y, we define confidence limits p_- and p_+ in analogy with equation (5.24)

$$p_-^c = \sum_{k=N^Y}^{N} P(k|N,p_-) \quad \text{and} \quad p_+^c = \sum_{k=0}^{N^Y} P(k|N,p_+) \qquad (5.27)$$

The interpretation is as before. For instance, the confidence interval $[p_-, p_+]$ corresponding to the standard error bar is again obtained with $p_-^c = p_+^c = 0.159$. Bounds calculated with $p_-^c = p_+^c = 0.023$ ensure the standard two error bar confidence level of 95.4%, and so on.

5.2 Signal Probability Distributions

Distribution functions (see Brandt 1983 or van der Waerden 1971) are defined as integrals over the corresponding probability densities $\rho(p')$

$$F(p) = \int_0^p \rho(p')\, dp'\,.$$

Here the lower integration bound already specializes to probability densities defined for $0 \le p \le 1$ and the fact that $\rho(p')$ is a probability density implies $0 \le F(p) \le 1$ with $F(0) = 0$ and $F(1) = 1$.

Equation (5.27), or of course (5.24) when applicable, can be used to sandwich the *a-posteriori* signal probability distribution $F(p) = F(p|N, N^Y)$ between lower and upper bounds. Namely, it is easy to see that

$$F_1(p) = 1 - p_+^c(p) = \sum_{k=N^Y+1}^{N} P(k|N,p) \le F(p) \le$$
$$F_2(p) = p_-^c(p) = \sum_{k=N^Y}^{N} P(k|N,p)\,. \qquad (5.28)$$

Using $N = 26000$ and $N_s = 130$, figure 4 depicts these functions for the examples $N^Y = 142$ and 194. Upper and lower bounds are seen close together, such that $F(p) = (F_1(p) + F_2(p))/2$ would be a reasonable working approximation. The corresponding probability densities are the derivatives with respect to p. Their numerical calculation is straightforward when analytical expressions for the derivatives of the binomial coefficients in equation (5.26) are used. Figure 5 exhibits the results, $P_1(p|N, N^Y)$ and $P_2(p|N, N^Y)$. At $p = 0$ the probability densities have δ-function contributions. Whereas in our example for good detection efficiency ($N^Y = 194$) the effect is practically negligible, it is quite remarkable for bad detection efficiency ($N^Y = 142$ in our example):

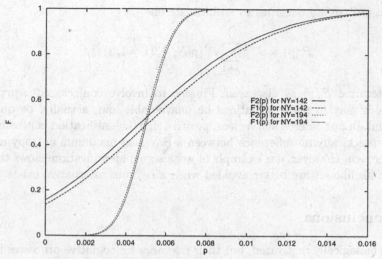

Fig. 4. *A-posteriori* signal probability distributions (upper and lower bounds).

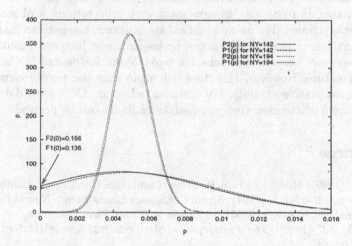

Fig. 5. *A-posteriori* signal probability densities (corresponding to upper and lower bounds of the distribution functions in figure 4).

$$P_i(p) = F_i(0)\,\delta(p) + \dots, \quad (i = 1, 2) \text{ with } F_1(0) = 0.136 \text{ and } F_2(0) = 0.156.$$

In typical applications the efficiencies F_s and F_b may not be known exactly either. Instead, a number of training sets ($j = 1, \dots, J$) may exist, each giving somewhat different efficiencies F_s^j, F_b^j. In this situation a bootstrap type of approach (see, e.g., Efron 1987) can be applied. We can linearly combine different probability densities to an ultimate one

$$P_i(p) = J^{-1} \sum_{j=1}^{J} P_i^j(p|N_j^Y) , (i = 1, 2) ,$$

and determine $P_i(p)$ as discussed. Finally, to involve conjectured *a-priori* likelihoods may in many situations be unavoidable and, actually, be quite successful. In our case: When a clear, positive signal identification is possible, we find practically no difference between a Bayesian maximum entropy and our approach. However, our example of weak signal identification shows that *a-priori* likelihoods are better avoided when a rigorous alternative exists.

6   Conclusions

NN are biologically motivated, but their relevance for cognitive processes has remained questionable. Essential ingredients for the latter may well be missing. On the other side, modelling of ANN has led to a number of interesting results. Statistical physics methods allow to demonstrate the existence of distinct phases. In particular, in agreement with expectations, a phase of retrieval states is found. For many practical applications, like pattern matching or data analysis in particle and astrophysics (to name just two), multilayer perceptrons have become indispensable tools. Many features of NN are of a statistical nature. However, this does not mean that the results cannot be used for quantitative analysis. For instance, when an ANN sorts data with exactly known efficiencies, rigorous confidence limits can be derived.

References

Amit, D.J. (1989): *Modeling Brain Function* (Cambridge Univ. Press, Cambridge)

Berg, B.A. and Riedler, J. (1997): Signal Confidence Limits from a Neural Network Data Analysis. hep-ex/9703001, to appear in Comp. Phys. Commun..

Boden, M., ed. (1990): *The Philosophy of Artificial Intelligence* (Oxford Univ. Press, Oxford)

Bechtel, W. and Abrahamsen, A. (1991): *Connectionism and the Mind* (Blackwell, Cambridge)

Brandt, S. (1983): *Statistical and Computational Methods in Data Analysis* (North-Holland, Amsterdam)

Churchland, P.M. (1989): *A Neurocomputational Perspective* (MIT Press, Cambridge)

Clopper, C.J. and Pearson, E.S. (1934): The Use of Confidence or Fiducial Limits Illustrated in the Case of the Binomial. Biometrika **26**, 404–413

Domany, E., van Hemmen, J.L. and Schulten, K., eds. (1994): *Models of Neural Networks* (Vols. 1 and 2, Springer, New-York)

Hertz, J., Krogh, A. and Palmer, R.G. (1991): *Introduction to the Theory of Neural Computation* (Addison-Wesley Pub. Comp.)

Efron, B. (1982): *The Jackknife, the Bootstrap and Other Resampling Plans* (Society for Industrial and Applied Mathematics, Philadelphia)

Kandel, E.K., Schwartz, J.H. and Jessel, T.M. (1995): *Essentials of Neuroscience and Behavior* (Appleton and Lange, Norwalk)

Kohonen, T. (1989): *Selforganization and Associative Memory* (Springer, Berlin)

Marchesini, G., Webber, B.R., Abbiendi, G., Knowles, I.G., Seymour, M.H. and Stanco, L. (1992): HERWIG 5.1 – A Monte Carlo Event Generator for Simulating Hadron Emission with Interfering Gluons. Comp. Phys. Commun. **67**, 465–508

McClelland, J.L., D.E. Rumelhart, D.E., eds. (1986): *Parallel Distributed Processing* (Vols. 1 and 2, MIT Press, Cambridge)

McCulloch, W.S., Pitts, W. (1943): A Logical Calculus of the Ideas Immanent in Nervous Activity, Bull. Math. Biophys. **5**, 115–133

Peterson, C., Rögnvaldsson, T and Lönnblad, L. (1994): JETNET 3.0 – A Versatile Artificial Neural Network Package. Comp. Phys. Commun. **81**, 185–220

Ratsch, U., Richter, and Stamatescu, I.-O., eds. (1997): *Aspects of Intelligence and Artificial Intelligence* (Springer, Heidelberg)

Rosenblatt, F. (1962): *Principles of Neurodynamics* (Spartan Books, Washington DC)

Stamatescu, I.-O. (1996): "The Neural Network Approach" in *Proceedings of the International Conference on the Philosophy of Biology, Vigo, 1996*

van der Waerden, B. (1971): *Mathematische Statistik* (Springer, Heidelberg)

Wasserman, Ph.D. (1993): *Advanced Methods in Neural Computing* (Van Nostrand Reinhold)

The Gross-Neveu Model and QCDs Chiral Phase Transition

Thomas Reisz *

Institut für Theoretische Physik, Universität Heidelberg, Philosophenweg 16, D-69120 Heidelberg, Germany

Abstract. Quantum chromodynamics has a rather complicated phase structure. The finite temperature, chiral phase structure depends on the number of flavours and to a large extent on the particular values of the fermion masses. For two massless flavours there is a true second order transition. It has been argued that this transition belongs to the universality class of the three-dimensional O(4) spin model. The arguments have been questioned recently, and the transition was claimed to be mean field behaved.

In this lecture we discuss this issue at the example of the three-dimensional, parity symmetric Gross-Neveu model at finite temperature, with a large number N of fermions. At zero temperature there is a phase where parity is spontaneously broken. At finite temperature, this model has a parity restoring second order transition. It reveals considerable similarity to the QCD chiral phase transition. There are related questions here concerning the universality class. We solve this problem essentially by means of the following methods: Large N expansion, dimensional reduction in the framework of quantum field theory, and high order convergent series expansions about disordered lattice systems.

1 Introduction - QCD and the Chiral Phase Transition

Quantum field theories provide the appropriate framework for the understanding and the quantitative description of high energy physics. In particular, quantum chromodynamics - QCD - is supposed to be the theory of strong interactions, the physics of quarks, mesons and hadrons. As such a theory it has to describe a large variety of high energy phenomena. For instance this concerns the confinement of quarks, the observed mesonic and hadronic particle mass spectrum, topological excitations. This theory evolves an enormous complexity. Computational techniques to be developed and used are rather complicated, both numerical and analytical ones.

There are a number of long-term projects running, with the aim of a quantitative understanding of particular properties of QCD from first principles. For example, one of these projects is the so-called alpha-collaboration (Lüscher, Sint, Sommer, Weisz and Wolff (1996)). One of the aims is to determine the flow of the (appropriately defined) renormalized coupling constant

---

* Supported by a Heisenberg Fellowship

$g_R(L)$ well above the perturbative short distance scale, that is on large spatial scale L. A remarkable strategy of this collaboration is the combination of the appropriate analytical and numerical (Monte-Carlo) techniques.

Other long-term projects concern the nature of the chiral phase transition in finite temperature QCD and the properties of the high temperature phase. The high temperature phase of QCD may be realized in heavy ion collisions, and it was certainly realized in the early universe. There are a large number of interesting physical questions related to this phase. For instance, screening of heavy quarks in the QCD plasma phase has been studied in detail and screening masses have been extracted, by combining analytical methods such as finite volume perturbation theory and dimensional reduction in the framework of quantum field theory, together with various Monte-Carlo methods. Other, yet open questions concern the (gauge-invariant) definition of a magnetic mass and its determination, the ability to disentangle infrared and ultraviolet singularities and their removal in the thermodynamic limit, that is to find appropriate resummation techniques.

In the limit of vanishing current quark masses, massless QCD has a chiral symmetry. This global symmetry is spontaneously broken at zero temperature, and it becomes restored in the high temperature phase. Whether this symmetry is restored by a 1st or 2nd order phase transition or just by a crossover phenomenon depends on the number of fermionic flavours. In addition, to a large extent it depends on the particular values of the fermion masses.

A detailed presentation of the QCD phase transitions according to our current knowledge can be found in the review of Meyer-Ortmanns (1996). Here we only give a very short outline of the state of the art. In Fig. 1 we show a qualitative plot of the supposed phase diagram for three flavours u, d, s in dependence on their masses. The u- and d-quark masses are of comparable size and are put to the same value. For the pure SU(3) Yang-Mills theory (right upper corner) the transition is the deconfinement transition, which is believed to be of 1st order. For the purely massless case (left lower corner) the transition refers to the chiral transition, which is supposed to be of first order in the case of 3 flavours. With increasing masses the chiral transition becomes weaker and finally turns over into a mere crossover. For 2 massless flavours, the chiral transition is of first order below a critical value m_s^* of the strange quark mass (corresponding to a tricritical point) and of second order above m_s^*.

The physical fermion masses are subject to a large uncertainty in relation to the phase boundaries. Even if there is no "true" phase transition in QCD, some correlation lengths might become large, generating large domains of chiral condensates.

In nature, two fermion flavours are realized as relatively light states. For the case of two flavours, both being massless, the chiral phase transition is supposed to be described by a critical point. What is the nature of this

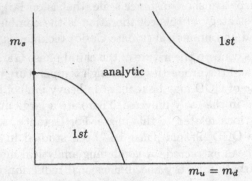

Fig. 1. A qualitative plot of the mass dependence of the QCD phase transition for three flavours. The dot identifies the tricitical strange quark mass, above which the transition is of 2nd order.

transition? It was argued by Pisarski and Wilczek that its universality class is that of the O(4) spin model in 3 dimensions. We will discuss this point of view in the next section. On the contrary, Kocic and Kogut argued that the transition is mean field behaved. So far, no definite answer has been given, neither by Monte Carlo simulations nor by analytical investigations.

In this lecture we discuss the issue of the finite temperature phase transition at the example of the 3-dimensional Gross-Neveu model. This model reveals considerable similarity to QCD. There are quite related questions concerning the nature of the transition, including as candidates both mean field and non-trivial spin model behaviour. However, in this case a definite answer can be given, mainly by analytical means.

1.1   Two-Flavour QCD

Before we turn over to the Gross-Neveu model, let us discuss in some more detail the effective scalar O(4) model for the QCD chiral phase transition. The order parameter is given by the chiral condensate $< \overline{\psi}(x)\psi(x) >$. It is argued that the transition belongs to the universality class of the scalar O(4) model in 3 dimensions (Pisarski and Wilczek (1984), Wilczek (1992), Rajagopal and Wilczek (1993)). An essential ingredient of this argument is the observation that under quite general conditions, for a field theory at high temperature fermionic degrees of freedom act only as composite states in the infrared. This fact is a consequence of the so-called dimensional reduction in quantum field theory and will be discussed in detail in the following sections.

Let us outline the way of reasoning towards the O(4) scenario. The idea is to build an effective model of the "localized" order parameter field, the condensate, and apply a Landau-Ginzburg type of argument. First of all, the order parameter field is a complex 2×2-matrix field in three dimensions $x \in \mathbf{R}^3$,

$$< \overline{\psi}_{Ri}(x)\psi_{Lj}(x) > \sim M_{ij}(x) \in M(2 \times 2, \mathbf{C}), \tag{1.1}$$

where R and L denotes right- and left-handed (relativistic) fermions, respectively. Close to the critical point the degrees of freedom that develop long-range fluctuations are assumed to be associated with the order parameter M, which is small in magnitude. Hence it is plausible to describe the effective model by the action

$$S_{eff}(M) = \int_x (\mathrm{tr}\, \partial_x M(x)^\dagger \partial_x M(x) + m^2 \, \mathrm{tr}\, M^\dagger(x)M(x) \tag{1.2}$$

$$+ g_1 \mathrm{tr}\, (M^\dagger(x)M(x))^2 + g_2 (\mathrm{tr}\, M^\dagger(x)M(x))^2).$$

The partition function becomes

$$Z = \int \mathcal{D}M \cdot \exp\left(-S_{eff}(M)\right), \tag{1.3}$$

where $\mathcal{D}M = \prod_x dM(x)$. We have implicitly assumed an ultraviolet cutoff, such as the Pauli-Villar cutoff or a lattice. At this stage then we can exploit the methods of renormalizable quantum field theory, such as the renormalization group and renormalized perturbation theory, and strong statements obtained by those methods. On the one hand, a posteriori this justifies a large freedom in the choice of the ultraviolet cutoff for the above (super-) renormalizable model. That is, a particular choice does not have influence on the low-momentum scale. On the other hand, the critical properties of the model can be studied.

Yet the model has too large a symmetry to describe a chiral phase transition. It is invariant under global $U(2)_L \times U(2)_R$ transformations

$$M \to UMV^\dagger, \quad U, V \in U(2). \tag{1.4}$$

Finally we want to reduce the symmetry to the smaller, physical symmetry group $SU(2)_L \times SU(2)_R \times U(1)$, (to be broken down to vector $SU(2) \times U(1)$ spontaneously at low temperature,) where $U(1)$ describes the vector-baryon symmetry. This restricted symmetry has to be implemented by imposing appropriate constraints on M. Whereas for larger number of flavours those constraints are to be non-linear in M, leaving the framework of renormalizable quantum field theory, for two flavours we can implement such constraints by restricting M to be a real multiple of a unitary matrix of unit determinant, $M \in c \cdot SU(2)$, $c \in \mathbf{R}$. Hence M can be parametrized linearly as

$$M = \pi_0 1_2 + \pi\sigma, \quad (\pi_0, \pi) \in \mathbf{R}^4, \tag{1.5}$$

where $\sigma = (\sigma_1, \sigma_2, \sigma_3)$ denotes the Pauli-matrices. It is now obvious that in terms of the π-fields we get a linear $O(4)$ model. This symmetry is broken at low temperature to $O(3)$.

If the universality arguments made so far are true, the QCD chiral phase transition for two massless flavours will belong to the universality class of the O(4) spin model in three dimensions. Of course, there are assumptions made in the course towards this conclusion. The most stringent ones are the following.

- The chiral phase transition is of 2nd order.
- The free energy density as a function of the order parameter field M is sufficiently smooth, allowing for a polynomial representation at small M. But we know this is problematic below the upper critical dimension, the dimension of strict renormalizability, which is 4 for the above model.
- In the language of statistical mechanics, the quantum critical system is equivalent to the corresponding classical critical system. In particular, the phase transition temperature is so high that the thermal fluctuations dominate the quantum fluctuations even at the transition. In the language of quantum field theory, this amounts to say that dimensional reduction is complete at the critical point, at least concerning the values of the critical exponents.

Recently Kocic and Kogut have criticised the above (σ-model) approach because the fermionic degrees of freedom are represented only via composite fields M (Kocic and Kogut (1994)). Actually, as we will see in the course of this lecture, this is a minor point of criticism. But even if fermions act only as composite states, there are several candidates for the universality class of the QCD chiral phase transition. Both the above argumentations mainly refer to conventional wisdom. In order to obtain a definite answer it is unavoidable to give strong reasonings, including explicit computations.

1.2 Why Studying the Gross-Neveu Model?

In this lecture the problem of the chiral phase transition will be discussed at the example of a much simpler model than QCD, which nevertheless bears the essential properties such as "chirality". Namely, I would like to discuss the 3-dimensional Gross-Neveu model at finite temperature, with a large number of fermionic flavours. It is a model with a four-fermion interaction, subject to a global $\mathbf{Z}_2$ symmetry (which is parity). This symmetry is known to be broken at zero temperature, and it is restored at high temperature by a 2nd order transition. We then ask for the universality class of this transition.

In quite analogy to QCD there are two opposite claims of the universality class. Adjusting the arguments of Pisarski and Wilczek, the transition is described by a $\mathbf{Z}_2$ symmetric scalar model in 2 dimensions that should belong to the Ising class. On the contrary, Kocic and Kogut claimed that the transition is that of mean field theory.

Beyond its similarity to QCD, the advantage of studying this model is that to a large extent we have a closed analytic control of it. This includes the finite temperature transition. As we will see, the above discrepancy can be resolved by analytical means. Also, in the course of the following sections it will become clear that the definite answer can be given only by explicit computations.

The answer will be given essentially by applying the following two techniques.

- Dimensional reduction in quantum field theory. Applied to the 3-dimensional, finite temperature Gross-Neveu model, it states that the infrared behaviour of the latter is described by a 2-dimensional scalar field theory at zero temperature. As we will see, this effective model is local and renormalizable. It should be emphasized that dimensional reduction provides an explicit computational device to compute the couplings of the effective model. This is important because a 2-dimensional scalar model does not necessarily belong to the Ising universality class.

- Convergent series expansions of high orders, applied to the free energy and connected correlations. In this way the critical region can be investigated, the order of the transition and eventually the universality class can be determined.

In outline, this lecture is organized as follows. Throughout we will use the path integral representation for quantum field theories. In Sect. 2 we describe the 3-dimensional Gross-Neveu model. The notions familiar from finite temperature quantum field theory are introduced. The large N representation and expansion of the model are discussed. The $N = \infty$ limit can be solved in closed form, including the phase structure. In Sect. 3 we outline the technique of dimensional reduction in the framework of renormalizable quantum field theory. In the recent past this technique became a very well developed, powerful computational machinery both for bosonic and fermionic field theories at finite temperature. It is applied to the Gross-Neveu model. Then we are prepared to study in Sect. 4 the phase structure of the model for finite number N of flavours. A first insight is obtained by the discussion of the strong coupling limit. In a subsection we shortly present convergent series expansion techniques, in particular linked cluster expansions, and the ideas behind their generation. They are then applied to determine the critical exponents, hence the universality class of the parity restoring transition. Summary and outlook for the case of QCD is given in the last Sect. 5.

2   The 3-Dimensional Gross-Neveu Model at Zero and Finite Temperature

2.1   The Model

For a detailed introduction to field theories at finite temperature, we refer to the books of Kapusta (1989) and of Rothe (1997). We first of all define the model under consideration. In 3 dimensions at finite temperature T (henceforth $3d_T$ in short), a field theory lives in a volume of the toroidal shape $T^{-1} \times \mathbf{R}^2$,

$$z = (z_0, z_1, z_2) \equiv (z_0, \mathbf{z}) \in \mathbf{R}/T \times \mathbf{R}^2, \tag{2.6}$$

cf. Fig. 2. The extension is set to infinity in both spatial directions, and the length of the torus component, the "temperature-direction", is fixed by the inverse temperature.

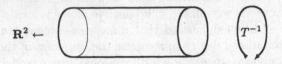

Fig. 2. The world of the $3d_T$ Gross-Neveu model. The fermionic degrees of freedom are subject to anti-periodic boundary conditions in the temperature direction.

We consider a model of N Dirac spinor fields, that is of N fermion species or flavours, with values in a Grassmann (exterior) algebra,

$$\psi(z) = (\psi_{\alpha i}(z))|_{\alpha=1,2;i=1,...,N},$$
$$\overline{\psi}(z) = (\overline{\psi}_{\alpha i}(z))|_{\alpha=1,2;i=1,...,N}. \qquad (2.7)$$

Due to the trace operation of the partition function, Grassmann-valued fields are subject to anti-periodic boundary conditions in the temperature direction,

$$\psi(z_0 \pm T^{-1}, \mathbf{z}) = -\psi(z_0, \mathbf{z}), \qquad (2.8)$$

and similarly for $\overline{\psi}$ (cf. e.g. Lüscher (1977)). As a first consequence of the anti-periodicity, the fermionic degrees of freedom always have an implicit infrared cutoff involved, which is of the order of the temperature T. This is easily seen by applying a Fourier transform to momentum space,

$$\psi(z) = \sum_{k_0=\pi T(2n+1),n\in\mathbb{Z}} e^{ik_0 z_0}\, \tilde{\psi}_n(\mathbf{z}). \qquad (2.9)$$

From this we infer that $|k_0| \geq \pi T$, which provides a lower bound on the total momentum.

The action is given by

$$S_{gn}(\psi,\overline{\psi}) = \int_z (\overline{\psi}(z)D\psi(z)) - \frac{\lambda^2}{N}\int_z \left(\overline{\psi}(z)\psi(z)\right)^2, \qquad (2.10)$$

where

$$\int_z \equiv \int_{\mathbf{R}^2} d^2\mathbf{z} \int_{-\frac{1}{2}T^{-1}}^{\frac{1}{2}T^{-1}} dz_0,$$

$$D = \sum_{i=0}^{2} \gamma_i \frac{\partial}{\partial z_i}, \qquad (2.11)$$

with $\gamma_0, \gamma_1, \gamma_2$ denoting the 3 basis elements of a Dirac representation. We normalize them according to

$$\gamma_i\gamma_j + \gamma_j\gamma_i = 2\delta_{ij} ; \quad i, j = 0, 1, 2. \tag{2.12}$$

Here and in the following for spinor fields $\overline{\psi}(z), \eta(z)$ we use the notations

$$\overline{\psi}(z)\eta(z) \equiv \sum_{\alpha=1,2} \sum_{i=1,\ldots,N} \overline{\psi}_{\alpha i}(z)\eta_{\alpha i}(z). \tag{2.13}$$

The partition function (generating functional of full correlation functions) is then given by

$$Z(\eta, \overline{\eta}) = \int \mathcal{D}\overline{\psi}\mathcal{D}\psi \, \exp\left(-S_{gn}(\overline{\psi}, \psi) + \int_z (\overline{\psi}(z)\eta(z) + \overline{\eta}(z)\psi(z))\right), \tag{2.14}$$

where

$$\int \mathcal{D}\overline{\psi}\mathcal{D}\psi = \prod_{z,\alpha,i} d\psi_{\alpha i}(z)d\overline{\psi}_{\alpha i}(z). \tag{2.15}$$

η and $\overline{\eta}$ are external sources introduced in order to derive correlation functions from $Z(\eta, \overline{\eta})$ by differentiation.

So far, the model is only formally defined by the action (2.10) and the partition function (2.14). In order to define it rigorously as a quantum field theory from the outset, we have to introduce an ultraviolet cutoff Λ. Among others, possible choices are a lattice, the Pauli-Villar cutoff, dimensional regularization, or just a simple momentum cutoff. The model becomes well defined for $p \leq \Lambda$, where p represents momentum. An essential point then is to prove the existence of this model in the large cutoff limit $\Lambda \to \infty$, and in this limit a sequence of axioms have to be satisfied (Osterwalder and Schrader (1975), Zinoviev (1995)). This is the issue of renormalization in the framework of quantum field theory. In many cases this is achieved by adjusting a finite number of bare parameters of the action, fixing corresponding renormalized coupling constants on a finite length scale. Renormalizability implies that the influence of the cutoff Λ on the low-lying momentum scale p is suppressed by some power of p/Λ. In particular, all the correlation functions of the model stay well-defined in the limit on finite scales.

One might argue that in the framework of statistical mechanics, one is not necessarily interested in the removal of the ultraviolet cutoff. For instance, the lattice spacing $a \sim 1/\Lambda$ is a finite physical length in solid state physics (except for the lattice spacing in the temperature direction, which still is to be sent to zero). However, concerning the critical regions of a model where the correlation length diverges in units of the lattice spacing, the motivation for renormalizability is quite similar as in quantum field theory. Instead of the cutoff being sent to infinity, the range of interest is where the momentum becomes arbitrarily small. Collective critical behaviour in this region to a large extent becomes independent of the cutoff scale, i.e. of the bare parameters. This identifies a universality class, that is universal critical behaviour. This would not be true for non-renormalizable models.

In the following we regularize the Gross-Neveu model by choosing a simple momentum cutoff Λ. This is done mainly for computational simplicity. Equally well we could have chosen a hybercubic lattice, with some care concerning the choice of lattice fermions.

There is some detailed knowledge of this model at zero temperature $T = 0$. First of all, applying power counting of ultraviolet divergencies, this 3-dimensional model appears to be non-renormalizable in the loop expansion. That is, as an expansion in powers of the coupling constant λ, we need more and more couterterms with increasing order of λ in order to render the ultraviolet limit finite. Fortunately, for large number N of fermionic flavours, the situation is different. Within the large N expansion, which will be explained in the very detail in the next section, the model becomes renormalizable, to every finite order in $1/N$ (Park, Rosenstein, Warr (1989a)). Actually, there is an even stronger statement. Namely, if N is sufficiently large, the 3-dimensional Gross-Neveu model can be shown to exist in the large cutoff limit (and in the infinite volume), that is, the correlation functions of the model stay finite. The proof is done in the framework of constructive field theory, using convergent expansions and the renormalization group (Faria da Veiga (1991), de Calan, Faria da Veiga, Magnen, Seneor (1991)). Remarkably enough, this is a construction about a non-Gaussian fixed point, leading to a non-trivial (interacting) quantum field theory.

Let us come back to the finite temperature. As already said in the introduction, the model (2.10) serves as a good example for studying a "chiral" phase transition because it reveals essential similarities to massless QCD. We seek for a symmetry that would be broken explicitly by a fermionic mass term of the form

$$m\overline{\psi}(z)\psi(z). \tag{2.16}$$

Actually, this will not be a true chiral symmetry. This would require the existence of a γ-matrix, the γ_5, which anticommutes with all generating elements $(\gamma_0, \gamma_1, \gamma_2)$ of a faithful representation of the associated Clifford algebra (cf. e.g. Chevalley (1997) or Porteous (1995)). It does not exist in three dimensions. Instead, the model is invariant under the global parity transformation

$$\psi(z) \to \psi(-z),$$
$$\overline{\psi}(z) \to -\overline{\psi}(-z). \tag{2.17}$$

Both the action (2.10) and the measure (2.15) stay invariant under this transformation. The symmetry under (2.17) is spontaneously broken at $T = 0$, and it becomes restored at high temperature by a second order phase transition, cf. below. The parity condensate

$$\frac{1}{N} < \overline{\psi}(z)\psi(z) >$$

$$= \frac{1}{Z(0,0)} \int \mathcal{D}\overline{\psi}\mathcal{D}\psi \, \frac{1}{N} \left(\overline{\psi}(z)\psi(z)\right) \exp\left(-S_{gn}(\overline{\psi}, \psi)\right) \qquad (2.18)$$

serves as an order parameter of the transition.

We remark in passing that, besides others, the model has a continuous global symmetry in flavour space, defined by

$$\psi_{\alpha i}(z) \rightarrow \sum_{j=1}^{N} U_{ij}\psi_{\alpha j}(z) , \quad \overline{\psi}_{\alpha i}(z) \rightarrow \sum_{j=1}^{N} U_{ji}^{\dagger}\overline{\psi}_{\alpha j}(z), \qquad (2.19)$$

with $U \in U(N)$. This symmetry is neither explicitly broken by a mass term (2.16) nor does it get broken spontaneously. In particular it yields fermion number conservation.

2.2   Many Flavours. The Large N Expansion

As mentioned in the previous section, much insight is obtained by representing the Gross-Neveu model in a form that is particularly well suited if the number of fermionic species is sufficiently large. The corresponding transformation is based on the resummation of so-called planar graphs according to Fig. 3.

Fig. 3. Resummation of planar graphs.

$$\frac{\lambda^2}{N} \left(1 - cN\frac{\lambda^2}{N} + \left(cN\frac{\lambda^2}{N}\right)^2 + \cdots\right) = \frac{1}{N} \left(\frac{1}{\lambda^2} + c\right)^{-1} \qquad (2.20)$$

Here c counts the combinatorical factor and the trace in spinor space over the single fermion bubble. Such a chain then is represented as an interaction mediated by an auxiliary scalar field σ.

We can do this resummation in closed form in the following way. First the local, quartic fermion interaction part of the action can be written as the result of a gaussian integration, using

$$\exp\left(\frac{\lambda^2}{N}\left(\overline{\psi}(z)\psi(z)\right)^2\right) = \left(\frac{N}{4\pi\lambda^2}\right)^{\frac{1}{2}} \cdot$$

$$\cdot \int_{-\infty}^{\infty} d\sigma(z) \exp\left(-\frac{1}{2}\left(\frac{N}{2\lambda^2}\right)\sigma(z)^2 \pm \sigma(z)\,\overline{\psi}(z)\psi(z)\right). \quad (2.21)$$

The action then becomes of the Yukawa-type and is bilinear in the fermionic fields. In turn, the integration over the Grassmann algebra valued fields can be done according to

$$\int \mathcal{D}\overline{\psi}\mathcal{D}\psi \, \exp\left(-\int_z\int_{z'} \overline{\psi}(z)\,Q(z,z')\psi(z') + \int_z(\overline{\psi}(z)\eta(z)+\overline{\eta}(z)\psi(z))\right)$$

$$= \det(-Q) \, \exp\left(\int_z\int_{z'} \overline{\eta}(z)Q^{-1}(z,z')\eta(z')\right). \quad (2.22)$$

In our case, Q is flavour diagonal, i.e. $Q = \mathbf{1}\otimes K$, with

$$K = D + \sigma\cdot\mathbf{1} \quad (2.23)$$

acting on the product of spinor and configuration space. D is defined by (2.11). As the remnant of the anti-periodic boundary conditions (2.8) imposed on the fermionic fields, the momentum space representation of D has nonvanishing components only for the discrete thermal momenta $k_0 = \pi T(2m+1)$, $m \in \mathbf{Z}$, cf. (2.9). In this way we obtain the following representation for the partition function (2.14) up to an unimportant normalization constant.

$$Z(\eta,\overline{\eta}) = \text{const} \int \mathcal{D}\sigma \, \exp\left(-N\,S_{eff}(\sigma)\right)$$

$$\cdot \exp\left(\int_z\int_{z'} \overline{\eta}(z)(1\otimes K^{-1})(z,z')\eta(z')\right), \quad (2.24)$$

with

$$\int \mathcal{D}\sigma = \prod_z d\sigma(z). \quad (2.25)$$

The effective action $S_{eff}(\sigma)$ is given by

$$S_{eff}(\sigma) = \int_z\left(\frac{\sigma(z)^2}{4\lambda^2} - \text{tr}_s\left(\log K\right)(z,z)\right), \quad (2.26)$$

with the trace taken over the spinor indices. A careful counting of the degrees of freedom (e.g. by imposing a lattice cutoff) shows that without loss of generality we may impose periodic boundary conditions on the auxiliary field $\sigma(z)$ in the temperature direction (it is "bosonic"). This is the natural and common choice and implies that constant field configurations of σ are to be identified as zero momentum configurations in the usual way. In turn the minima of the effective action for large N belong to the latter ones.

It should be emphasized that up to now no approximation is involved. The partition function has only been rewritten in a form that is most convenient if the number N of fermions becomes large. The only place where the number N enters in (2.24) is as the prefactor of the action. The explicit form (2.24) reveals the standard saddle point expansion as the convenient method to study the Gross-Neveu model for large N. The parity condensate (2.18) becomes the expectation value of $-\mathrm{tr}\,_s K^{-1}(z, z)$, and similar for the condensate correlations. It is straightforward to show to all orders of $1/N$ that a vanishing condensate (2.18) is equivalent to a vanishing expectation value of the auxiliary field σ. The auxiliary field $\sigma(z)$ itself is parity odd. The parity symmetry of the originial model becomes invariance under

$$\sigma(z) \;\to\; -\sigma(-z). \tag{2.27}$$

The large N expansion proceeds along the following steps.
- Localization of the translation invariant minima of the effective action S_{eff}. This yields the so-called gap equation

$$\frac{\delta S_{eff}(\sigma)}{\delta\sigma(z)}\Big|_{\sigma(z)\equiv\mu} \;=\; 0\,! \tag{2.28}$$

- Minimization requires positivity of

$$\frac{\delta^2 S_{eff}}{\delta\sigma(z)\delta\sigma(z')}\Big|_{\sigma(z)\equiv\sigma(z')\equiv\mu}. \tag{2.29}$$

- Expansion about the solutions according to

$$\sigma(z) \;=\; \mu \;+\; \frac{\phi(z)}{N^{\frac{1}{2}}}. \tag{2.30}$$

Existence of stable solutions of the gap equation and their properties will be discussed in the next section. For further reference we write down the model as obtained by inserting the ansatz (2.30) into the effective action (2.26) and expand for large N. Towards this end we first introduce some notations used in the following. Fermionic and bosonic thermal momenta are defined as the sets

$$\mathcal{F} = \{\frac{\pi}{\beta}(2m+1) \mid m \in \mathbf{Z}\},$$

$$\mathcal{B} = \{\frac{\pi}{\beta}2m \mid m \in \mathbf{Z}\}. \tag{2.31}$$

For the inverse temperature we write $\beta = T^{-1}$. For $x, \beta \geq 0$ and $n = 0, 1, 2, \ldots$ let

$$J_n(x,\beta) = \frac{1}{\beta} \sum_{q_0 \in \mathcal{F}} \int' \frac{d^2\mathbf{q}}{(2\pi)^2} \frac{1}{(q^2+x)^n},$$

$$Q_n(x,\beta) = \frac{1}{\beta} \sum_{q_0 \in \mathcal{F}} \int' \frac{d^2\mathbf{q}}{(2\pi)^2} \frac{q^2}{(q^2+x)^n}. \tag{2.32}$$

The prime at the integral sign denotes that the sum and integral is confined to the region with

$$q^2 \equiv q_0^2 + \mathbf{q}^2 \leq \Lambda^2. \tag{2.33}$$

For $\beta < \infty$ the $J_n(x, \beta)$ are always infrared finite for every $x \geq 0$, whereas

$$J_n(x, \infty) = \int_{q^2 \leq \Lambda^2} \frac{d^3 q}{(2\pi)^3} \, (q^2 + x)^{-n} \tag{2.34}$$

is infrared singular as $x \to 0+$ for $n \geq 2$. Furthermore, the ultraviolet singularities of the J_n are temperature independent, which means that

$$\lim_{\Lambda \to \infty} (J_n(x, \beta) - J_n(x, \infty)) < \infty, \tag{2.35}$$

for each $x \geq 0$. This is a property which is shared by a large class of 1-loop Feynman integrals. They can be rendered ultraviolet finite by imposing the appropriate zero temperature normalization conditions.

Inserting (2.30) into the effective action (2.26), we obtain the following large N representation. Let

$$K_0 = D + \mu. \tag{2.36}$$

The partition function then becomes

$$Z(\eta, \overline{\eta}) = \mathcal{N} \int \mathcal{D}\phi \, \exp\left(-S(\phi) + \int_z \int_{z'} \overline{\eta}(z)(1 \otimes \widetilde{K}^{-1})(z, z')\eta(z') \right), \tag{2.37}$$

with

$$\mathcal{N} = \text{const} \, \exp\left(-N \int_z (4\lambda^2 \mu^2 (J_1(\mu^2, \beta))^2 - \text{tr}_s \log(K_0)(z, z)) \right), \tag{2.38}$$

$$\widetilde{K} = K_0 + \frac{\phi}{N^{\frac{1}{2}}}, \tag{2.39}$$

and

$$\begin{aligned} S(\phi) &= \int_z \left(\frac{\phi(z)^2}{4\lambda^2} - N \, \text{tr}_s \left(\log\left(1 + \frac{1}{N^{\frac{1}{2}}}\phi K_0^{-1}\right) - \frac{1}{N^{\frac{1}{2}}}\phi K_0^{-1} \right)(z, z) \right) \\ &= \sum_{n \geq 2} S_n(\phi). \end{aligned} \tag{2.40}$$

Here,

$$S_2(\phi) = \int_z \left(\frac{\phi(z)^2}{4\lambda^2} + \frac{1}{2} \, \text{tr}_s \left(\phi K_0^{-1} \right)^2 (z, z) \right),$$

$$S_n(\phi) = \frac{(-1)^n}{N^{n/2-1}} \frac{1}{n} \int_z \text{tr}_s \left(\phi K_0^{-1} \right)^n (z, z), \quad n \geq 3. \tag{2.41}$$

$S_n(\phi)$ is of the order of $O(N^{1-n/2})$. For later convenience we write the quadratic part of the action in its momentum space representation obtained by Fourier transform. With

$$\phi(x) = \frac{1}{\beta} \sum_{k_0 \in \mathcal{B}} e^{ik \cdot x} \, \widetilde{\phi}(k) \tag{2.42}$$

we obtain after some algebra

$$S_2(\phi) = \frac{1}{2} \frac{1}{\beta} \sum_{k_0 \in \mathcal{B}} \int' \frac{d^2 k}{(2\pi)^2} \, |\widetilde{\phi}(k)|^2 \Big\{ 2 \Big(\frac{1}{4\lambda^2} - J_1(\mu^2, \beta) \Big)$$
$$+ \frac{1}{\beta} \sum_{q_0 \in \mathcal{F}} \int' \frac{d^2 q}{(2\pi)^2} \Big[\frac{k^2 + 4\mu^2}{(q^2 + \mu^2)((q+k)^2 + \mu^2)} \tag{2.43}$$
$$- \Big(\frac{1}{(q+k)^2 + \mu^2} - \frac{1}{q^2 + \mu^2} \Big) \Big] \Big\}.$$

The wave function renormalization constant Z_R and the renormalized mass m_R (the inverse correlation length) are defined by the small momentum behaviour of the (connected) 2-point function

$$\widetilde{W}^{(2)}(k) = \, < \widetilde{\phi}(k)\widetilde{\phi}(-k) > \tag{2.44}$$

according to

$$\widetilde{W}^{(2)}(k_0 = 0, \mathbf{k}) = \frac{1}{Z_R} \left(m_R^2 + \mathbf{k}^2 + O(\mathbf{k}^4) \right) \quad \text{as } \mathbf{k} \to 0. \tag{2.45}$$

2.3 Phase Structure in the $N = \infty$ Limit

We discuss the phase structure of the Gross-Neveu model in the limit of infinite number of fermionic flavours first (Park, Rosenstein, Warr (1989b)). The gap equation (2.28) reads

$$\mu \left(\frac{1}{4\lambda^2} - J_1(\mu^2, \beta) \right) = 0, \tag{2.46}$$

the solution of which yields the condensate $\mu(\beta)$. We are mainly interested in the phase structure at finite temperature, but it is instructive to include the case $\beta = \infty$ of zero temperature. $\mu = 0$ is always a solution of (2.46). For fixed β, $J_1(\mu^2, \beta)$ is monotonically decreasing with μ^2. Hence, for $1/(4\lambda^2) > J_1(0, \beta)$, $\mu = 0$ is the only solution, and it is straightforward to show that it is a minimum of the action. On the other hand, for $1/(4\lambda^2) < J_1(0, \beta)$ there are two non-trivial stable solution $\sigma(z) = \pm\mu(\beta)$ of (2.28) with $\mu(\beta) > 0$ satisfying (2.46).

Now suppose that the bare coupling constant λ is so large that the parity-broken phase is realized at zero temperature. We define a zero temperature mass scale M by

$$J_1(M^2, \infty) \equiv \frac{1}{4\lambda^2} < J_1(0, \infty). \tag{2.47}$$

For fixed μ^2, as a function of β, $J_1(\mu^2, \beta)$ is monotonically increasing in β. Decreasing β down from $\beta = \infty$, there is a unique critical value β_c where the broken phase becomes unstable, cf. Fig. 4. This β_c is given by

$$J_1(M^2, \infty) = J_1(0, \beta_c) \quad \text{or} \quad \beta_c = \frac{2 \ln 2}{M}. \tag{2.48}$$

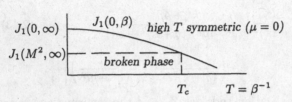

Fig. 4. Phase structure of the $3d_T$-dimensional Gross-Neveu model at $N = \infty$. For given mass scale M, the symmetric high temperature phase becomes unstable at $T_c = M/(2 \ln 2)$, and the model undergoes a second order phase transition to the broken phase.

In order to show that the symmetry gets actually broken spontaneously for $\beta > \beta_c$, before the infrared cutoff is removed (the volume is sent to infinity) we introduce an explicit parity breaking term (2.16) into the original fermionic action (2.10). Without loss we take $m > 0$. Then we remove the infrared cutoff and finally let $m \to 0+$. The only modification the fermionic mass term implies for the derivation of the large N representation of the partition function as done in the last section is the replacement of (2.23) by

$$K = D + (m + \sigma) \cdot 1. \tag{2.49}$$

The gap equation (2.46) is replaced by

$$(\mu - m) \frac{1}{4\lambda^2} - \mu J_1(\mu^2, \beta) = 0. \tag{2.50}$$

For small m and $\beta > \beta_c$ there are two stable solutions of (2.50) given by $\mu(m, \beta) = \pm\mu(\beta) + \delta$, where $\mu(\beta) > 0$ is the "unperturbed" solution of (2.46) and

$$\delta = \frac{1}{8\lambda^2 \mu(\beta)^2 J_2(\mu(\beta)^2, \beta)} \, m + o(m). \tag{2.51}$$

Inserting this into (2.38), we see that the negative solution of the gap equation gets exponentially suppressed in the thermodynamic limit for positive m. We are left with a unique absolut minimum of the action, breaking the parity symmetry.

The phase transition is of second order. The correlation length m_R^{-1} diverges at β_c. We define critical exponents in the standard way, in particular with χ_2 as the 2-point susceptibility,

$$\chi_2 \equiv \widetilde{W}^{(2)}(k = 0)^{-1} \simeq (\beta_c - \beta)^{-\gamma},$$
$$m_R \simeq (\beta_c - \beta)^\nu, \qquad \text{as } \beta \to \beta_c - \qquad (2.52)$$
$$Z_R \simeq (\beta_c - \beta)^{\nu\eta},$$

and similarly primed exponents ν', γ' by approaching the phase transition from the broken phase. To leading order in N^{-1} the 2-point function is given by

$$\widetilde{W}^{(2)}(k_0 = 0, \mathbf{k}) = 2\left(J_1(M^2, \infty) - J_1(0, \beta)\right)$$
$$+ \mathbf{k}^2 \, 2Q_3(0, \beta) + O(\mathbf{k}^4) \qquad (\text{as } \mathbf{k} \to 0) \quad (2.53)$$
$$= \frac{\ln 2}{\pi \beta \beta_c}(\beta_c - \beta) + \mathbf{k}^2 \frac{\beta_c}{16\pi} + O(\mathbf{k}^4)$$

in the symmetric phase, and

$$\widetilde{W}^{(2)}(k_0 = 0, \mathbf{k}) = 4\mu^2 J_2(\mu^2, \beta)$$
$$+ \mathbf{k}^2 \, 2\left(J_2(\mu^2, \beta) - 2\mu^2 J_3(\mu^2, \beta) + 4\mu^2 Q_4(\mu^2, \beta) - Q_3(\mu^2, \beta)\right)$$
$$+ O(\mathbf{k}^4) \qquad (\text{as } \mathbf{k} \to 0) \qquad\qquad (2.54)$$
$$\simeq \frac{2\ln 2}{\pi \beta_c^2}(\beta - \beta_c) + \mathbf{k}^2 \frac{\beta_c}{16\pi} + O(\mathbf{k}^4) \qquad (\text{as } \beta \to \beta_c-)$$

in the broken phase. We thus obtain $\gamma = \gamma' = 1$, $\nu = \nu' = 1/2$ and $\eta = 0$. Other critical exponents are determined in the straightforward way, with the result that $\alpha = \alpha' = 0$, $\delta = 3$ and $\beta = 1/2$.

2.4   What Happens for Finite N?

So far we know the phase structure of the $3d_T$ Gross-Neveu model in the limit of infinite number of flavours. As next, we are interested in the physics of the model if there is a large but still finite number of fermionic species. The parity symmetry still remains broken at small temperature and will be restored at sufficiently high temperature. Concerning the universality class of the proposed second order transition two opposite claims have been made.

On the one hand, one might follow the reasoning proposed by Pisarski and Wilczek. The length of the temperature torus is given by the inverse temperature β. In the high temperature regime this length is rather small, hence the geometry is essentially two-dimensional on length scales large compared to β. Furthermore, due to the anti-periodic boundary conditions to be imposed on the fermionic degrees of freedom, the latter are subject to an infrared cutoff, cf. (2.9). Hence, in the far infrared, only the purely scalar modes survive. It might be natural to assume that this provides a model that belongs to the

Ising universality class in two dimensions. This would imply the following values for the critical exponents: $\nu = 1$, $\gamma = 1.75$, $\eta = 0.25$. This is to be contrasted to the case of $N = \infty$.

On the contrary, the reasoning proposed by Kocic and Kogut roughly goes as follows. The model is subject to a Z_2 symmetry. This symmetry does not depend on the number of fermionic flavours. For $N = \infty$ the model is gaussian ($\nu = 1/2$, $\gamma = 1$), hence it might be so for finite N. A basic assumption made in this way of reasoning is that it is the symmetry pattern which determines the universality class of a phase transition. But we know that this can be completely misleading. For instance, let us consider a classical spin model in 3 dimensions as described by an action of the type $\phi^2 + \lambda \phi^4$. It has a global symmetry $\phi \to -\phi$ that does not depend on a particular value of the quartic coupling λ. For $\lambda = 0$ the model is purely Gaussian. But this is in sharp contrast to the critical behaviour at finite λ. Similarly, for the Gross-Neveu model, the large N behaviour is not predicitve for finite N, at least not without closer inspection.

Which scenario is the right one? The reduction argument as applied above is purely geometrical. Does it generalize to field theory, and if so, how does the effective 2 dimensional model look like? Renormalization group studies reveal an infinite number of fixed points of those models, likely to be related to various different universality classes. It is not just the symmetry pattern alone that determines the critical behaviour. To decide the nature of the transition requires more sophisticated methods known from quantum field theory.

Below we shall combine two techniques to resolve the puzzle of the Gross-Neveu model. The first one is dimensional reduction in the framework of quantum field theories. This has become a well established machinery in the recent past. It provides a computational device to derive effective models for the quantitative description of the infrared behaviour at high temperature. The second method concerns convergent high temperature series. Under very general conditions, the high order behaviour of those series allows for the study of the critical region, including the measurement of critical exponents.

3   Dimensional Reduction in Quantum Field Theory

Field theories for realistic systems and statistical models with a large number of degrees of freedom evolve considerable complexity. From the point of view of high energy physics, good examples are quantum field theories such as quantum chromodynamics, which is supposed to be the microscopic theory of the strong interaction, and the electroweak standard model. The investigation of the large scale properties becomes very demandingful. In many instances we are forced to use methods that imply appropriate simplifications. On the one hand we will learn for the more complicated situation. On the other hand it is desirable to have a systematic approach, such as convergent or asymptotic expansions. This allows for a quantitative investigation and statements to be made for the full theory.

For field theories at finite temperature, one of those systematic techniques is dimensional reduction. Originally it served as a more or less qualitative reasoning for the description of the infrared behaviour at high temperature. It is easy to understand and works well for classical field theories. The generalization to quantum field theories, however, is neither simple nor obvious. The first attempts were based on a naive generalization, by just assuming that decoupling properties of heavy masses that are well known for zero temperature generalize to finite temperature. It took some time until it was made definite that this is not the case (Landsman (1989)).

A systematic way in which dimensional reduction is realized in renormalizable quantized field theories was given by the author (Reisz (1992)). Meanwhile it has become a very powerful technique. In particular it has been used to study the infrared properties of the QCD high temperature phase, so as to extract screening masses of the quark-gluon plasma (Kärkkäinen, Lacock, Petersson and Reisz (1993)). For a summary and further references, cf. Lacock and Reisz (1993). Later, the success of this well developed technique and the experiences obtained inspired its application to the electroweak standard model of the early universe (Kajantie, Laine, Rummukainen and Shaposhnikov (1996)). In the following we will shortly outline the ideas of dimensional reduction and then will apply it to the Gross-Neveu model.

3.1 A Short Survey of Dimensional Reduction

We consider a local, renormalizable quantum field theory in thermal equilibrium by the interaction with a heat bath, in $D \geq 3$ dimensions. In the functional integral formalism such a field theory lives in a D-dimensional volume with one dimension being compactified to a torus of length given by the inverse temperature (Fig. 2). There are boundary conditions imposed on the physical fields in this direction due to the trace operation of the partition function. Bosonic degrees of freedom are subject to periodic boundary conditions, fermionic degrees of freedom to anti-periodic boundary conditions.

The idea behind dimensional reduction is based on the geometrical picture that at high temperature T the length of the temperature torus given by T^{-1} becomes small. So as the temperature increases, the D-dimensional system becomes more and more $D - 1$-dimensional on spatial length scales L large to T^{-1}. In the following we discuss the implications for a renormalizable quantum field theory.

Consider, for simplicity, a scalar model with field content $\phi(z)$, described by the action

$$S(\phi) = \int_0^{T^{-1}} dz_0 \int d^{D-1}\mathbf{z} \left(\frac{1}{2} \, \phi(z) \, (-\Delta_D \phi) \, (z) + V(\phi(z)) \right), \quad (3.55)$$

where $\Delta_D = \sum_{i=0}^{D-1} \partial^2/(\partial z_i^2)$ denotes the D-dimensional Laplacian with periodic boundary conditions in z_0, and V the self-interaction of the field. Periodicity implies that the corresponding momentum components in this direction are discrete. We apply a so-called Matsubara decomposition, which is nothing but a Fourier transform with respect to this momentum component,

$$\phi(z) = T^{1/2} \sum_{k_0 = 2n\pi T, n \in \mathbf{Z}} e^{ik_0 z_0} \, \widetilde{\phi}_n(\mathbf{z})$$

$$= T^{1/2} \, \phi_{st}(\mathbf{z}) + \phi_{ns}(z), \tag{3.56}$$

where the static field ϕ_{st} denotes the $n = 0$ component and the non-static field ϕ_{ns} sums all components with $n \neq 0$. The kinetic part of the action becomes

$$\int dz_0 \int d^{D-1}\mathbf{z} \, \frac{1}{2} \, \phi(z) \, (-\Delta_D \phi) \, (z)$$

$$= \int d^{D-1}\mathbf{z} \, \frac{1}{2} \, \widetilde{\phi}_0(\mathbf{z}) \left(-\Delta_{D-1}\widetilde{\phi}_0 \right)(\mathbf{z}) \tag{3.57}$$

$$+ \sum_{n \neq 0} \int d^{D-1}\mathbf{z} \, \frac{1}{2} \, \widetilde{\phi}_n(\mathbf{z}) \left(\left(-\Delta_{D-1}\widetilde{\phi}_n \right)(\mathbf{z}) + (2\pi n T)^2 \widetilde{\phi}_n(\mathbf{z}) \right).$$

Apparently the field components with $n \neq 0$ acquire a thermal mass proportional to T. This provides an infrared cutoff to these non-static modes mediated by the interaction with the heat bath. The only modes which do not acquire a T-dependent mass term are the $\widetilde{\phi}_0$. They are called static modes because they are precisely those field components that do not fluctuate in T-direction. They are the only degrees of freedom that survive dynamically on scales $L \geq T^{-1}$. All the other modes decouple. We call this scenario *complete dimensional reduction*.

So far, the consideration is done on the level of the action, that is, for the classical field theory. How does this picture change under quantization? Now we have to investigate in full the partition function given by

$$Z = \int \mathcal{D}\phi \, \exp\left(-S(\phi)\right) \quad ; \quad \mathcal{D}\phi = \prod_z d\phi(z), \tag{3.58}$$

with the action $S(\phi)$ as in (3.55). From the Matsubara-decomposition (3.56) it appears that the finite temperature field theory if the $\phi(z)$ in D dimensions can be viewed as a $D - 1$-dimensional field theory of the $\widetilde{\phi}_n(\mathbf{z})$ at zero temperature. There are multiple fields $\widetilde{\phi}_n$ now almost all of which become massive. The only effect of the temperature is to renormalize the interaction.

One might guess that on the infrared scale the heavy modes decouple as in the classical case, the only degrees of freedom left being the static fields $\widetilde{\phi}_0(\mathbf{z}) \equiv \phi_{st}(\mathbf{z})$. There are well known statements in quantum field theory on large mass behaviour. The Ambjørn-Appelquist-Carazonne theorem predicts the decoupling of massive fields from the low-energy scale of the theory under quite general conditions. (It was proposed originally by Appelquist and Carazonne (1975). The proof of this theorem uses the BPHZ renormalization scheme and is due to Ambjørn (1979).) In many cases it applies, such as for the simple scalar field theory above. The statement is very specific for the renormalization scheme used. All the coupling constants of the model have to be defined on the low momentum scale. In other instances the theorem does not apply. This concerns models with an anomaly or those subject to a non-linear symmetry, in which case the correlation functions are constrained by Ward identities.

The obstruction that prevents the application of this theorem in our case is peculiar. The number of heavy modes is infinite, whereas the above theorem does not make any prediction uniform on the number of modes. Anyway it would be a surprise if this theorem applies to field theories at finite temperature. For there is a strong interplay between renormalization and decoupling properties. Its validity would allow to circumvent the issue of ultraviolet renormalization of the theory, which is an intrinsic D-dimensional problem.

Whether or not the quantum fluctuations destroy complete dimensional reduction must be decided by other means. The answer was given by Landsman (1989) by means of a careful perturbative renormalization group analysis. Apart from some exceptional cases (such as QED), the answer is negative. To understand this fact, we have to account for renormalization theory. Let $\widetilde{\Gamma}(\mathbf{p}, T)$ denote a generic connected or 1-particle irreducible correlation function, $\mathbf{p}$ a generic spatial momentum. $\widetilde{\Gamma}$ achieves contributions both from the static and the non-static modes. The analogy of the classical decoupling property of the latter reads

$$\widetilde{\Gamma}(p, T) = \widetilde{\Gamma}_{st}(p, T) \left(1 + O(\frac{p}{T}, \frac{\mu}{T})\right), \tag{3.59}$$

where $\widetilde{\Gamma}_{st}$ denotes the purely static contribution to $\widetilde{\Gamma}$. μ denotes some finite mass parameter. For renormalizable field theories we can always achieve this in the first instance, by imposing normalization conditions on some low momentum scale. These conditions define the physical, renormalized coupling constants of the theory. The bare parameters of the model, that is the coupling constants of the action, are functions of the physical coupling constants and the ultraviolet cutoff. They are defined in such a way that the correlation functions $\widetilde{\Gamma}$ stay finite in the large cutoff limit and become functions of the renormalized coupling constants and the temperature only. This is the issue of renormalization.

The point to be made now is that the bare parameters do not depend on the temperature T. The latter only enters through the torus of length T^{-1}. In order to obtain (3.59), the subtractions must be temperature dependent, and so are the renormalized parameters. In particular, the renormalized mass μ that enters (3.59) becomes of the order of T in the generic case. Hence the correction term becomes of the order of 1. This shows that complete dimensional reduction breaks down.

In accordance to the decomposition (3.56) of the fields into static and non-static components, let us write the partition function as

$$Z = \int \mathcal{D}\phi_{st} \, \exp\left(-S_{eff}(\phi_{st})\right) \tag{3.60}$$

with

$$\exp\left(-S_{eff}(\phi_{st})\right) = \int \mathcal{D}\phi_{ns} \, \exp\left(-S(T^{1/2}\phi_{st} + \phi_{ns})\right) \tag{3.61}$$

obtained by integrating out the non-static modes. Formally we can write this as

$$S_{eff}(\phi_{st}) = S_0(\phi_{st}) + (\delta_{ns}S)(\phi_{st}), \tag{3.62}$$

where S_0 is the classically reduced action of the temporal zero modes ϕ_{st}. This action is completely super-renormalizable. For the example above, it is given by

$$S_0(\phi_{st}) = \int d^{D-1}\mathbf{z} \left(\frac{1}{2} \, \phi_{st}(\mathbf{z}) \, (-\Delta_{D-1}\phi_{st}) \, (\mathbf{z}) + V(\phi_{st}(\mathbf{z}))\right). \tag{3.63}$$

The quantum fluctuations of the non-static modes are completely described by the correction term $\delta_{ns}S$. This induced interaction is non-local. We know from the discussion above that it cannot be neglected even at very high temperature. Nevertheless one would still think of an appropriate high temperature expansion. Not any expansion will work because we have to avoid the same obstruction leading to (3.59).

The solution is to generate the high temperature reduction as the low momentum expansion (Reisz (1992)). $\delta_{ns}S$ is made up of purely non-static modes. Thus it has an intrinsic mass gap. Its momentum representation is analytic at zero momentum $\mathbf{p}$. We are thus allowed to expand about $\mathbf{p} = 0$. In a renormalizable field theory, the high temperature behaviour of the renormalized n-point function $\widetilde{\Gamma}_{ns}^{(n)}$ is proven to be

$$\widetilde{\Gamma}_{ns}^{(n)}(\mathbf{p},T) = T^{\rho_n} P_{\rho_n}(\frac{\mathbf{p}}{T}) + o(1) \text{ as } T \to \infty. \tag{3.64}$$

P_m denotes a (multidimensional) polynomial of degree m for $m \geq 0$, $P_m = 0$ for $m < 0$, and

$$\rho_n = (D-1) - n\frac{D-3}{2}. \tag{3.65}$$

The degree ρ_n depends only on n (and D), but is independent of the order to which $\widetilde{\Gamma}_{ns}^{(n)}$ has been computed in a perturbative (e.g. weak coupling) expansion.

Let us write $T_{\mathbf{p}}^{\delta} f(\mathbf{p})$ for the (multidimensional) Taylor expansion about $\mathbf{p} = 0$ to the order δ of the function $f(\mathbf{p})$. To keep the finite T contributions on the infrared scale amounts to replace for those n with $\rho_n \geq 0$

$$\widetilde{\Gamma}_{ns}^{(n)}(\mathbf{p}, T) \rightarrow T_{\mathbf{p}}^{\rho_n - q_n} \widetilde{\Gamma}_{ns}^{(n)}(\mathbf{p}, T) \tag{3.66}$$

and put all others to 0. Here, q_n denotes some integer number with $0 \leq q_n \leq \rho_n$. We notice that the expansion does not generate ratios of the form μ/T. The expansion is applicable even for T-dependent normalization conditions where the renormalized masses μ increase with T.

Only non-static correlations $\widetilde{\Gamma}_{ns}^{(n)}$ with $\rho_n \geq 0$ contribute to $\delta_{ns} S$ at high temperature. The contribution is polynomial in momentum space, that is, $\widetilde{\Gamma}_{ns}^{(n)}$ generates a local n-point interaction of $\delta_{ns} S$. Hence we end up with a locally interacting model of the fields $\phi_{st}(\mathbf{z})$.

If we put $q_n = 0$ above, i.e. keep the complete high T part of the non-static amplitudes, the effective model becomes a (power counting-) renormalizable field theory in $D - 1$ dimensions.

We may be interested to keep only the leading term for every correlation function. In this case we put $q_n = \rho_n$ in (3.66) and obtain

$$\widetilde{\Gamma}_{ns}^{(n)}(\mathbf{p}, T) \rightarrow T_{\mathbf{p}}^{0} \widetilde{\Gamma}_{ns}^{(n)}(\mathbf{p}, T) \equiv \widetilde{\Gamma}_{ns}^{(n)}(\mathbf{0}, T) \tag{3.67}$$

for all amplitudes $\widetilde{\Gamma}_n$ with $\rho_n \geq 0$. This restriction is meaningful at sufficiently large spatial separations compared to the inverse temperature T^{-1}, for instance for the investigation of universal critical behaviour at a second order phase transition. Corrections are suppressed by the order of $\mathbf{p}/T$.

We should emphasize at this point that (3.67) is not equivalent to keeping only those effective interactions that stay super-renormalizable (except for the reduction from $D = 3$ to $D - 1 = 2$ dimensions, cf. below). In general, in addition this would require to omit the contributions from the non-static correlation functions $\widetilde{\Gamma}_{ns}^{(n)}$ with $\rho_n = 0$. For example, for the reduction from 4 to 3 dimensions this would imply to ignore interactions of the form ϕ^6. This is justified only if the temperature is sufficiently large. Without further justification this omission is a potential danger for the investigation of the finite temperature phase transition.

If only the effective interactions mediated by the non-static modes are kept which do preserve super-renormalizability of the (D-1)-dimensional effective model, the latter is made ultraviolet finite by the D-dimensional counterterm of the original theory, of course projected in the same way as in (3.67). In the more general case, supplementary normalization conditions have to be imposed. They are obtained by matching conditions to the D-dimensional theory.

Let us summarize the state of the art for a local renormalizable quantum field theory at finite temperature. Its infrared behaviour on spatial length scales $L \geq T^{-1}$ is quantitatively described by a local, generically renormalizable or even super-renormalizable field theory of the purely static modes

$\phi_{st}(\mathbf{z})$. This is a field theory in one less dimension and at zero temperature. The non-static degrees of freedom ϕ_{ns}, which account for the oscillations along the temperature torus, have an intrinsic mass gap of the order of T and hence do not survive as dynamical degrees of freedom in this region. However, they induce local interactions of the static fields that inevitably have to be taken into account in order to describe the infrared properties in a correct way. This concerns both the properties of the high temperature phase as well as finite temperature phase transitions. From the point of view of statistical mechanics, precisely these interactions are responsible for the phenomenon that quantum critical behaviour can be different from the critical behaviour of the corresponding classical model. Which interactions actually have to be kept depends on the situation at hand, as explained above.

A remarkable property of dimensional reduction is the fact that fermions do not survive the reduction process as dynamical degrees of freedom. As discussed in the last chapter, the anti-periodic boundary conditions imposed on the fermion fields in the T-direction provide a mass gap, identifying fermions as purely non-static modes. The "only" effect they have on the infrared scale is mediated by renormalization of the interaction of the otherwise bosonic fields.

The actual computation of the effective action of the static fields ϕ_{st} can be done by various means. For four-dimensional gauge theories, QCD and the electroweak standard model weak coupling renormalized perturbation theory has been used extensively. For the Gross-Neveu model below we will carry out the dimensional reduction by means of the large N expansion.

3.2 Application to the Three-Dimensional Gross-Neveu Model at Finite Temperature

We are back now to the Gross-Neveu model. This model is renormalizable for a large number N of flavours. Hence, dimensional reduction as presented in the last section applies. We have $D = 3$, so by (3.65) we get $\rho_n = 2$ for all n. All non-vanishing correlation functions $\widetilde{\Gamma}_{ns}^{(n)}(\mathbf{p}, T)$ grow at most as T^2 at high temperature.

We are interested in the parity restoring phase transition and apply the high temperature expansion in the form (3.67) to all non-static effective vertices. That is, we keep the leading high temperature behaviour of $\widetilde{\Gamma}_{ns}^{(n)}(\mathbf{p}, T)$ for all n. It is convenient to rescale the twodimensional fields by a factor of $(16\pi)^{1/2}/\beta$. Then the effective action is given by

$$S_{eff}(\phi) = \int \frac{d^2\mathbf{z}}{(2\pi)^2} \left(\frac{1}{2}m_0^2\phi(\mathbf{z})^2 + \frac{1}{2}\sum_{i=1}^{2}\left(\frac{\partial}{\partial z_i}\phi(\mathbf{z})\right)^2 \right.$$

$$\left. + \sum_{m\geq 2}(-1)^m\frac{c_m}{N^{m-1}\beta^2}\phi(\mathbf{z})^{2m} \right). \tag{3.68}$$

Here and in the following we write ϕ instead of ϕ_{st} for the static fields. Only even powers of ϕ occur in the effective action. The bare mass squared m_0^2 and the constants c_m for $m \geq 2$ are given by

$$m_0^2 = \frac{16 \ln 2}{\beta^2 \beta_c} (\beta_c - \beta) + \frac{\overline{m}^2}{\beta^2 N} + O(\frac{1}{N^2})$$

$$c_m = \frac{1}{m} \frac{J_m(0, \beta)}{\beta^{2m-3}} (16\pi)^m (1 + O(\frac{1}{N})). \tag{3.69}$$

The c_m do not depend on the number of flavours. Also, they are temperature independent non-negative numbers because of $0 \leq J_m(0, \beta) = \text{const } \beta^{2m-3}$, with m-dependent constant. In particular, we have

$$c_2 = 8\,\pi,$$

$$c_3 = \frac{32}{9} \pi^2, \tag{3.70}$$

$$c_4 = \frac{128}{75} \pi^3.$$

If $N = \infty$, only the quadratic part of the action survives and the model is purely Gaussian. On the other hand, for large but finite N, which we are interested in, we obtain an interacting model. The interactions are alternating in sign. For every $m \geq 2$, the $2m$-point coupling constant is of the order of $N^{1-m/2}$. For large N interactions $\phi(z)^{2m}$ with large m are highly suppressed.

The $\mathbf{Z}_2$-symmetric model described by the action (3.68) is super-renormalizable. The $1/N$ expansion is an expansion in the number of loops. The only ultraviolet divergence occurs for the 2-point function to the one loop order, which is the order N^{-1}. There is only one corresponding, logarithmically divergent, mass counterterm included. This is the second term for m_0^2 of (3.69). Its precise value does not play a role for the universality class of the phase transition, as will be seen in the next chapter. All other correlation functions stay ultraviolet finite and are not subject to subtractions.

4   Phase Transition for Finite Number of Flavours N

We are prepared now to investigate the parity restoring phase transition of the Gross-Neveu model. This will be done by determining the phase structure of the associated 2-dimensional effective model described by the action (3.68).

Two-dimensional field theories typically evolve a rich and complicated phase structure. For instance, renormalization group studies of hierarchical models reveal that $\mathbf{Z}_2$ symmetric scalar models have complicated fixed point manifolds in field space (Pinn, Pordt and Wieczerkowski (1994)). We have to expect more than just one (Ising) universality class for 2d scalar models, depending to a large extent on their interactions and the coupling constant strengths.

The interactions and the values of the coupling constants of the effective model (3.68) have been computed by dimensional reduction. They are given functions of the number N of flavours and of the temperature $T = \beta^{-1}$. The model is superrenormalizable, and there is only one logarithmically cutoff dependent mass counterterm. This is the only term that depends on the particular cutoff chosen. For the following non-perturbative investigation we put the model on the lattice.

First of all, we truncate the action beyond the interaction ϕ^8. Terms of the order of N^{-4} are omitted. This is legitimate for finite, but large N, which we are interested in. On the lattice, the action becomes

$$
a^2 \sum_{\mathbf{x} \in \Lambda_2} \left\{ \left(\frac{m_0^2}{2} \phi(\mathbf{x})^2 + \frac{1}{2} \sum_{\mu=1,2} (\widehat{\partial}_\mu \phi)(\mathbf{x})^2 \right) \right.
$$
$$
\left. + T^2 \left(\frac{c_2}{N} \phi(\mathbf{x})^4 - \frac{c_3}{N^2} \phi(\mathbf{x})^6 + \frac{c_4}{N^3} \phi(\mathbf{x})^8 \right) \right\}. \tag{4.71}
$$

Here Λ_2 denotes the two-dimensional hypercubic lattice with lattice spacing a, and $\widehat{\partial}$ is the lattice derivative

$$
(\widehat{\partial}_\mu \phi)(\mathbf{x}) = \frac{1}{a} \left(\phi(x + a\widehat{\mu}) - \phi(\mathbf{x}) \right), \tag{4.72}
$$

$\widehat{\mu}$ as unit vector in the positive μ-direction.

The choice of the lattice cutoff is consistent. For, the lattice model is (super-)renormalizable. Applying well known power counting theorems on the lattice (Reisz (1988)), it can be shown that the continuum limit where the lattice spacing is sent to zero exists, to all orders of the $1/N$-expansion. That is, all correlation functions of the basic field ϕ stay ultraviolet finite. Furthermore, in this large cutoff limit they agree with those of the model with e.g. the simple momentum cutoff, which we have chosen so far.

For further use we write the action in a form that is well known for lattice spin models. Furthermore, the lattice spacing is put to 1 in the following. We rescale the fields by introducing the so-called hopping parameter κ and define further ultralocal coupling constants λ, σ and ω by

$$
\phi(\mathbf{x}) = (2\kappa)^{1/2} \phi_0(\mathbf{x}),
$$
$$
\kappa(4 + m_0^2) = 1 - 2\lambda + 3\sigma - 4\omega,
$$
$$
4c_2 \kappa^2 \frac{T^2}{N} = \lambda - 3\sigma + 6\omega, \tag{4.73}
$$
$$
-8c_3 \kappa^3 \frac{T^2}{N^2} = \sigma - 4\omega,
$$
$$
16c_4 \kappa^4 \frac{T^2}{N^3} = \omega.
$$

Then, up to an irrelevant constant normalization factor, we obtain the partition function in the form

$$Z = \int \prod_{x \in \Lambda_2} \exp\left(-S_0(\phi_0)\right) \tag{4.74}$$

with the action

$$S_0(\phi_0)) = \sum_{x \in \Lambda_2} \left(S^{(0)}(\phi_0(x)) - (2\kappa) \sum_{\mu=1,2} \phi_0(x)\phi_0(x + \widehat{\mu}) \right),$$

$$S^{(0)}(\phi_0(x)) = \phi_0(x)^2 + \lambda(\phi_0(x)^2 - 1)^2 \tag{4.75}$$
$$+\sigma(\phi_0(x)^2 - 1)^3 + \omega(\phi_0(x)^2 - 1)^3.$$

The coupling constants λ, σ and ω are of the order of N^{-1}, N^{-2} and N^{-3}, respectively, and are given by

$$\lambda = 36 \left(1 - \frac{4}{\mathcal{R}} + \frac{288}{25\mathcal{R}^2}\right) \bar{l},$$

$$\sigma = -\frac{48}{\mathcal{R}} \left(1 - \frac{144}{25\mathcal{R}}\right) \bar{l}, \tag{4.76}$$

$$\omega = \frac{1728}{25} \frac{1}{\mathcal{R}^2} \bar{l},$$

with

$$\bar{l} = \frac{8\pi\kappa^2 T^2}{9N} \quad \text{and} \quad \mathcal{R} = \frac{3N}{2\pi\kappa}. \tag{4.77}$$

The particular values where simultaneoulsy $\bar{l} = 0$ and $\mathcal{R} = \infty$ correspond to the case of infinite number of flavours, $N = \infty$. The model then is a free Gaussian model. For all other values we introduce the ratios

$$\widehat{\alpha} = \frac{\lambda}{\omega} = \frac{25}{48}\mathcal{R}^2 \left(1 - \frac{4}{\mathcal{R}} + \frac{288}{25\mathcal{R}^2}\right),$$

$$\widehat{\beta} = \frac{\sigma}{\omega} = -\frac{25}{36}\mathcal{R} \left(1 - \frac{144}{25\mathcal{R}}\right). \tag{4.78}$$

As we will see, the phase structure can be parametrized as function of the ratios $\widehat{\alpha}$ and $\widehat{\beta}$.

The finite temperature phase transition and its properties will be determined non-perturbatively by means of convergent high temperature series. We will apply the linked cluster expansion (LCE). It provides hopping parameter series for the free energy density and the connected correlation function of the model about completely disordered lattice systems. It should be emphasized that LCE generates convergent power series, convergence being uniform in volume. This implies that correlations can be measured with arbitrary precision within the domain of convergence, the precision depending on the order of computation. Even more, under quite general conditions, the phase transition is identified with the radius of convergence. The critical behaviour of the model is encoded in the high order coefficients of the series. This concerns both the location of the transition as well as the critical exponents and amplitudes.

The hopping parameter series have been generated recently for two-, four-, and six-point functions up to the 20th order, for arbitrary coupling constants (Reisz (1995a)). The series are available for general $O(N)$ scalar field models on hypercubic lattices in $D \geq 2$ dimensions. The lattices are unbounded in all directions or they have one compact dimension, representing the temperature torus. Also, the expansion has been generalized to finite lattices in order to allow for the determination of the order of a transition by finite size scaling analysis (Meyer-Ortmanns and Reisz (1997)). This LCE technique allows for a rapid generation of the series representation for all coupling constants. We will outline the idea behind this in the next section.

4.1   Convergent Series Expansion - LCE

Hopping parameter expansions of classical lattice spin systems have a long tradition in statistical physics. Good references are Wortis (1974) and Itzykson and Drouffe (1989). In contrast to generic perturbation theory about eigenstates of harmonic oscillators or free fields, which are normally at best asymptotically convergent, such expansions provide convergent series about disordered lattice systems.

Linked cluster expansion (LCE) is a hopping parameter expansion applied to the free energy density, connected correlation functions and susceptibilities. Providing absolutely convergent power series, under quite general conditions they allow for the determination of the critical properties of a statistical model. This includes critical couplings as well as critical exponents and amplitudes. These numbers are encoded in the high order behaviour of the series coefficients. Their determination requires a computation to high orders.

The realization of such expansions by convenient algorithms and with the aid of computers have been poineered by Lüscher and Weisz (1988). They were the first to perform such expansions to the 14th order of the hopping parameter. Recent progress and generalizations have been made in various directions. They have been generalized to field theories at finite temperature respectively quantum spin systems, where one dimension is compact with torus length given by the inverse temperature (Reisz (1995)). To the finite volume, allowing for finite size scaling analysis in order to determine the nature of a phase transition (Meyer-Ortmanns and Reisz (1997)). In the course of this construction it was possible to prolong the length of the series to the order 20. Other progresses concern classical fixed-length spin systems (Butera and Comi (1997)) or higher moments of two-dimensional scalar models (Campostrini, Pelissetto, Rossi and Vicari (1996)).

Below we only give a short outline of the ideas behind the expansion technique. Details will be given only to the extent they are needed for the application in the next section.

Let Λ_D denote a D-dimensional hypercubic lattice. For simplicity we work with units where the lattice spacing becomes 1. We consider a class of O(N) symmetric scalar fields living on the sites of the lattice. The partition function is given by

$$Z(H,v) = \int \prod_{x \in \Lambda_D} d^N \Phi(x) \, \exp\left(-S(\Phi,v) + \sum_{x \in \Lambda} H(x) \cdot \Phi(x)\right). \qquad (4.79)$$

Here $\Phi = (\Phi_1, \ldots, \Phi_N)$ denotes a real, N-component scalar field, H are external (classical) fields, and

$$H(x) \cdot \Phi(x) = \sum_{a=1}^{N} H_a(x) \Phi_a(x).$$

The action is given by

$$S(\Phi,v) = \sum_{x \in \Lambda_D} S^0(\Phi(x)) - \frac{1}{2} \sum_{x \neq y \in \Lambda_D} \sum_{a,b=1}^{N} \Phi_a(x) v_{ab}(x,y) \Phi_b(y). \qquad (4.80)$$

The "ultra-local" (0-dimensional) part S^0 of the action describes the single spin measure. We assume that it is $O(N)$ invariant and that it ensures stability of the partition function (4.79). For instance,

$$S^0(\Phi) = \Phi^2 + \lambda(\Phi^2 - 1)^2 + \sigma(\Phi^2 - 1)^3 + \omega(\Phi^2 - 1)^4, \qquad (4.81)$$

with $\omega > 0$. Dependence on the bare ultra-local coupling constants λ, σ and ω is suppressed in the following. Fields at different lattice sites are coupled by the hopping term $v(x,y)$. For the case of nearest neighbour interactions,

$$v_{ab}(x,y) = \delta_{a,b} \begin{cases} 2\kappa & , \text{ x, y nearest neighbour,} \\ 0 & , \text{ otherwise.} \end{cases} \qquad (4.82)$$

δ is the Kronecker symbol.

The generating functional of connected correlation functions (free energy) is given by

$$W(H,v) = \ln Z(H,v), \qquad (4.83)$$

$$W^{(2n)}_{a_1 \ldots a_{2n}}(x_1, \ldots, x_{2n}) = \langle \Phi_{a_1}(x_1) \cdots \Phi_{a_{2n}}(x_{2n}) \rangle^c$$
$$= \left. \frac{\partial^{2n}}{\partial H_{a_1}(x_1) \cdots \partial H_{a_{2n}}(x_{2n})} W(H,v) \right|_{H=0}. (4.84)$$

In particular, the connected 2-point susceptibility χ_2 and the moment μ_2 are defined according to

$$\delta_{a,b}\, \chi_2 = \sum_x\, <\Phi_a(x)\Phi_b(0)>^c,$$

$$\delta_{a,b}\, \mu_2 = \sum_x \left(\sum_{i=0}^{D-1} x_i^2\right)<\Phi_a(x)\Phi_b(0)>^c. \qquad (4.85)$$

The Fourier transform $\widetilde{W}(p)$ of the 2-point function is defined by

$$W^{(2)}_{a_1 a_2}(x_1, x_2) = \delta_{a_1, a_2} \int_{-\infty}^{\infty} \frac{d^D p}{(2\pi)^D}\, e^{ip\cdot(x_1 - x_2)}\, \widetilde{W}(p). \qquad (4.86)$$

The standard definitions of the renormalized mass m_R (as inverse correlation length) and the wave function renormalization constant Z_R are by the small momentum behaviour of $\widetilde{W}$,

$$\widetilde{W}(p)^{-1} = \frac{1}{Z_R}\left(m_R^2 + p^2 + O(p^4)\right) \quad \text{as } p \to 0. \qquad (4.87)$$

From (4.85) we obtain the relations

$$m_R^2 = 2D\frac{\chi_2}{\mu_2},\ \ Z_R = 2D\frac{\chi_2^2}{\mu_2}. \qquad (4.88)$$

The critical exponents γ, ν, η are defined by the leading singular behaviour at the critical point κ_c,

$$\begin{aligned}
\ln \chi_2 &\simeq -\gamma \ln(\kappa_c - \kappa),\\
\ln m_R^2 &\simeq 2\nu \ln(\kappa_c - \kappa), \quad \text{as } \kappa \nearrow \kappa_c, \\
\ln Z_R &\simeq \nu\eta \ln(\kappa_c - \kappa),
\end{aligned} \qquad (4.89)$$

such that $\nu\eta = 2\nu - \gamma$.

Let us assume for the moment that the interaction part (4.82) of the action between fields at different lattice sites is switched off, i.e. $v = 0$. Then the action decomposes into a sum of the single spin actions over the lattice sites, $S(\Phi, v = 0) = \sum_x S^0(\Phi(x))$, and the partition function factorizes according to

$$Z(H, v = 0) = \prod_{x \in \Lambda_D} Z^0(H(x)). \qquad (4.90)$$

It follows that

$$W(H, v = 0) = \sum_{x \in \Lambda_D} W^0(H(x)), \qquad (4.91)$$

with $W^0 = \ln Z^0$. In particular, the connected correlation functions (4.84) vanish except for the case that all lattice sites agree, i.e. for $x_1 = x_2 = \cdots = x_{2n}$. In that case,

$$W^{(2n)}_{a_1 \ldots a_{2n}}(x_1, \ldots, x_{2n}) = \frac{\overset{0\,c}{v_{2n}}}{(2n-1)!!} C_{2n}(a_1, \ldots, a_{2n}). \qquad (4.92)$$

The C_{2n} are the totally symmetric O(N) invariant tensors with $C_{2n}(1, \dots, 1)$ $= (2n-1)!! \equiv 1 \cdot 3 \cdots \cdots \cdot (2n-1)$. We have introduced the vertex couplings as defined by

$$\overset{oc}{v}_{2n} = \frac{\partial^{2n}}{\partial H_1^{2n}} \, W^0(H) \bigg|_{H=0}. \tag{4.93}$$

The vertex couplings $\overset{oc}{v}_{2n}$ depend only on the ultralocal part S^0 of the action, that is on the single spin measure. They are obtained from the relation $W^0(H) = \ln Z^0(H)$ or

$$W^0(H) = \sum_{n \geq 1} \frac{1}{(2n)!} \, \overset{oc}{v}_{2n} \, (H^2)^n = \ln\big(1 + \sum_{n \geq 1} \frac{1}{(2n)!} \, \overset{o}{v}_{2n} \, (H^2)^n\big), \tag{4.94}$$

where

$$\overset{o}{v}_{2n} = \frac{\int d^N \Phi \, \Phi_1^{2n} \exp\left(-S^0(\Phi)\right)}{\int d^N \Phi \exp\left(-S^0(\Phi)\right)}, \tag{4.95}$$

by comparing coefficients of the Taylor expansions about $H = 0$. The resulting relations of the $\{\overset{oc}{v}_{2n}\}$ and the $\{\overset{o}{v}_{2n}\}$ is one-to-one. To some extent, the $\overset{o}{v}_{2n}$ encode universality classes, as we will see below.

The linked cluster expansion for W and connected correlations $W^{(2n)}$ is the Taylor expansion with respect to the hopping couplings $v(x,y)$ about this completely decoupled case,

$$W(H,v) = \left(\exp \sum_{x,y} \sum_{a,b} v_{ab}(x,y) \frac{\partial}{\partial \widehat{v}_{ab}(x,y)} \right) W(H, \widehat{v}) \bigg|_{\widehat{v}=0}. \tag{4.96}$$

The corresponding expansions of correlation functions are obtained from (4.83) and (4.84). Multiple derivatives of W with respect to $v(x,y)$ are computed by the generating equation

$$\frac{\partial}{\partial v_{ab}(x,y)} W(H) = \frac{1}{2} \left(\frac{\partial^2 W}{\partial H_a(x) \partial H_b(y)} + \frac{\partial W}{\partial H_a(x)} \frac{\partial W}{\partial H_b(y)} \right). \tag{4.97}$$

The expansion (4.96) is convergent under the condition that the interaction $v(x,y)$ is sufficiently local and weak (cf. Pordt (1996) and Pordt and Reisz (1996)). Convergence is uniform in the volume. In all that follows we restrict attention to the case of (4.82), that is to euclidean symmetric nearest neighbour interaction of strength κ. By inspection, for the susceptibilities such as (4.85) the coefficients of the power series of κ are of equal sign. This remarkable property implies that the singularity closest to the origin lies on the positive real κ-axis. This singularity then is the phase boundary or the critical point. In turn it is possible to extract quantitative information on behaviour even close to criticality from the high order coefficients of the series.

We know that e.g. the 2-point susceptibility (4.85) is represented by the convergent series expansion

$$\chi_2 = \sum_{L \geq 0} a_{L,2} \, (2\kappa)^L \qquad (4.98)$$

with real coefficents $a_{L,2}$. From the high order behaviour of the coefficients of the series we extract the critical point κ_c and the singular critical behaviour of χ_2 close to κ_c. As

$$\chi_2 \simeq \left(1 - \frac{\kappa}{\kappa_c} \right)^{-\gamma}, \qquad (4.99)$$

we have

$$\frac{a_{L,2}}{a_{L-1,2}} = \frac{1}{2\kappa_c} \left(1 + \frac{\gamma - 1}{L} + o(L^{-1}) \right). \qquad (4.100)$$

Analogous relations hold for the other correlations, as for the correlation length m_R^{-1} and for Z_R^{-1}. Similarly, critical amplitudes are extracted. There are more sophisticated and also more involved methods than (4.100) to obtain the critical numbers to high precision. Our application below is to distinguish Ising and mean field behaviour in two dimensions, which have rather different critical exponents. For this the ratio criterion (4.100) turns out to be sufficiently precise.

The explicit computation of the series becomes rapidly complex with increasing order. It is mainly a combinatorical problem and is conveniently handled by graph theoretical techniques. Correlation functions are represented as a sum over equivalence classes of connected graphs, each class being endowed with the appropriate weight it contributes. For instance, the coefficients $a_{L,2}$ are obtained as the sum over the set $\mathcal{G}_{L,2}$ of connected graphs with 2 external and L internal lines,

$$a_{L,2} = \sum_{\Gamma \in \mathcal{G}_{L,2}} w(\Gamma), \qquad (4.101)$$

with $w(\Gamma)$ the weight of the graph Γ. Let $\mathcal{V}_\Gamma$ denote the set of vertices of Γ, and for every $v \in \mathcal{V}_\Gamma$ let $l(v)$ be the sum of internal and external lines of Γ attached to the vertex v. The weight $w(\Gamma)$ then is given by

$$w(\Gamma) = \mathcal{R}(\Gamma) \cdot \prod_{v \in \mathcal{V}_\Gamma} \frac{\overset{\circ c}{v}_{l(v)} (S^0)}{(l(v) - 1)!!}. \qquad (4.102)$$

The first factor $\mathcal{R}(\Gamma)$ is a rational number. It is the product of the (inverse) topological symmetry number of Γ, the internal O(N) symmetry factor, and the lattice embedding number of the graph Γ. This is the number of ways Γ can be put onto the lattice under the constraints imposed by the graph topology, appropriately supplemented with possible geometrical weights, as e.g. for μ_2 of (4.85). Two vertices have to be put onto nearest neighbour lattice sites if they have at least one internal line in common. Beyond this there are no further exclusion constraints. In particular, a lattice site may be covered by more than just one vertex (free embedding).

The second factor of (4.102) is the product over all vertices v of Γ of the vertex couplings $\overset{oc}{v}_{l(v)}$ (S^0). This is the only place where the ultralocal coupling constants of S^0 enter. Generally this factor is a real number, it becomes rational only in exceptional cases. An example of a weight $w(\Gamma)$ is given in Fig. 5.

$$\left(\tfrac{1}{3!3!}\right)(2D)^2\left(9(N+2)(N+4)\right)\left(\tfrac{1}{3}\overset{oc}{v}_4\right)^2\left(\tfrac{1}{15}\overset{oc}{v}_6\right)$$

Fig. 5. A graph contributing to the connected 2-point function of the D-dimensional O(N) model, and its weight.

For given dimension, lattice topology and internal symmetry, $\mathcal{R}(\Gamma)$ is a fixed rational number. Its computation is expensive, but needs to be done only once. Changing the parameters of the action (that is, κ and S^0) does not require to compute $\mathcal{R}(\Gamma)$ a new. S^0 only enters the vertex couplings $\overset{oc}{v}_{2n}$. Once we have computed all the $\mathcal{R}(\Gamma)$, the computation of the LCE series for a new S^0 requires only a minimal amount of computing time.

The number of graphs Γ and in turn of the $\mathcal{R}(\Gamma)$ is large. To the 20th order there are more than 10^8 numbers $\mathcal{R}(\Gamma)$. Actually, it is not necessary to keep the $\mathcal{R}(\Gamma)$ for all graphs Γ of interest. The idea to circumvent this is to introduce the so-called vertex structures. Roughly speaking, a vertex structure accounts for the numbers of vertices with particular numbers of lines attached. More precisely, a vertex structure w is a sequence of non-negative integers,

$$w = (w_1, w_2, \ldots) \quad, \quad w_i \in \{0, 1, 2, \ldots\}, \tag{4.103}$$

with only finitely many $w_i \neq 0$. We write $\mathcal{W}$ for the set of all vertex structures. For a graph Γ the vertex structure $w = w(\Gamma) \in \mathcal{W}$ associated to Γ is defined by (4.103) with w_i equal to the numbers of vertices of Γ that have precisely i lines attached. With the definition

$$\mathcal{R}(w) = \sum_{\substack{\Gamma \in \mathcal{G}_{L,2} \\ w(\Gamma)=w}} \mathcal{R}(\Gamma) \tag{4.104}$$

we obtain instead of (4.101) and (4.102)

$$a_{L,2} = \sum_{w \in \mathcal{W}} \mathcal{R}(w) \prod_{i \geq 1}\left(\frac{\overset{oc}{v}_i}{(i-1)!!}\right)^{w_i}. \tag{4.105}$$

The sum is finite because most of the $\mathcal{R}(w)$ vanish. Actually there are only a couple of hundreds of non-trivial $\mathcal{R}(w)$'s, to be compared to 10^8 $\mathcal{R}(\Gamma)$'s.

In order to make high orders in the expansion feasible it is necessary to introduce more restricted subclasses of graphs such as 1-line and 1-vertex irreducible graphs and renormalized moments. The correlations are then represented in terms of the latter. The major problems to be solved are algorithmic in nature. Examples are:

– unique algebraic representation of graphs
– construction and counting of graph
– computing weights, in particular lattice embedding numbers.

High orders are feasible by largely exploring simplifications. On the hypercubic lattice, only graphs matter that have an even number of lines in each loop. For $O(N)$ symmetric models, vertices with an odd number of lines attached vanish. The most important series are available to the 20th order yet.

4.2   The Nature of the Parity Transition

We determine the phase transition of the model (4.74), (4.75). Before we put the coupling constants to their prescibed values given by (4.76), it is instructive to discuss the strong coupling limit first.

As we have mentioned in the last section, universality classes can be read off to some extent from the n-dependence of the vertex couplings $\overset{\circ}{v}_{2n}$ as defined in (4.95), without having to go through the complete analysis of the LCE series. For $\mathbf{Z}_2$-symmetric models, the two cases we are interested in are the following. Let z be any positive real number. Then

$$\overset{\circ}{v}_{2k} = \frac{(2k-1)!!}{2^k}\, z^k \tag{4.106}$$

for all $k = 0, 1, 2, \ldots$ implies that the model belongs to the universality class of the Gaussian model. The critical hopping parameter κ_c is given by $\kappa_c z = 1/(2D)$ in D dimensions, and the critical exponents are $\gamma = 1$, $\nu = 1/2$ and $\eta = 0$. On the other hand, let the vertex couplings be given by

$$\overset{\circ}{v}_{2k} = \frac{(2k-1)!!}{2^k}\, z^k\, \frac{\Gamma(\frac{1}{2})}{\Gamma(\frac{1}{2}+k)}. \tag{4.107}$$

Then the universality class is that of the Ising model. In two dimensions, $\kappa_c z = (1/4)\ln{(2^{1/2}+1)}$, $\gamma = 1.75$, $\nu = 1$ and $\eta = 0.25$.

The strong coupling limit of (4.74), (4.75) is defined by $\omega \to \infty$ with the ratios $\widehat{\alpha} = \lambda/\omega$ and $\widehat{\beta} = \sigma/\omega$ hold fixed. The behaviour of the vertices $\overset{\circ}{v}_{2k}$ is obtained by the saddle point expansion. As function of $\widehat{\alpha}$ and $\widehat{\beta}$ the model evolves a rather complicated phase structure. We don't need to discuss it in detail, but the following regions are of particular interest to us.

There are regions where the behaviour is Gaussian, and regions where it is Ising like. For $\widehat{\alpha}/\widehat{\beta}^2 > 1/4$ we obtain (4.107) with $z = 1$. In this case,

$\kappa_c = (1/4)\ln(2^{1/2}+1)$ independent of $\widehat{\alpha}$ and $\widehat{\beta}$. For $0 < \widehat{\alpha}/\widehat{\beta}^2 < 1/4$ and $\widehat{\beta} < 0$ we again get (4.107), with $z = 1 - (3\widehat{\beta}/8)\left(1 + \left(1 - \frac{32}{9}\frac{\widehat{\alpha}}{\widehat{\beta}^2}\right)^{1/2}\right) > 0$. In both cases the transition is Ising like. On the other hand, in the region $0 < \widehat{\alpha}/\widehat{\beta}^2 < 1/4$ and $\widehat{\beta} > 0$, $\widehat{\alpha} < \min\left(\frac{3}{2}\widehat{\beta} - 2, \widehat{\beta} - 1\right)$ implies (4.106), with $z \sim \omega^{-1}$. The model is Gaussian behaved or completely decouples over the lattice because of $z \to 0$. In Fig. 5 we show a qualitative plot of the universality domains.

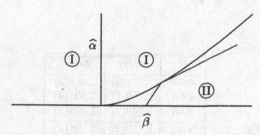

Fig. 6. Part of the phase structure of the effective model in the strong coupling limit. In the regions I the critical exponents are those of the 2-dimensional Ising model. Region II is the Gaussian domain or decoupled across the lattice.

In our case the ratios $\widehat{\alpha}$ and $\widehat{\beta}$ are largely fixed. They are given by (4.78), and

$$\frac{\widehat{\alpha}}{\widehat{\beta}^2} = \frac{27}{25}\frac{1 - \frac{4}{\mathcal{R}} + \frac{288}{25\mathcal{R}^2}}{\left(1 - \frac{144}{25\mathcal{R}}\right)^2}. \tag{4.108}$$

Because $\mathcal{R}$ is proportional to N, for sufficiently large N we have both $\widehat{\beta} < 0$ and $\widehat{\alpha}/\widehat{\beta}^2 > 1/4$. This inevitably selects the Ising universality class.

Next we consider the case of finite coupling constants λ, σ and ω. We invoke the LCE to construct the large order series of the susceptibilities χ_2 and μ_2 as defined in (4.85) for the various coupling constants. From the series coefficients then we extract the critical coupling κ_c and the critical exponents ν and γ, according to (4.99) and (4.100) and in analogy for the correlation length m_R^{-1}.

We proceed in the following way. We choose $\mathcal{R} = O(N)$ sufficiently large. Then $\widehat{\alpha}$ and $\widehat{\beta}$ are fixed. For every $\mathcal{R}$ we vary the 8-point coupling ω from $\omega = \infty$ down to small values, the other couplings running according to $\lambda = \widehat{\alpha}\omega$ and $\sigma = \widehat{\beta}\omega$. From the LCE series of χ_2 we obtain the critical hopping parameter $\kappa_c(\lambda, \sigma, \omega)$. This then determines the precise value of N corresponding to the transition point by

$$N = \frac{2\pi\kappa_c}{3}\mathcal{R}. \tag{4.109}$$

Then the critical exponents γ and ν are measured. As a result, their values are independent of ω, and they are equal to the Ising model ones. The exception is a small region about the Gaussian model with $\lambda = \sigma = \omega = 0$, where there is a smooth crossover to the Gaussian numbers. This is to be ascribed as a truncation effect of the finite order series. The above process is repeated for various values of $\mathcal{R}$ so that N varies over a large region.

In Table 1 and Fig. 7 we have collected some numbers on the critical exponents γ and ν obtained in this way. For all N the critical exponents are those of the 2-dimensional Ising model. For small N this might be not trustworthy because we have obtained the effective model within the large N expansion. On the other hand, for large N this is predictive.

N	$\gamma(N)$	$2\nu(N)$
46.1	1.751(2)	2.001(2)
24.3	1.752(3)	2.001(4)
4.68	1.749(2)	1.999(2)
1.41	1.750(2)	2.001(2)

Table 1. The critical exponents γ and ν of the three-dimensional finite temperature Gross-Neveu model for various numbers N of flavours. They agree to those of the two-dimensional Ising model.

As a result, we conclude that the 3-dimensional Gross-Neveu model, with large but finite number N of fermionic flavours, has a finite temperature phase transition that belongs to the universality class of the 2-dimensional Ising model.

5 Summary

We have investigated the finite temperature phase structure of the three-dimensional Gross-Neveu model. It is a parity symmetric, four-fermion interacting model with N fermion species, and we kept N large but finite.

From the point of view of high energy physics, the interest in this model is based on the observation that it reveals remarkable similarities to the chiral phase transition of QCD with two massless flavours. Being a three-dimensional model, chirality is replaced by parity, but both symmetries would be broken explicitly by a fermionic mass term.

Beyond its similarity to QCD, the advantage of this model is that to a large extent we have a closed analytic control of it. This includes the finite temperature transition.

At zero temperature, parity symmetry is spontaneously broken. It becomes restored at some critical temperature T_c by a second order transition.

There are at least two candidates of universality classes this transition might belong to. These are the two-dimensional Ising model and mean field behaviour. Both scenarios could be true, following different conventional wisdom arguments. It became clear in this lecture that the definite answer can be given only by explicit computations.

Our result is that for large but finite number N of fermions, the transition belongs to the universality class of the two-dimensional Ising model, with critical exponents $\gamma = 1.75$, $\nu = 1$ and $\eta = 0.25$. This result is obtained by the combination of three computational techniques: Large N expansion, dimensional reduction in quantum field theory and high order linked cluster expansions.

The large N representation of the Gross-Neveu model amounts to a resummation of planar graphs, that is, of chains of fermion bubbles, cf. Fig. 3. In the path integral approach this can be achieved in closed form by representing the four-fermi interaction as the result of a Yukawa-type interaction with an auxiliary scalar field. Whereas the weak (quartic) coupling expansion is not renormalizable, the model becomes (strictly) renormalizable for large N, in particular in the expansion in powers of N^{-1}.

Dimensional reduction is to be considered as a special high temperature expansion that is applied in the course of integrating out the degrees of freedom that fluctuate along the temperature torus. It is applied to a quantum field theory (rather than a classical field theory) to describe its infrared properties. Applied to the Gross-Neveu model, the latter is mapped onto a two-dimensional effective field theory. Due to the anti-periodicity of the fermionic fields, the model is purely bosonic. It is local and super-renormalizable. The interactions of this model are computed within the large N expansion. Both because the transition at the critical temperature T_c is of second order and N is supposed to be sufficiently large, dimensional reduction will work down to T_c.

Finally, the relevant phase structure of the dimensionally reduced effective model is determined. This is done by applying the linked cluster expansion. This technique allows us to compute convergent hopping parameter series for the free energy and connected correlation functions to very high orders. It provides the series also for the correlation length and the wave function renormalization constant. The high order behaviour of the series coefficients allow for the determination of critical couplings, including critical exponents.

The effective model has a complex phase structure. In particular there are both regions in coupling constant space with mean field and with Ising behaviour. However, the coupling constants are not free but have been computed already by dimensional reduction. They identify the Ising region as that part which describes the finite temperature Gross-Neveu model.

It would be interesting to supplement these considerations by other means. In the large N expansion, N^{-1} acts as a real coupling constant. The parity symmetry of the model does not depend on N. This promises an investigation by means of the renormalization group, with N^{-1} kept as a running coupling constant. This investigation should reveal a non-trivial infrared fixed point at some finite $N^* < \infty$. For instance, one could try an ϵ-expansion about $1 + 1$ dimensions.

What do we learn for the QCD chiral phase transition? The situation is similar to the Gross-Neveu model. The effective model now is a three-dimensional, pure bosonic theory. Fermions do not act as dynamical degrees of freedom on the infrared scale $R \geq T^{-1}$, but they determine the bosonic interactions. The σ-model scenario of the two flavour case is not ruled out, with critical exponents of the three-dimensional $O(4)$ model. This is also supported by recent numerical investigations (Boyd, Karsch, Laermann and Oevers (1996), cf. also the contribution of Frithjof Karsch to these proceedings).

Dimensional reduction has been proved to be a powerful method to study infrared properties of the QCD plasma phase. The reduction step from four to three dimensions is normally done by means of renormalized perturbation theory, that is as an expansion with respect to the renormalized gauge coupling constant $g_{ren}(R, T)$ at temperature T and length scale R. In this way there is analytic control of the cutoff and volume dependence. On lattices with moderate extension in the temperature direction, these effects are rather large. This implies a considerable complication for numerical investigations. Their analytic control allows to remove these effects. In this way it is possible to determine screening masses of the quark gluon plasma already on small lattices.

This works fine for temperatures at least above twice the phase transition temperature T_c. Compared to the Gross-Neveu model we have one more spatial dimension. From dimensional power counting this implies that interactions of higher operators are suppressed by inverse powers of the temperature. We do not have to rely on a large number of flavours any more to truncate those terms. However, approaching the phase transition, the renormalized gauge coupling constant becomes large, and we can no longer rely on perturbation theory. The relevant coupling constants of the effective model have to be computed by non-perturbative means. Therefore, the universality class of the two-flavour chiral transition in QCD is still an open question.

References

Ambjørn, J. (1979): On the decoupling of massive particles in field theory, Comm. Math. Phys. **67**, 109-119

Appelquist, T. and Carazzone, J. (1975): Infrared singularities and massive fields, Phys. Rev. **D11**, 2856-2861

Boyd, G., Karsch. F., Laermann, E. and Oevers, M. (1996): Two flavour QCD phase transition, talk given at the 10th International Conference on Problems of Quantum Field Theory, Alushta, Ukraine, 13-17 May 1996, Univ. Pisa preprint IFUP-TH-40-96, and e-Print archive: hep-lat@xxx.lanl.gov 9607046

Butera, P. and Comi, M. (1997): 2n-point renormalized coupling constants in the three-dimensional Ising model: Estimates by high temperature series to order β^{17}, Phys. Rev. **E55**, 6391-6396

Campostrini, M., Pelissetto, A., Rossi, P. and Vicari, E. (1996): A strong coupling analysis of two-dimensional O(N) sigma models with $N \geq 3$ on square, triangular and honeycomb lattices, Phys. Rev. **D54**, 1782-1808

Chevalley, C. (1997): *The Algebraic Theory of Spinors and Clifford Algebras* (Springer, Berlin, Heidelberg)

de Calan, C., Faria da Veiga, P.A., Magnen, J., Seneor, R. (1991): Constructing the three-dimensional Gross-Neveu model with a large number of flavor components. Phys. Rev. Lett. **66**, 3233-3236

Faria da Veiga, P.A., (1991): PhD thesis, University of Paris (unpublished).

Itzykson, C. and Drouffe, J.-M. (1989): *Statistical field theory II* (Cambridge University Press, Cambridge)

Kärkkäinen, L., Lacock, P., Petersson, B. and Reisz, T. (1993): Dimensional Reduction and Colour Screening in QCD, Nucl. Phys. **B395**, 733-746

Kajantie, K., Laine, M., Rummukainen, K. and Shaposhnikov, M. (1996): The electroweak phase transition: a nonperturbative analysis, Nucl. Phys. **B466**, 189-258

Kapusta, J. I. (1989): *Finite-temperature field theory* (Cambridge University Press, Cambridge)

Kocic, A., Kogut, J. (1994): Can sigma models describe the finite temperature chiral transitions?, Phys. Rev. Lett. **74**, 3109-3112

Lacock, P. and Reisz, T. (1993): Dimensional reduction of QCD and screening masses in the quark gluon plasma, Nucl. Phys. (Proc. Suppl.) **B30**, 307-314

Landsman, N.P. (1989): Limitations to dimensional reduction at high temperature, Nucl. Phys. **B322**, 498-530

Lüscher, M. (1977): Construction of a selfadjoint, strictly positive transfer matrix for euclidean lattice gauge theories, Comm. Math. Phys. **54**, 283-292

Lüscher, M. and Weisz, P. (1988): Application of the linked cluster expansion to the n-component ϕ^4 theory, Nucl. Phys. **B300**, 325-359

Lüscher, M., Sint, S., Sommer. R., Weisz, P. and Wolff, U. (1996): Nonperturbative O(a) improvement of lattice QCD, Nucl. Phys. **B491**, 323-343

Meyer-Ortmanns, H. (1996): Phase Transitions in Quantum Chromodynamics, Rev. Mod. Phys. **68**, 473-598

Meyer-Ortmanns, H. and Reisz, T. (1997): Critical phenomena with convergent series expansions in a finite volume, Jour. Stat. Phys. **87**, 755-798

Osterwalder, K., Schrader, R. (1975): Axioms for Euclidean Green's functions II, Comm. Math. Phys. **42**, 281-305

Park, S., Rosenstein, B., Warr, B. (1989a): Four-fermion theory is renormalizable in 2+1 dimensions. Phys. Rev. Lett. **62**, 1433-1436

Park, S., Rosenstein, B., Warr, B. (1989b): Thermodynamics of (2+1) dimensional four-fermion models. Phys. Rev. **D39**, 3088-3092

Pisarski, R. D. and Wilczek, F. (1984): Remarks on the chiral phase transition in chromodynamics. Phys. Rev. **D29**, 338-341

Pinn, K., Pordt, A. and Wieczerkowski, C. (1994): Computation of hierarchical renormalization group fixed points and their ϵ-expansion, Jour. Stat. Phys. **77**, 977-1005

Pordt, A. (1996): A convergence proof for linked cluster expansions, Univ. Münster preprint MS-TPI-96-05, and e-Print archive: hep-lat@xxx.lanl.gov 9604010

Pordt, A. and Reisz, T. (1996): Linked cluster expansions beyond nearest neighbour interactions: Convergence and graph classes (1996): Univ. Heidelberg preprint HD-THEP-96-09 and Univ. Münster preprint MS-TPI-96-06, and e-Print archive: hep-lat@xxx.lanl.gov 9604021, to appear in Int. J. Mod. Phys. A

Porteous, I. R. (1995): *Clifford algebras and the classical groups* (Cambridge University Press, Cambridge)

Rajagopal, K. and Wilczek, F. (1993): Static and dynamic critical phenomena of a second order QCD phase transition, Nucl. Phys. **B399**, 395-425

Reisz, T. (1988): A power counting theorem for Feynman integrals on the lattice, Comm. Math. Phys. **116**, 81-126, and: Renormalization of Feynman integrals on the lattice, Comm. Math. Phys. **117**, 79-108

Reisz, T. (1992): Realization of dimensional reduction at high temperature, Z. Phys. **C53**, 169-176

Reisz, T. (1995a): Advanced linked cluster expansion: Scalar fields at finite temperature, Nucl. Phys. **B450**[FS], 569-602

Reisz, T. (1995b): High temperature critical O(N) field models by LCE series, Phys. Lett. **B360**, 77-82

Rothe, H. J. (1997): *Lattice gauge theories, an introduction*, 2nd edition (World Scientific, Singapore)

Wilczek, F. (1992): Application of the renormalization group to a second-order QCD phase transition. Int. Jour. Mod. Phys. **A7**, 3911-3925

Wortis, M. (1974): Linked cluster expansion, in *Phase Transitions and Critical Phenomena* Vol. 3, Domb, C. and Green, M. S., eds. (Academic Press, London)

Zinoviev, Yu. M. (1995): Equivalence of Euclidean and Wightman field theories, Comm. Math. Phys. **174**, 1-28

The TBA, the Gross-Neveu Model, and Polyacetylene

Alan Chodos [1] and Hisakazu Minakata [2]

[1] Yale University, Center for Theoretical Physics, P.O. Box 208120
New Haven, Connecticut 06520-8120, USA
[2] Tokyo Metropolitan University, Department of Physics
1-1 Minami-Osawa, Hachioji, Tokyo 192-03, Japan

Abstract. We summarize recent work showing how the Thermodynamic Bethe Ansatz may be used to study the finite-density first-order phase transition in the Gross-Neveu model. The application to trans-polyacetylene is discussed, and the significance of the results is addressed.

In this contribution, rather than simply repeat what is contained in the literature, we should like to attempt to place this work in context, and to emphasize those points that are either the most novel or the most inviting of further investigation. Details of calculation can be found in the published references (Chodos and Minakata 1994, 1997).

This work brings together three disparate elements: the Gross-Neveu (GN) model, the thermodynamic Bethe Ansatz (TBA), and the phenomenology of polyacetylene $((CH)_x)$.

Polyacetylene is essentially a linear chain which comes in two forms, dubbed *trans* and *cis*. The *trans* form, which is the more stable, has a doubly-degenerate ground state. It is this circumstance that allows for the existence of topological excitations, or solitons, and these lead in turn to a rich phenomenology. An excellent review of polyacetylene and other similar polymers is the article in Reviews of Modern Physics by Heeger, Kivelson, Schrieffer and Su (1988).

It was realized long ago, by Takayama, Lin-liu and Maki (1980) and by Campbell and Bishop (1982), that in the continuum limit, and in an approximation that ignores the dynamics of the lattice vibrations, the insulator-metal transition in polyacetylene can be described by the $N = 2$ Gross-Neveu model, i.e. by the Lagrangian density

$$\mathcal{L} = \bar{\psi} i \not{\partial} \psi - g\sigma\bar{\psi}\psi - \frac{1}{2}\sigma^2. \tag{0.1}$$

This is a $1 + 1$ dimensional model in which ψ is a 2-component Dirac spinor with an internal index that runs over two values (hence $N = 2$). It describes the electronic degrees of freedom of $(CH)_x$, while the auxiliary field $\sigma(x)$ represents the lattice distortion. It is interesting to note, however, that the spin degrees of freedom of the electron are described by the internal index of ψ, while the spinor index of ψ takes account of the right-moving and left-moving electrons near the Fermi surface. It is also remarkable that polyacetylene can be described, albeit approximately, by a relativistic model. Of course the limiting velocity in the Lorentz group is not the speed of light but rather the Fermi velocity v_F.

Our original motivation in this work was to understand an observed phase transition in polyacetylene, as a function of the concentration of dopants. At a critical concentration of .06 dopants per carbon atom, the data are consistent with a first-order phase transition to a metallic state. We attempt to describe this by computing the effective potential of the $GN-$model in the presence of a chemical potential h. The role of h is to account for the extra density of electrons that is supplied to the system by the dopants. It is introduced by adding a term $h\psi^\dagger\psi$ to the Hamiltonian of the system.

The behavior of the GN model at finite temperature and density has received considerable attention over the years. We note in particular the work of Wolff (1985), of Karsch, Kogut and Wyld (1987), of Osipov and Fedyanin (1987) and of Klimenko (1988). In our case, we are interested in the effect of the chemical potential (or electron density) with the temperature set to zero. We recall that, in leading order in the $1/N$ expansion, the zero-chemical-potential GN model exhibits spontaneous breaking of discrete chiral symmetry (as well as other interesting phenomena like asymptotic freedom and dimensional transmutation). This corresponds nicely to the degenerate ground state of $trans - (CH)_x$.

As the chemical potential is increased from zero, one finds that at a certain value, $h = m/\sqrt{2}$, where m is the (dynamically-generated) mass of the fermion, the system undergoes a first order phase transition in which the symmetry is restored. The dynamical mass goes to zero above the transition point.

Remarkably enough, the critical value of h that is found above corresponds to a dopant concentration of .06. Furthermore, the transition from the soliton-dominated broken phase to the unbroken phase in which the fermions behave like a free gas is consistent with the observed transition in which the electronic properties resemble those of a metal. And the data also suggest that the transition is first-order (Tsukamoto 1992; Chen, et al. 1985; Moraes, et al. 1985).

The puzzling aspect of all this is why the agreement between theory and experiment should be so good. Not only are we using the GN model, which is an approximation to the lattice Hamiltonian of Su, Schrieffer and Heeger (1980), but the results have been obtained in leading order in the $1/N$ expansion, and, as we have already remarked, the correspondence between the GN model and polyacetylene requires $N = 2$. (In fairness, one should note that $N = 2$ refers to 2 Dirac fermions; in the literature one often calls this the $N = 4$ model, because there are 4 Majorana fermions. Thus a simple change of notation would appear to improve the validity of the approximation significantly.)

Ideally, one would like to solve the thermodynamics of the $N = 2$ GN model exactly. This is not totally out of the question, because the S-matrix, which as we shall see below is the required input for the TBA, is known exactly (Karowski and Thun 1981), and in fact work is in progress to solve the TBA (at zero temperature) for the $N = 2$ GN model numerically (Chodos, Klümper and Minakata, 1997). However, in the remainder of this note we shall discuss the problem of extending the analysis to the next-to-leading order in $1/N$. This should at least give us some indication as to whether the corrections might significantly affect the location, or even the existence, of the phase transition we have already found.

The straightforward way to go beyond leading order would be to compute the effective potential for the GN model to next-to-leading order in $1/N$, including the corrections due to a non-vanishing chemical potential. The analogous computation at zero chemical potential was performed long ago by Schonfeld (1975) and by Root (1975), and an inspection of their work reveals that it is already at a level of complexity that the labor required to incorporate $h \neq 0$ seems excessive. We choose instead to attack the problem through the use of the TBA.

Perhaps the most interesting feature of this investigation is simply the discovery of how to establish the correspondence between the results obtained directly from the effective potential and the information contained in the TBA. We shall find that, with the exception of a single dimensionless constant, we can recover all the information about the phase transition from the TBA. Furthermore, the unknown constant can be obtained from the effective potential evaluated at zero chemical potential. Once this has been understood and the correspondence has been verified to leading order in $1/N$, going to next order is simply a matter of making use of results already in the literature and performing one integral numerically. It remains an interesting question whether even the missing constant can be extracted from the TBA, so that, given the S-matrix the TBA would provide a self-contained and complete description of the thermodynamics of the GN system.

The thermodynamic Bethe Ansatz, in a non-relativistic setting, appears in the classic paper by Yang and Yang (1969). The zero-temperature version can be found in the earlier work of Lieb and Liniger (1963). More recent work, including the extension to relativistic field theory, can be found in the papers of Thacker (1981), Zamolodchikov (1990), Klassen and Melzer (1991), and Forgacs, et al. (1991a; 1991b).

At zero temperature, the essence of the TBA is a linear integral equation for the dressed single-particle excitation energy $\epsilon(\theta)$. Here θ is the rapidity of the particle: $E = m cosh\theta$, $p = m sinh\theta$. The equation reads:

$$\epsilon = h - mcosh\theta + \int_{-B}^{B} d\theta' K(\theta - \theta')\epsilon(\theta') \tag{0.2}$$

where the kernel K is the logarithmic derivative of the S-matrix:

$$K(\theta) = \frac{1}{2\pi i}\frac{d}{d\theta}lnS(\theta). \tag{0.3}$$

[The TBA applies to theories in $1 + 1$ dimension, such as the GN model, where there is only two-body scattering and the S-matrix is therefore only a phase factor depending on the relative rapidities of the two particles.] The parameter B is determined by the condition $\epsilon(\pm B) = 0$ (one is implicitly assuming here first, that $\epsilon(-\theta) = \epsilon(\theta)$, and second, that ϵ is positive for $-B < \theta < B$ and negative for $| \theta |> B$).

Once ϵ has been obtained from the solution to this equation, one can compute the free energy density of the system as (Forgacs, et al. 1991a; 1991b)

$$f(h) - f(0) = \frac{-m}{2\pi}\int_{-B}^{B} d\theta\epsilon(\theta)cosh\theta. \tag{0.4}$$

The constant $f(0)$, on dimensional grounds, must have the form $f(0) = -bm^2$, where b is a dimensionless constant. Our notation anticipates the fact that b will turn out to be positive.

One notes that in the GN model, the expansion of $K(\theta)$ begins in order $1/N$, and therefore to leading order eqn. (2) is extremely simple:

$$\epsilon(\theta) = h - mcosh\theta \tag{0.5}$$

with B determined by $coshB = h/m$. We see that this is only possible (for real B) if $h \geq m$, and so we have

$$f(h) - f(0) = \frac{-m^2}{2\pi}\theta(h - m)[\frac{1}{2}sinh2B - B]. \tag{0.6}$$

As was shown in (Chodos and Minakata 1997), one finds that the free energy given above can be interpreted as the value of the effective potential $V(\sigma)$ (properly normalized) at the point $\sigma = \sigma_0$ corresponding to the broken vacuum; the computation of the effective potential is, however, considerably more involved than the manipulations described above for the TBA, so the TBA does indeed permit a much more efficient evaluation of the free energy.

There is, however, an essential point which demands resolution. Where is the evidence for a phase transition? From the effective potential, one learns that there is a first order transition at $h = m/\sqrt{2}$, but the free energy obtained from the TBA is absolutely flat at $h = m/\sqrt{2}$, having non-trivial functional dependence on h only for $h \geq m$. For $h > m/\sqrt{2}$, the expression for the free energy obtained from the effective potential that agrees with the TBA is not actually the value from the true minimum. Rather, as stated above, it is from the point representing the broken vacuum, which ceases to be the global minimum for $h > m/\sqrt{2}$.

To resolve this difficulty, one must recognize that there are really two phases involved: the massive phase characteristic of spontaneously broken chiral symmetry, and described by eqn. (6), and a massless phase corresponding to the restoration of this symmetry. One can obtain the free energy for this phase simply by taking the limit as $m \to 0$ (with h fixed) of eqn. (6), recognizing that $f(0)$ vanishes in this limit. The result is

$$f_0(h) = \frac{-h^2}{2\pi} . \tag{0.7}$$

The system will choose to be in whichever of the two phases has the lower free energy, and if there is a value of h for which $f(h) = f_0(h)$, the system will undergo a phase transition at that point.

In order to make this comparison, we need to know the constant b that appears in the formula

$$f(0) = -bm^2. \tag{0.8}$$

If $0 < b < 1/2\pi$, then the phase transition will occur at $h/m = \sqrt{2\pi b}$.

According to our present understanding, the TBA itself does not determine b. However, within the effective potential formalism, $f(0)$ is the difference $V_0(\sigma_0) - V_0(0)$, where $V_0(\sigma)$ is the effective potential computed at zero chemical potential (up to a normalization factor of $1/N$, this is just the effective potential computed originally by Gross and Neveu) and σ_0 is the value of V_0 at its minimum. By consulting Gross and Neveu's original work (1974), one obtains

$$b = 1/4\pi \tag{0.9}$$

and hence at the transition $h = m/\sqrt{2}$, reproducing the result from the effective potential.

It is now straightforward to extend this reasoning to next order in $1/N$. The steps are the following:

(a) One makes use of the results of Schonfeld and Root (in particular, in (Chodos and Minakata 1997) we employed an integral formula due to Schonfeld (1975)) to obtain the correction, at zero chemical potential, to $V_0(\sigma)$ of Gross and Neveu. One finds thereby that $b \to b + \Delta b$, with $\Delta b = (-\frac{1}{4\pi})\frac{2.12}{3N}$.

(b) One inserts into the TBA the first non-vanishing contribution to K, which occurs at order $1/N$. Following the work of Forgacs, et al. (1991a), we assume that only the S-matrix describing the scattering of fundamental fermions is relevant. [If this is not the case, there will be further, presumably smaller, corrections to this order, but we should still obtain a result that is qualitatively correct.]

(c) One takes the massless limit of the free energy obtained in part (b), giving $f_0(h) = \frac{-1}{2\pi}(1 + \delta)h^2$, with $\delta = (\frac{.232}{N})$.

One then makes the same comparison between $f(h)$ and $f_0(h)$ that one did in leading order, to see to what extent the phase transition point has been shifted.

The result for the new critical h is

$$h = \frac{m}{\sqrt{2}} [1 + (\frac{-.47}{N})] \tag{0.10}$$

which for $N = 2$ amounts to about a 20% correction. Since, in the massless phase, the dopant concentration is directly proportional to the chemical potential (see Chodos and Minakata, 1994), this implies that the critical concentration of dopants is about 20% lower, not quite as good as the leading order result, but still comfortably compatible with experiment.

Let us conclude with a few remarks concerning a class of models for which the TBA is exactly solvable. These models a priori have nothing to do with the GN model that is our principal concern in this paper, but being exactly solvable they can provide some insight concerning the behavior and mathematical properties of eqn. (2).

The simplest such model has a kernel given by

$$K(\theta) = \lambda cosh\theta . \tag{0.11}$$

It is not hard to show that the corresponding S-matrix is $S(p_1, p_2) = e^{i\Phi}$, $\Phi = \frac{2\pi\lambda}{m^2}\epsilon_{\mu\nu}p_1^\mu p_2^\nu$ where $\theta = \theta_1 - \theta_2$, and p_1 and p_2 are the 4-momenta of the scattered particles. This S satisfies the appropriate restrictions imposed by analyticity, unitarity and crossing, but it is not polynomially bounded, and hence it is not clear what kind of underlying degrees of freedom are being described.

In any case, solving the TBA equation one finds

$$\epsilon(\theta) = h - \tilde{m}cosh\theta \tag{0.12}$$

where

$$\tilde{m} = \frac{m}{1 + \lambda(\frac{1}{2}sinh2B - B)} \tag{0.13}$$

with $h = \tilde{m}coshB$. The free energy becomes

$$f(h) - f(0) = -\theta(h - \tilde{m})\frac{m^2}{2\pi}\frac{1}{\lambda + [\frac{1}{2}sinh2B - B)]^{-1}} \qquad (0.14)$$

One can see for $\lambda > 0$ that h is bounded above, and in fact $h \to 0$ as $B \to \infty$. For $\lambda < 0$, h can take on all values, in particular $h \to \infty$ when $B \to B_c$ given by $1 + \lambda(\frac{1}{2}sinh2B_c - B_c) = 0$, but one finds that $\lim_{m \to 0} f(h) = 0$, so there is no "asymptotically free" behavior in which the free energy has the form $f(h) \to -\kappa h^2$ for sufficiently large h.

If we study the next simplest model,

$$K(\theta) = \lambda cosh3\theta \qquad (0.15)$$

[the case $K = \lambda cosh2\theta$ is excluded by crossing symmetry] we again find that this is represented by a non-polynomially bounded S-matrix, and that there is a marked difference in behavior between positive and negative λ. The solution for ϵ takes the form

$$\epsilon(\theta) = h + \epsilon_1 cosh\theta + \epsilon_3 cosh3\theta . \qquad (0.16)$$

The coefficient ϵ_1 is negative, but for $\lambda > 0$, ϵ_3 is positive. This means that the condition $\epsilon(B) = 0$ is not sufficient to guarantee a solution, because there exists a $B' > B$ such that $\epsilon(\theta) > 0$ for $| \theta |> B'$. We conclude that no solution exists for $\lambda > 0$. For $\lambda < 0$, solutions do exist that are qualitatively similar to the $\lambda < 0$ solutions of the previous example; i.e., they do not exhibit the asymptotic freedom that would allow for the existence of a massless phase such as is found in the GN model.

Some of the questions regarding these models are: (a) is it possible to discover what the underlying degrees of freedom are? (b) can one find an example that is more "realistic" (in the sense that it shares the important features of the GN model)? (c) can one use these models as a laboratory for understanding how to derive the constant $f(0)$ directly from the TBA?

Finally, we conjecture that it may be possible to expand a physically interesting (i.e. one derived from a polynomially bounded S-matrix) kernel $K(\theta)$ in a series

$$K(\theta) = \sum_{n=0}^{\infty} C_n cosh(2n + 1)\theta . \qquad (0.17)$$

This would reduce the original TBA equation to a matrix equation for the coefficients ϵ_n in the expansion

$$\epsilon(\theta) = h + \sum_{n=0}^{\infty} \epsilon_n cosh(2n + 1)\theta \qquad (0.18)$$

which in turn might bring the problem to a more tractable mathematical form.

Our overall assessment of the status of this work is as follows: we have shown that the Gross-Neveu model appears to give a remarkably good description of the finite-density phase transition observed in polyacetylene, and that the result is stable against higher-order $1/N$ corrections. Numerical work is in progress to solve the TBA integral equation exactly for the case of interest. Deeper questions remain, such as whether the TBA can provide a complete account of the thermodynamics of the GN model, and whether models can be found which permit the exact solution of the TBA equation while at the same time making unambiguous physical sense.

Acknowledgements: A.C. wishes to thank H. Meyer-Ortmanns and A. Klümper for the opportunity to attend such a stimulating workshop. This work was performed under the Agreement between Yale University and Tokyo Metropolitan University on Exchange of Scholars and Collaborations (May 1996). A.C. was supported in part by the DOE under grant number DE-FG02-92ER-40704. H.M. was partially supported by Grant-in-Aid for Scientific Research #09640370 of the Japanese Ministry of Education, Science and Culture, and by Grant-in-Aid for Scientific Research #09045036 under the International Scientific Research Program, Inter-University Cooperative Research.

References

Campbell, D.K., Bishop, A.R. (1982): Nucl. Phys. **B200**, 297.

Chen, J., et al. (1985): Solid State Commun. **53**, 757.

Chodos, A., Klümper, A. , Minakata, H. (1997): work in progress.

Chodos, A., Minakata, H. (1994): Phys. Lett. **A191**, 39.

Chodos, A., Minakata, H. (1997): Phys. Lett. **B490**, 687.

Forgacs, P., Niedermayer, F., Weisz, P. (1991a): Nucl. Phys. **B367**, 123.

Forgacs, P., Niedermayer, F., Weisz, P. (1991b): Nucl. Phys. **B367**, 144.

Heeger, A.J., Kivelson, S., Schrieffer, J.R., Su, W.P. (1988): Rev. Mod. Phys. **60**, 781.

Karowski, M., Thun, H.J (1981): Nucl. Phys. **B190**, 61.

Karsch, F., Kogut, J., Wyld, H.J. (1987): Nucl. Phys. **B280**, 289.

Klassen, T.R., Melzer, E. (1990): Nucl. Phys. **B338**, 485.

Klassen, T.R., Melzer, E. (1991): Nucl. Phys. **B350**, 635.

Klimenko, K.G. (1988): Teor. Mat. Fiz. **75**, 226.

Lieb, E.H., Liniger, W. (1963): Phys. Rev. **130**, 1605.

Moraes, F.M. et al. (1985): Synth. Met. **11**, 271.

Osipov, V.A., Fedyanin, V.K. (1987): Theor. Math. Phys. **73**, 393 (in Russian); 1296 (in English).

Root, R.G. (1975): Phys. Rev. **D11**, 831.

Schonfeld, J. (1975): Nucl. Phys. **B95**, 148.

Su, W.P., Schrieffer, J.R., Heeger, A.J., (1980): Phys. Rev. **B22**, 2099.

Takayama, H., Lin-liu, Y.R., Maki, K., (1980): Phys. Rev. **B21**, 2388 .

Thacker, H.B. (1981): Rev. Mod. Phys. **53**, 253.

Tsukamoto, J. (1992): Adv. Phys. **41**, 509.

Wolff, U. (1985): Phys. Lett. **B157**, 303.

Yang, C.N., Yang, C.P (1969): J. Math. Phys. **10**, 1115.

Zamolodchikov, Al.B. (1990): Nucl. Phys. **B342**, 695.

Solitons in Polyacetylene

Siegmar Roth

Max-Planck-Institut für Festkörperforschung, Heisenbergstr. 1, 70569 Stuttgart, Germany

1 Introduction

In the context of this workshop it is not necessary to introduce solitons. The previous contributions have done this thoroughly. We can restrict ourselves to the minimum requirements:
1. In mathematics solitons are analytical solutions of non-linear differential equations (This is in contrast to the wave the equation, which is a linear differential equation and the solutions of which are harmonic waves).
2. In physics solitons are "particle-like" objects. They conserve their form in motion and at collisions.

From the pragmatic point of view there are two families of solitons:

a) "Pulses", or the Korteweg-deVries family.

b) "Steps", or the Sine-Gordon family.

These families are schematized in Fig. 1.

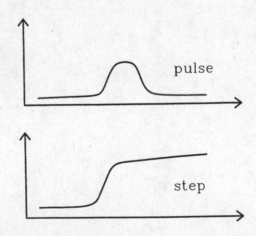

Fig. 1. Two families of solitons: Pulses (Korteweg-deVries family) and Steps (Sine-Gordon family)

Tsunamis, or flood waves, have often been evoked as examples of soliton-like excitations in hydrodynamics. Fig. 2 shows a non-linear water wave in the famous wood carving of Hokusai, overlaid by a histogram of the annual number of publications on solitons. We see that the literature on solitons by itself is a bibliographic wave of solitonic character (Roth (1995)).

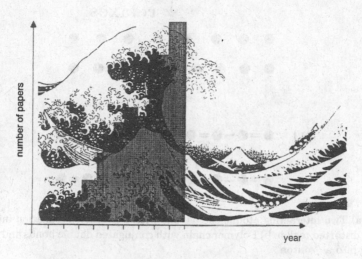

Fig. 2. Wood carving of water wave by Hokusai, overlaid by bibliographic soliton wave in the literature on solitons.

An early example of solitons in solid state physics is Seegers treatment of two interacting atomic chains (Seeger (1948)), as shown in Fig. 3a. One chain is perfect, the other has an atom missing (vacancy). This second chain has to find a compromise between perfect superposition ("register") over the first chain and equidistant arrangement of its atoms. So the vacancy relaxes into a region of enlarged interatomic distances. In polymer physics a similar defect can be realized in one single chain - if this is a chain with conjugated double bonds. This is illustrated in Fig. 3b. In a conjugated chain double bonds alternate regularly with single bonds. The soliton-like defect is a misfit where two single bonds meet (or two double bonds). To see the "step" we plot the "bond alternation parameter". This is the difference in length between two adjacent bonds - Fig. 4. At the misfit this parameter changes sign. Does a misfit behave like a particle? Does it move? Does it conserve its shape during motion? One can simulate a conjugated chain on a computer and one can show that, indeed, a misfit has soliton properties. Moreover, one can even treat a polymer chain in the continuum model, in which the atoms or molecules are not discrete entities sitting on lattice sites with spaces in-between, but where they are continuously smeared out all over the chain ("jellium"). In the continuum limit misfits in conjugated chains turn into true solitons (Takayama et al. (1980)). 2

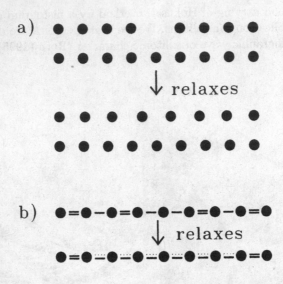

Fig. 3. a) Two interacting atomic chains, one with a vacancy. Relaxation into soliton-like distorted region. b) Polymer chain with conjugated double bonds and misfit, relaxing into a "soliton".

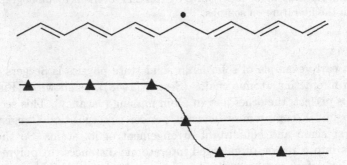

Fig. 4. Bond alternation parameter changing sign at a misfit on a conjugated chain, to illustrate soliton-like step.

2   Conjugated Polymers

Polyacetylene is the prototype conjugated polymer. It consists of a linear chain of CH-radicals, i.e. of carbon atoms, each bound by so-called σ-bonds to two carbon neighbors and to a hydrogen atom. The fourth valence forms the system of conjugated double bonds (π-bonds). We have already used this chain in Figs. 3 and 4 (where we have ignored the hydrogen atoms). In Fig. 3 we have drawn the chain straight, which is an oversimplification. The correct chemical structure is the zigzag configuration used in Fig. 4. Polyacetylene is not the only conjugated polymer, some others are shown in Fig. 5, using the conventional chemical structure formulae, and in Fig. 6, using computer-generated space-filling models.

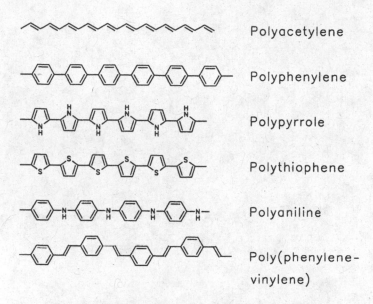

Polyacetylene

Polyphenylene

Polypyrrole

Polythiophene

Polyaniline

Poly(phenylene-vinylene)

Fig. 5. Chemical Structure of conjugated polymers.

3   Solitons and Conductivity

Conjugated polymers have attracted considerable attention because they become electrically conducting after a chemical treatment called "doping" and which bears some similarity to the doping of semiconductors. For these substances the word "Synthetic metals" has been coined and many topical conferences (Proc. ICSM (1996)) and monographs (Nalwa (1997)) have been devoted to this topic. Figure 7 shows the increase in conductivity when polyacetylene is doped by iodine. We note that the doping concentrations are much higher than we are used to from inorganic semiconductors. But also the change in conductivity is much higher. At the doping concentration of about 7 molar percents the conductivity appears to saturate, and from several other experiments, notably from the magnetic susceptibility, there is evidence of an "insulator-to-metal" transition at this point. If the doping concentration is expressed in "electrons transferred from the polymer to the dopant" the critical concentration is fairly independent of the type of conjugated polymer and the type of dopant involved. Figure 8 shows a comparison of the conductivity of conducting polymers to those of other materials. The

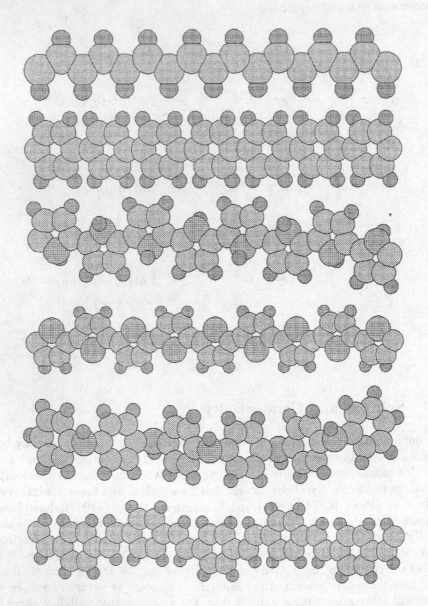

Fig. 6. Space-filling model for the conjugated polymers shown in Fig. 5. (The small spheres correspond to hydrogen atoms).

hatched area of the arrow corresponds to the highest values obtained in some special types of polyacetylene, and we see that, indeed, these conductivities are already in the range of the best metallic conductors.

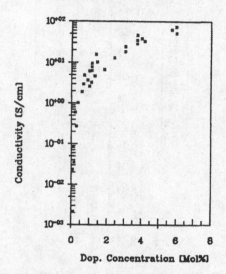

Fig. 7. Conductivity of polyacetylene as a function of doping concentration (Type S polyacetylene, room temperature, iodine doping).

The doping-induced conductivity change in polyacetylene was discovered in 1977 (Chiang et al. (1977)). Soon afterwards it was speculated that the conductivity might be caused by mobile charged solitons. In Fig. 9 we show neutral and charged solitons. The neutral soliton we have already seen in Figs. 3 and 4: a misfit where two single bonds meet. Then there is a non-bonding electron left over, a "dangling bond", or a "radical", as the chemists call it. As computer simulations show, this defect can move along the chain. If in doping electrons are withdrawn from the chain, the non-bonding electron will be the first one to go. There is still a defect, still two single bonds meeting, but now the defect is charged - it is positively charged, the positive charge of the carbon nuclei is looking through the hole left by the missing electron. If an additional electron is injected into the chain, the defect will become negatively charged. Such electron injection occurs when the polymer is doped by alkali metals. Formally, a negative soliton corresponds to a misfit in which two double bonds meet.

It is very difficult to prove experimentally that the conductivity of poly-acetylene is due to solitons. In "real" polyacetylene the polymer chains have only a finite length, often not more than some hundred monomeric units. So the solitons cannot move very far before they have to hop to a neighbor

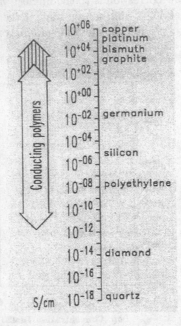

Fig. 8. Comparison of the room temperature conductivity values of conducting polymers with conductivities of other materials.

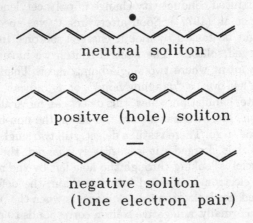

Fig. 9. Neutral and charged solitons in polyacetylene. Originally it was speculated that the conductivity of doped polyacetylene might be due to the motion of charged solitons along the polymer chains.

chain, and all the experiments show is the evidence of hopping. Whether the hopping entities are solitons or not cannot be decided. In particular it is not possible to demonstrate conservation of shape during motion. Moreover, at high doping concentrations the solitons overlap and cannot be treated as individual particles any more (See Fig. 3, where we intended to indicate that the dangling bond relaxes into a more spread-out distorted region).

Usually some solitons are present in polyacetylene as an unintended by-product from the chemical synthesis. These are neutral solitons. By doping they are converted into charged solitons. If all neutral solitons are used up further doping creates additional solitons by breaking double bonds. Solitons can also be created by absorption of light. Light breaks double bonds and thus creates solitons. It turns out that not only bonds are broken, but also charges are separated, so that most light generated solitons are not neutral but charged. An alternative way of thinking is to assume that, as in a conventional semiconductor, light first creates an electron-hole pair by lifting an electron from the valence to the conduction band, and then the electron and the hole relax into negative and positive solitons. Photogeneration of charged solitons leads to photoconductivity, and from the photoconductivity the lifetime of the solitons can be estimated. It turns out that this lifetime is only in the range of some picoseconds (Reichenbach et al. (1993)), consistent with the limited soliton mobility in short polymer chains. More important for technical applications is the inverse process: electroluminescence by soliton injection from external electrodes and radiative recombination of soliton pairs. This is the basis for organic light emitting diodes (Miyata and Nalwa (1997)), which are candidates for large flat and flexible video screens. And for optimizing the efficiency of these devices the relaxation behavior of charge carriers injected into the polymer is of utmost importance. Thus solitons in polyacetylene is not only an academic topic.

4   Solitons: Optical Fingerprints

Several times we have referred to the soliton as a "non-bonding" electron. We know that electrons in a solid cannot just have any energy, there are allowed energy bands and forbidden energy gaps. In a semiconductor the most important energy bands are the valence band and the conduction band. In the un-doped case the valence band is completely filled and the conduction band is empty. The valence band consists of the "bonding states" and the conduction band of the "anti-bonding states". In polyacetylene the situation is very similar. Here the valance and conduction band are usually called π and $\pi*$ band. Non-bonding states are neither bonding nor anti-bonding, and it is not difficult to guess that they must sit in the energy gap. Optical spectroscopy is a classical method to determine the gap in a semiconductor: There is an absorption edge which corresponds to the width of the gap. Figure 10 shows the optical absorption of polyacetylene at various doping levels. We clearly see the absorption edge in the undoped sample. Upon doping a peak arises below the absorption edge. In first approximation the "soliton peak" should be exactly at midgap. Due to higher order effects it can shift towards one of the band edges.

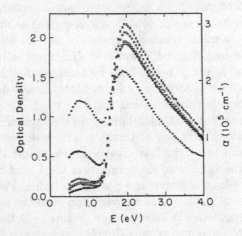

Fig. 10. Optical absorption of polyacetylene showing the emergence of a midgap state upon doping. Note that the absorption curves cross at an isosbestic point, because the midgap peak grows at the expense of the band states.

5   Metal-Insulator Transition

Semiconductors and insulators are solids in which the electronic energy bands are either completely filled or completely empty. By contrast, metals have partially filled bands. In Fig. 10 we have seen that solitons are accommodated in the band gap, in the ideal case exactly at midgap. If there are many solitons the levels will interact, the midgap peak will widen, and finally it will fill the gap completely. Then there is no more gap separating the filled valence band from the empty conduction band, but there is one wide overall band filled to 1/2: An insulator-to-metal transition has occurred. Solitons can be created by doping, by photoexcitation, and by thermal excitation. In principle, all these three processes can induce an insulator-to-metal transition in polyacetylene. It is an interesting gedankenexperiment to follow this transition from the metallic side. For this we start with a polyacetylene chain at the hypothetical temperature of 10 000 K (real polyacetylene decomposes at some 600 K). At 10 000 K there are so many solitons that practically all double bonds are broken - there is just a linear chain of equidistant carbon atoms, each carrying a hydrogen atom (which does not matter in this context) and a dangling bond. This high temperature polyacetylene chain is very similar

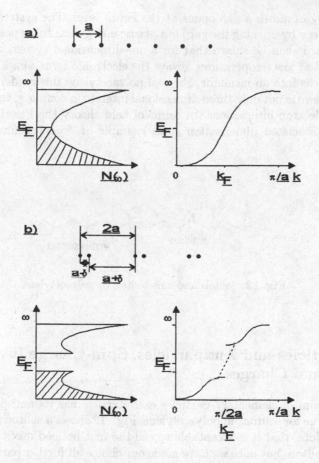

Fig. 11. Peierls distortion in an isolated chain of equidistant monatomic sodium. Crystal lattice, electronic density of states and electron dispersion relation without distortion (a) and with distortion (b)

to a monoatomic alkali metal chain. For the alkali chain Peierls has shown in his famous 1955 text book (Peierls (1955)) that at low temperatures the monoatomic metallic chain is unstable, it will "dimerize" into a bi-atomic insulating chain. The Peierls transition is illustrated in Fig. 11. The upper part shows the high temperature situation: a monoatomic chain, the density of states function with a half-filled band, and a continuous electron dispersion relation. In the lower part the atoms approach pairwise. This leads to a doubling of the elementary cell of the crystalline lattice and consequently in reciprocal space to a reduction of the Brillouin zone by a factor of 2. The electrons are scattered at the "superstructure" (doubled elementary cell), the dispersion relation becomes discontinuous at the new zone end, and in the density of states a gap opens at the Fermi level. The system wins electronic energy by opening the gap, but elastic energy has to be paid to distort the lattice. It can be shown that for a one-dimensional system with a half-filled band at low temperatures always the electronic term wins and that the system turns into an insulator. For real polyacetylene this model is oversimplified: There is not only three-dimensional interchain coupling, there are also electron-electron interactions. In terms of field theory the Peierls transition and the associated dimerisation is an example of "spontaneous symmetry breaking".

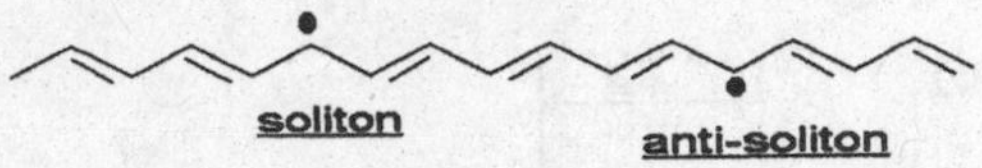

Fig. 12. Soliton and anit-soliton in polyacetylene.

6   Particles and Antiparticles, Spin-Charge Inversion, Fractional Charges

In the realm of elementary particles each particle has its anti-particle. The same is true for solitons in polyacetylene. Fig. 12 shows a soliton and an anti-soliton. Note, that it is acceptable to call the first noticed misfit a soliton or an anti-soliton, but once we have made our choice all further particles on the same chain are defined either as solitons or as anti-solitons. As a by-product of the chemical synthesis there will be some solitons in any polyacetylene sample. Further solitons can only be created as soliton/anti-soliton pairs. We can call this fact "conservation of the particle number". From a chemical point of view it is evident that solitons have to be created as particle/anti-particle pairs: If a bond is broken two dangling bonds are left. A curious consequence of bond conjugation is "soliton motion by double-steps". If a soliton moves it links a bond to the next site and leaves a dangling bond at a position two sites further. The situation is illustrated in Fig. 13. A further surprising feature of solitons in polyacetylene is "spin-charge inversion", which is indicated in Fig. 14. If a soliton has a spin it has no charge, and vice versa. Of course, a neutral soliton has a spin: It is an unpaired electron. A neutral soliton has no charge, because the charge of the electron is compensated by the positive background charge of the carbon nuclei. If the electron is removed, evidently the spin will also be removed, and the defect is positively charged because now the nucleonic charge is uncompensated. If, however, a second electron is accommodated on the defect, the nucleonic charge is overcompensated and

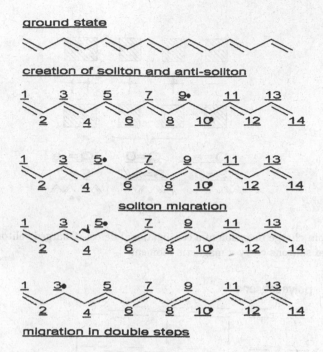

Fig. 13. Solitons and anti-solitons. The presence of a soliton on a polyene chain allows the classification of the lattice sites as even and odd, and of all further conjugational defects as solitons and anti-solitons.

the defect becomes negative. The two electrons will have opposite spin and the net spin will be zero.

In a gedankenexperiment we can even construct solitons with fractional charges, not in polyacetylene, but in the hypothetical substance "polyfractiolene", shown in Fig. 15. In polyacetylene there is an alternation of double and single bonds, in polyfractiolene the double bonds would be separated by two single bonds (This substance cannot be synthesized; we want the bond system to be mobile, so that is can slide as "charge density wave" along the chain; the hydrogen atoms - not shown in the figure - would block such a motion). In the upper part of Fig. 15 two solitons are created. This leaves the chain unchanged outside the soliton region. The electron count at the soliton region reveals that the chain bears two positive elementary charges, which corresponds to one elementary charge per soliton. In the lower part, in fractiolene, three solitons are needed to bring the chain back to its original configuration. Again the chain bears two positive charges, but now they are shared by three solitons, leading to the fractional charge 2/3.

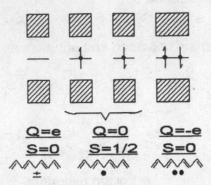

Fig. 14. Spin charge inversion of a conjugational defect. Charged solitons are spinless; neutral solitons carry a magnetic moment.

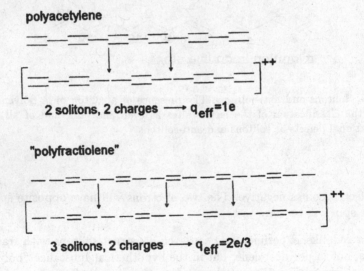

Fig. 15. Electron counting to demonstrate fractional charges on one-dimensional chains.

7   Polarons, Confinement

In Figs. 5 and 6 we have shown that polyacetylene is not the only polymer with conjugated double bonds. It is, however, the conjugated polymer with the highest symmetry. The importance of symmetry becomes evident if we look at Fig. 16, where a soliton in polyacetylene is compared to a soliton in polyparaphenylene. In polyacetylene a soliton separates two parts of a chain with equal energy. In polyparaphenylene there are phenylene rings on the left hand side of the soliton, but on the right hand side the rings are quinoidal - and the energy at the quinoidal part is considerably higher. Therefore the soliton is driven to the right hand side, and on passing it converts high energy quinoid rings into low energy phenylene rings (very much like a Bloch wall in a ferromagnet exposed to an external magnetic field). Solitons are only stable

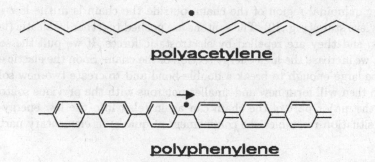

polyacetylene

polyphenylene

Fig. 16. A soliton is free to move in polyacetylene whereas in polyphenylene it is pushed to the chain end by lattice forces.

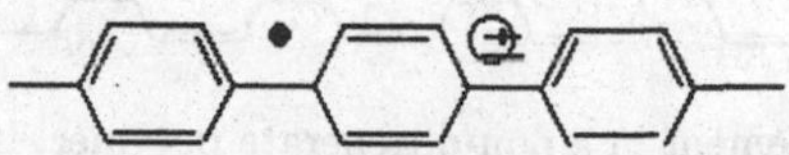

Fig. 17. Polaron in polyphenylene.

in polyacetylene. In all other conjugated polymers they are driven to the chain ends. A complex particle, consisting of two solitons, however, is stable in low symmetry polymers also. Such complex particles are called polarons. Figure 17 shows a polaron in polyparaphenylene. It consists of a neutral and a charged soliton. The name is derived from polarons in semiconductors, which are electrons strongly coupled to the crystalline lattice. Polarons are packages of charge, spin, and lattice distortion. Polarons are the relevant charge carriers in polymer light emitting devices and therefore polarons are of large technological importance.

Figure 18 shows a bi-polaron. This is a complex of two positive solitons. One can also construct a bi-polaron by combining two ordinary polarons. Then the neutral parts of the polarons annihilate to form a double bond and

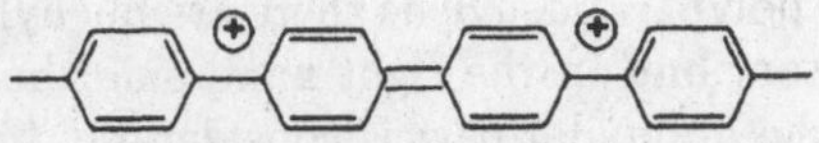

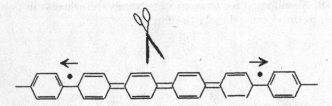

Fig. 18. Bipolaron in polyphenylene.

the charged parts are left over. A bi-polaron is a nice object to demonstrate what is called "soliton confinement". Between the solitons there is a high energy quinoidal region of the chain, outside the chain is in the low energy phenylene state (Fig. 19). The solitons are pushed together by elastic (lattice) forces, and they are repelled by electrostatic forces. If we pull the solitons apart we increase the high energy section of the chain. Soon the elastic energy will be large enough to break a double bond and to create two new solitons, which then will form new and smaller polarons with the previous solitons, so that the major part of the chain can again relax into the low energy form. This situation resembles the confinement of quarks in elementary particles.

Fig. 19. Soliton confinement in a non-degenerate polymer. If the solitons forming a polaron are pulled apart, a high-energy section of the chain will be created between the solitons. Soon the energy stored will be large enough to break a double bond and create two new solitons.

In Fig. 20 we present a compilation of the various conjugational defects, which can be constructed from solitons in conjugated polymers. For simplicity we demonstrate the defects in polyacetylene, although the complex defects are more appropriate for the low symmetry polymers (in polyacetylene there are no elastic forces pushing the components of a polaron together, unless we assume interchain coupling in a rope of chains). This compilation can be called "physical-chemical dictionary", although the chemical terms on the right hand side do not necessarily imply mobility of the defects, and the physical terms on the left hand side do strictly apply only to the continuum model of conjugated polymers.

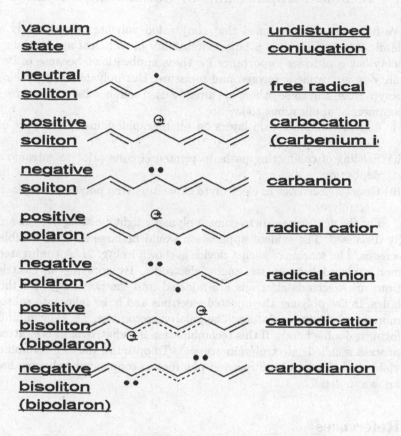

vacuum state		**undisturbed conjugation**
neutral soliton		**free radical**
positive soliton		**carbocation (carbenium i**
negative soliton		**carbanion**
positive polaron		**radical catior**
negative polaron		**radical anion**
positive bisoliton (bipolaron)		**carbodicatior**
negative bisoliton (bipolaron)		**carbodianion**

Fig. 20. Complex conjugated defects constructed from solitons: a "physical-chemical dictionary".

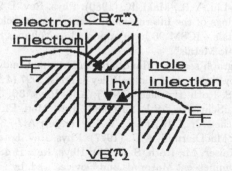

Fig. 21. Soliton generation by charge injection. In a first step electrons and holes are injected from the electrodes. These then relax to form solitons and anti-solitons.

8   Technical Applications of Conducting Polymers

We have already mentioned that conjugated polymers are not only of academic interest. There is a large potentiality of technical applications. Polyacetylene is of lesser importance for these applications, because of its instability to atmospheric oxygen and moisture. But polyaniline, polythiophene, polypyrrole, and some others are already used commercially. The three major commercial applications today are

i) Conductive protection layers on photographic films to dissipate electrostatic charges
ii) Seeding of conducting paths in printed circuits prior to galvanic copper deposition
iii) Counterelectrodes in electrolyte capacitors (the polymer is replacing the electrolyte).

For the next few years to come, polymeric light emitting devices are heavily discussed. The evident application would be large flat and flexible video screens. The scheme of such a device is shown in Fig. 21. A conjugated polymer is placed between two metal electrodes. By applying an electric field, from one electrode electrons are injected into the polymer, from the other holes. In the polymer the injected electrons and holes relax into solitons - or more precisely, into polarons. The polarons recombine, i.e. they annihilate by forming double bonds. If this recombination is radiative light is emitted. This process is called "electroluminescence". To optimize the electroluminescence yield the behavior of solitons and polarons in conjugated polymers has to be known in detail.

References

Roth, S.(1995): One-Dimensional Metals, VCH, Weinheim.

Seeger, A. (1948): Thesis, University of Stuttgart.

Takayama, H., Lin-Liu, Y.R., Maki, K. (1980): Phys. Rev. B **21**, 2388.

The latest proceedings of the International Conference on Science and Technology of Synthetic Metals - ICSM 96 have appeared as Volumes 84, 85, and 86 of the journal "Synthetic Metals".

For a recent monograph see e.g. "Handbook on Organic Conductive Molecules and Polymers", ed. by Nalwa, H.S., John Wiley & Sons, 1997 (4 Volumes), For review articles see e.g. S. Roth, H. Bleier (1987): Adv. Phys. **36**, 385 and A.J. Heeger, S. Kivelson, J.R. Schrieffer, W.P. Su (1988): Rev. Mod. Phys. **60**, 781.

Chiang, C.K., Fincher, Jr.,C.R., Park, Y.W., Heeger, A.J., Shirakawa, H., Louis, E.J., Gau, S.C., MacDiarmid, A.G. (1977): Phys. Rev. Lett. **39**, 1098.

Reichenbach, J., Kaiser, M., Roth, S. (1993): Phys. Rev. B **48**, 14104.

"Organic Electroluminescent Materials and Devices", ed. by Miyata, S. and Nalwa, H.S., Gordon and Breach, (1997).

Peierls, R.E.: "Quantum Theory of Solids", Oxford, Clarendon Press, (1955).

Subject Index

σ-model   106, 196

aging   88
Aizenman random walk   28
asymmetric exclusion process   89, 97

backpropagation   179, 185
Bethe ansatz   88, 94, 233
bi-polaron   253
bilocal   104
biopolymerization   97
bosonization   103, 105

Cartan decomposition   111
Casimir amplitude   18
chiral symmetry   103, 193, 232
compactification   144, 146
configurational average   7
conformal   16
contact-interacting (CIW) model   27
cumulant average   7

diffusion-limited
 – annihilation   93
 – coagulation   99
 – pair-annihilation   91
dimerize   249
Dirichlet condition   11
domain
 – growth   90, 92
 – wall   90, 91
Domb-Joyce model   28
dopants   232, 236

enantiodromy   94
Euclidean   60
excitons   91, 99

exclusion interaction   50, 86

Feynman rules   75
Fisher–de Gennes–amplitude   18
Fokker-Plank equations   79
foldicity   147
Fourier acceleration   136

gel-electrophoresis   95
Glauber dynamics   91, 177
gluon   104, 124
Grassmann (exterior) algebra   124, 198
Gross-Neveu model   194, 231

Haar measure   53
hadronization   116
Hamiltonian formalism   82, 85
Hausdorff dimension   4
Hebb rule   180
Heisenberg   86, 97
HMC   122, 136
Hopfield model   178
Hubbard-Stratonovich transformation 103
Hybrid Monte Carlo (HMC)   126, 133

integrable   90, 95, 96
interacting polymers   70
Ising
 – model   48, 90
 – representation   177

kink-antikink pair   167, 169
Kirkwood-Salsburg equations   46, 57
Krylov-subspace   122

Landau–Ginzburg–Wilson   5, 8

Lecture Notes in Physics

For information about Vols. 1–469
please contact your bookseller or Springer-Verlag

New Series m: Monographs